RNA SILENCING

With an Appendix on Gene Therapy
by Eithan Galun

RNA SILENCING

With an Appendix on Gene Therapy
by Eithan Galun

Esra Galun

Department of Plant Genetics
The Weizmann Institute of Science

World Scientific

NEW JERSEY · LONDON · SINGAPORE · BEIJING · SHANGHAI · HONG KONG · TAIPEI · CHENNAI

Published by

World Scientific Publishing Co. Pte. Ltd.

5 Toh Tuck Link, Singapore 596224

USA office: 27 Warren Street, Suite 401-402, Hackensack, NJ 07601

UK office: 57 Shelton Street, Covent Garden, London WC2H 9HE

British Library Cataloguing-in-Publication Data
A catalogue record for this book is available from the British Library.

RNA SILENCING

ISBN 981-256-206-0

Typeset by Stallion Press
Email: enquiries@stallionpress.com

Printed by FuIsland Offset Printing (S) Pte Ltd, Singapore

Contents

Preface

וַיֹּאמֶר הַמֶּלֶךְ קְחוּ-לִי חֶרֶב
גִּזְרוּ אֶת הַיֶּלֶד הַחַי לִשְׁנָיִם

"And the king said, bring me a sword …
and divide the living child in two"
(King Solomon's trial, Hebrew Bible, Kings I, Chap. 3.)

Τόδε περί ἁμάξης ἐμυθεύετο, ὅστις λύσειε τοῦ ζυγοῦ τῆς ἁμάξης τόν δεσμόν, τοῦτον χρῆναι ἄρξαι τῆς Ἀσίας….Ἀλέξανδρος δέ…παίσας τῷ ξίφει διέκοψε τόν δεσμόν…

"This was the myth about the coach, that whoever would undo the knot on its yoke, he would become the ruler of Asia. Alexander then drew his sword and cut the knot."
(The myth of the Gordian Knot and Alexander the Great in *Alexandrou Anabasis Book B*, by Arrianos.)

****

In the above two mottos the **Sword** was instrumental in resolving intriguing problems. In the trial of King Solomon, the wise King revealed who the mother of the living child was. In the cutting of the Gordian Knot, Alexander quickly resolved an ancient problem and marched to his conquests.

How are these mottos related to RNA Silencing? Scholars of RNA silencing also used a **sword** although they termed it *Dicer* and the cutting, *Dicing*. Most RNA silencings are initiated with *Dicing* double-stranded RNA by the *Dicer*. We shall see that these scholars followed the wisdom of King Solomon and made ample use of *Dicing* to study major problems in genetics, in the regulation of gene expression and in the development of eukaryotes.

Scientific revolutions may be sensed in advance by scholars with an exceptional intuition but the *site* where such revolutions take place is frequently a complete surprise. Not all major revolutions in science took place in main towns. Who would have predicted that the revolution in cosmology, initiated by Copernicus, would take place in a tiny town as Frauenburg (then in Poland, thereafter in Eastern Prussia and now called Formbork in Poland)? In genetics, the first revolution, initiated by Gregor Mendel, happened in a modest Augustinian monastery near Brunne, in Moravia. A further revolution, initiated by Barbara McClintock, took place in a very tranquil surrounding: the Cold Spring Harbor Laboratory of the Carnegie Institution of Washington. This latter revolution heralded the *Transposable Elements* and was the subject of my previous book (Galun, 2003). Another department of the Carnegie Institution of Washington (in Baltimore, MD) was the site of a further genetic revolution: the awareness of the profound role of small dsRNA sequences in gene expression, defense against invaders, remodeling chromatin and regulation of development in eukaryotic organisms.

For brevity I shall adhere to the term *RNA Silencing* as the name of this book. But there are several types of RNA sequences that affect gene expression (e.g. small interfering RNA, microRNA). As in the cases of the revolutions initiated by Mendel and McClintock, the *RNA silencing* was also preceded by information that by retrospect could lead to novel molecular-genetic understanding. But the "birth" of this field of endeavor can be traced to one specific publication by Mello and Fire (Fire *et al.*, 1998) that heralded the silencing of RNA.

Although *RNA silencing* is a young field of investigation it has already solved several genetic/biological riddles and has started to be instrumental to further genetic and developmental studies as well as to lead to practical applications as gene therapy. As in my previous books, *Transgenic Plants* (Galun and Breiman, 1997), *Manufacture of Medical and Health Products by Transgenic Plants* (Galun and Galun, 2001) and the aforementioned book *Transposable Elements* (Galun, 2003), this book is targeted to a wide range of readers. Therefore, some basic subjects that are prerequisites for the understanding of *RNA silencing* will be discussed briefly as a kind of introduction to the main theme of this book. While most of this book is intended for the novice in this field, it also includes an Appendix that deals with the use of gene silencing for gene therapy. This Appendix was added for the benefit of those who are considering using this approach in their investigations. Consequently, the Appendix is written in a manner that will render it useful for such an endeavor.

While writing this book, I have received generous help from many people and I am very grateful for this help. Professor Dan Segal of the Tel Aviv University invested a great deal of efforts to correct my mistakes in the drafts of the chapters on the nematodes and on *Drosophila* and mosquitoes. My wife, Professor Margalith Galun, read parts of the manuscript and made useful remarks. My son, Professor Eithan Galun, not only wrote the Appendix of this book, but also read and corrected parts of the book that deal with the medical aspects.

My gratitude goes also to several colleagues in the Department of Plant Sciences of the Weizmann Institute of Science: Dr Yuval Eshed, Professor Robert Fluhr, Professor Gad Galili and Professor Avraham Levy who provided useful remarks on specific chapters. I owe thanks to Professor Shulamit Michaeli of Bar–Ilan University who read and corrected Chap. 6. I also thank Dr Izhak Bentwich of the Rosetta–Genomics company in Nes–Ziona for reading Chap. 10 and to Professor Aglaia Athannassiadou of the Medical School of the University of Patras, Greece, for the Greek text of the *Myth of the Gordian Knot*.

I appreciate very much the work of the Secretary of our Department, Mrs Renee Grunebaum, who typed and retyped the manuscript and handled the hundreds of references almost daily during many months and with endless patience and for being able to bear with me during my moody moments. Thanks are due to the Graphic Arts

Department of the Weizmann Institute of Science for performing very professional work on the figures of this book.

I acknowledge the outstanding collaboration of the Publisher, Dr K.K. Phua, Chairman of the World Scientific Publishing Co. (WSPC) in Singapore, and Ms Magdalene Ng, senior editor at WSPC, who was instrumental in rendering my draft into a real book.

I acknowledge the permissions to use figures and tables that were granted by the respective publishers. I provided the source below each table or figure and the full citations are listed in the references. These citations should mean that I am grateful for the consents of the respective publishers to use their copyright material in this book.

To Dr Nillie Weinstein I endow my final and very special gratitude. She encouraged me to commence this book and provided continuous support during my writing period. I benefited immensely from her wisdom and knowledge especially but not only in philosophical issues. Dr Weinstein took an active part in the integration of these issues into the book and carefully reviewed the appropriate phrasings. For the above and for additional contributions, she became an inseparable constituent of my endeavor.

# Acronyms and Abbreviations

AGO	ARGONAUTE
CP	viral coat protein
DCL	dicer-like (enzymes)
GFP	green fluorescent protein
GUS	β-glucuronidase
HR	hypersensitive response
miR…	genes coding for miRNAs
miRNA	microRNA
ORF	open reading frame
PAZ	PIWI, argonaute & zwille (a protein interaction domain)
PCR	polymerase chain reaction
PKR	protein kinase, dsRNA dependent
PSTV	potato spindle tuber viroid
PTGS	post transcriptional gene silencing
PVK	potato virus X
PVY	potato virus Y
RdDM	RNA-directed DNA methylation
RdRP	RNA-dependent RNA polymerase
RISC	RNA-induced silencing complex
RT	reverse transcription
TE	transposable element
TEV	tobacco etch virus
TGMV	tomato golden mosaic virus
TGS	transcriptional gene silencing
TMV	tobacco mosaic virus
TncRNA	3′-untranslated region (of transcripts)
VIGS	virus induced gene silencing

Introduction

Philosophical Contemplations

The motto of my previous book, *Transposable Elements — Guide to the Perplexed and the Novice* (Galun, 2003), was a quote from Heraclitus of Ephesus (500 BC): "Many fail to grasp what they have seen and cannot judge what they have learned, although they tell themselves they know". In the preface of the above mentioned book I then reiterated the wisdom of several old masters, stating that only the combination of keen observations with rationalization and logic thinking will lead to meaningful new knowledge. This old wisdom is also relevant to the emerging field of RNA Silencing. I shall present a brief history of RNA Silencing below. We shall see that an important step in the awareness of RNA silencing was the study by Mello and Fire (Fire *et al.*, 1998) on the impact of short RNA fragments on the nematode *Caenorhabditis elegans*. These investigators found that if they introduced *C. elegans* into fragments of a single-stranded mRNA or the antisense of this mRNA, there was a specific genetic interference; but when they used the double-stranded RNA (dsRNA), composed of sense and antisense mRNA, the interference was several-fold greater. Albeit, further studies showed that the single-stranded RNA (ssRNA) effects were *artifacts*. The preparations of the ssRNAs were contaminated by dsRNA and it was only the latter RNA which caused the interference. The verification of the effect of the dsRNA provided the accelerated endeavor of RNA silencing. One additional example to support the notion that you believe what you see can be misleading is from my own research that led to the publication of an artifact. In collaboration with the laboratory professor Nathan Sharon (also of the Weizmann Institute of Science) to study the impact of a lectin on fungal growth,

I observed that the lectin wheat-germ-agglutinin (WGA) caused the explosion of the hyphal tips of *Trichoderma*, arresting the growth of the fungal hyphae. We published it in the journal *Nature* (Mirelman *et al.*, 1975) and this publication was then amply cited... but later it was found that because we used "home-made" WGA, this WGA contained traces of chitinase. Why was there a contamination of chitinase in our WGA? Now we know. The Sharon laboratory devised an efficient method to isolate and purify WGA. A wheat-germ homogenate was passed through an affinity column that contained oligo N-acetyl glucosamine. The WGA was bound to the oligomer and subsequently washed out of the column. But the same oligomer also retained the trace amounts of chitinase in the wheat-germ homogenate so that this chitinase was washed out with the WGA. These chitinase traces caused the antifungal effect.

Let us briefly carry the question of the difference between what the investigator "sees" and what the "real" nature is, to the philosophical level. Since the classical Greek philosophical period (i.e. 500–300 BC) there is debate whether or not humans are capable of perceiving the real nature. Could it be that by using his senses man perceives only an *illusion* of the real nature? If the latter is true, is there a way by which the real nature can be perceived by human rationalism? Note that during this classical period, there were only observations; experimentation was not applied yet. Plato illustrated this enigma by his famous *Analogy of the Cave* (The Republic, book VII, Plato, and see the Epilogue in Galun, 2003). Maimonides (1135–1204) claimed, on the basis of logical arguments, that there must be forces (or laws) of nature that humans cannot comprehend but these laws do exist. In his *Guide to the Perplexed* (Part I, Chap. 73) Maimonides claimed that it is agreed that the earth is ball-shaped (a globe) and that there are people on the opposite side of the globe. These people are also standing on the globe but with their heads away from us; still they do not fall "down". Hence, there must be a force (law of nature) that keeps them from falling. Maimonides claimed that the people at the other side (the lower side opposite us) should fall only by our imagination; if we consider the situation by logical thinking we shall recognize that the latter people were also standing on earth, from their point of view. Maimonides did not mention gravity. Well, with respect to this example of Maimonides,

Isaac Newton (1642–1727) revealed the Laws of Universal Gravitation. These laws could explain the enigma of Maimonides. They were since amended although the real nature of gravitation is still enigmatic. Shall we ever reveal and understand the real nature? Immanuel Kant (1724–1804) added a decisive contribution to the question of apparent versus real nature. In brief he claimed that there is a real nature (world), but humans cannot perceive this noumenon by direct cognition. Humans are only able to perceive the real world when rational recognition is added to direct observations. In a way Kant argued what Heraclitus of Ephesus and Maimonides claimed 2300 years and 500 years earlier, respectively. Kurt Gödel (1906–1978) who was born in Brünne, the Moravian town near which Gregor Mendel made his revolutionary genetic discoveries, developed an important mathematical-logic theorem. He stated that certain branches of mathematics are based, in part, on propositions that are not provable *within* the system itself, although they may be proven by means of logical systems outside of (pure) mathematics. Does this lead to the philosophical claim that a full understanding of the real nature of a system can be achieved only if one is positioned *outside* of the system? Hence, for our deliberation: outside of the known world. If this claim is correct a full understanding of the world will never be achieved by humans. Does this mean we should give up? The answer is negative. Being aware of the logical possibility that a certain goal cannot be achieved should not deter us from trying. The *way* and the *effort* to achieve a goal could be more rewarding than *reaching* the goal. This was clearly presented by Albert Camus (1913–1960) in his book, *Le Mythe de Sisyphe.*

Not equipped with adequate philosophical background to contribute convincing arguments in favor or against the claim that humans can comprehend the nature of the *real* world, I shall thus leave this question open.

There is a claim, expressed especially by mathematicians and physicists that esthetically beautiful solutions to problems are commonly the correct solutions. This was the notion of Watson and Crick when they derived an elegant solution to the double helix structure of DNA: the elegant solution must be the correct one. Let us now turn our attention to biology. We shall assume that a certain biological phenomenon is being investigated and performed from several angles and in a step-wise manner. When all the results are combined, the phenomenon appears clarified; moreover, the now clarified phenomenon is a sound basis for further investigations of related phenomena, until an esthetic, elegant and beautiful general-picture of biological phenomenon is reached. Do we mind if this picture may not represent the "real" nature in the philosophical sense? Obviously, a step-wise endeavor to understand biological phenomena, and in general terms any phenomenon in nature, is rewarding by itself (i.e. Sisyphus of Camus). For the scholars themselves this reward would suffice. Baruch Spinoza (1632–1677) went even further and in his Ethics claimed that the revelation of the nature of Nature is a source of joy and happiness. Albeit, this may be a sufficient reward to satisfy the investigator, it could pose a problem with respect to the recognition of the investigators' achievement by his peers. The Nobel committees for prizes in the sciences have an additional requirement. These committees usually wait with the endowment of prizes until the discoveries find applications in further studies. This waiting could take many years. As prizes are awarded only to living people, longevity is of advantage to potential Nobel awardees.

We shall see that components of the RNA silencing systems, such as *micro RNA*, may affect the differentiation of Eukaryotic organisms and thus be decisive for the final shape of their organs. Here is a pitfall that is commonly overlooked by naïve biologists. When the latter are told about the gene that causes additional petals in a flower or additional fingers in a hand, they may assume that these genes contain all the information to shape a flower or a hand, respectively. This is obviously wrong. We still have no notion as to how, from the *linear* information that is stored in the DNA, a three-dimensional structure can emerge. Well, with a few exceptions. The genes in some bacterial phages are now known and their expression leads to proteins that interact to produce the final shape of the respective phage. Also, the genes that code for the histone components are known and the derived structure of the nucleosome is now understood at the molecular level. We are therefore beginning to understand, at the chemical/molecular level, how euchromatin, in which the nucleosomes are separated, is converted to heterochromatin in which there is a dense packaging of nucleosomes. But these examples are the exception; we still do not know how the final structure of organs (e.g. of a specific petal) is derived. True enough, it is obviously structured by orchestration of cell division, cell enlargement and arrest of cell division, at very specific locations in the organ. But for that to happen, there must be a coordination based on sensing the space in which this orchestration of differentiation takes place. We shall see below that there are beginnings to the understanding the structuring of animal and plant organs and that this understanding is assisted by changes in gene-expression. RNA silencing mechanisms are involved in these changes and therefore the basic questions of control of differentiation are relevant to the theme of this book. For correct differentiation a given cell should either stay idle or divide and/or enlarge. For that the cell should have a correct orientation of its location in "space" — relative to the other cells in the specific organ. We can carry our question further. When we look from a certain distance, at a sycamore tree and a pine tree, the silhouettes of each of these trees are very different and typical to each of them. How do the cells in these trees know where and how to divide and enlarge to reach the same general shape of the tree all the time? Do the individual cells attain a perception of space? If they indeed have a perception of space, their perception could be fundamentally different from ours. Here we return to Immanuel Kant who claimed, already in the 18th century, that humans have an inherited and intuitive sense of Euclidian geometry and perceive space accordingly. We now know that there are other geometries. Could it be that cells of animals and plants use different space-orientations from those used by the intuitive human mind? All the above contemplations are intended to draw the attention of the reader to the rather complicated issue of differentiation. A comprehensive and beautiful analysis of the problem of differentiation was presented in the book of Enrico Coen (1999).

A Short History of RNA Silencing

History is commonly meant to start with written records. Take, for example, the *history* of the Middle East started with the first writings that were found in Uruk (South-Eastern Mesopotamia) and in Memphis (Egypt). Both were dated to about 3000 BC. Anything that happened before this time is considered *Pre-history*. We shall adopt the same approach to RNA silencing, meaning that the history of RNA silencing started with the first written reports on this phenomenon. Still, there are differences. First, in the early reports on what we now term RNA silencing, the term Gene silencing was used. Albeit, Gene silencing included not only the phenomenon of RNA silencing. This will be clarified below. Another difference is that for the pre-history of RNA silencing there are still living witnesses (as the author of this book) who remember this period, while for the common pre-history of countries and nations, our "witnesses" are merely archeological artifacts.

Remarks on the "pre-history" of gene (RNA) silencing

Summarizing "unrecorded" records is ambiguous. Still, the following remarks and information are relevant to our deliberations. The first remark is that these "unrecorded-records" concerned primarily plants (in this context — actually angiosperms). Investigators dealing with disease resistance of plants and especially with viroid and viral diseases were faced with a situation in which plants that were infected with a mild pathogen (i.e. a mild virus) showed various degrees of resistance to a second infection by the same or a similar pathogen. In a few cases such cross protections were reported in reviewed publications or in lectures in scientific meetings (see: Niblett *et al.*, 1978; Sherwood, 1987; Lomonossoff, 1995). During this period, that is, *prehistoric* with respect to RNA silencing, there were numerous written records on cross protection. In several of these records the authors even furnished suggestions for the mechanisms of these protections, mentioning the involvement of proteins and/or RNA. Albeit, the specific role of dsRNA was not put forward until the mid 1990s. It is noteworthy that such protections were not always reproducible but they did appear "real" and horticulturalists swiftly utilized this phenomenon. Take, for example, certain *Citrus* cultivars were infected with a mild virus (that occasionally caused some stunting) to immunize the trees against virulent viruses. Whole orchards were thus immunized for a double purpose: to reduce the size of the trees, for easier management of the orchard and to prevent the severe virus diseases.

In later investigations on viral cross-protection in plants, the involvement of dsRNA came to light. These latter investigations will be discussed in some detail in this book.

Another case of *pre-historical* silencing concerns transposable elements (TEs) (see: Galun, 2003). Since the pioneering studies of Barbara McClintock on "controlling" TEs it became evident that the maize transposons can undergo a reversible change from active to inactive (silent) elements. Note that the two first transposable element systems, *Ac/Ds* and *Spm* were established already in 1948 and 1953, respectively (*Ac/Ds* by McClintock and *Spm* independently by McClintock and Peterson; see Galun, 2003). In some cases McClintock even revealed cycled changes from active to inactive phases of the transposable elements. This was observed in the *Suppressor-mutator (Spm)* elements as well as in the *Activator (Ac)* element of the *Ac/Ds*, and later also in the *Mu* transposable elements of maize. These changes that led to active and to silenced phases of *Ac* and *Spm* were revealed by genetic studies but the molecular meaning of silencing the activity of a TE was not known for 30 years after the discovery of these elements. Only after molecular studies identified genes for transposases in the TEs, the silencing of the respective transcripts could be followed. A detailed account on this subject, with emphasis on the *Spm* transposable elements, was provided by Fedoroff (1995).

Efficient genetic transformation of plants was achieved by the use of *Agrobacterium*-mediated transformation and properly engineered plasmids. This happened in the early 1980s (see: Galun and Breiman, 1997, for a detailed review). Very hectic activity of plant-genetic transformation was then initiated in many laboratories. All those engaged in these transformation activities witnessed the same general phenomenon: the expression of the transgene in the transformed plants was very variable even among the plants that resulted from the same transformation experiment. Moreover, when the transgene was inserted into a plant genome, in more than a single site, the expression of the transgene was frequently *lower* than when introduced only into one site. This silencing of the transgenes had an epigenetic character, commonly carried to the next sexual generations. Most of this silencing was not reported. The investigators were interested in the highest expression of the transgene. So they just focused on one or a few transgenic plants for their further research and ignored the transgenic plants that had a low expression of the transgene. There could be various reasons for the low expression of transgenes, such as position effect (i.e. insertion into heterochromatic regions of the chromosome) and rearrangement of the coding and cis-regulatory sequences of the transgene. Obviously, RNA silencing, through dicing, imposed by dsRNA, was not taken into account because it was unknown during this period.

If indeed written records mark the border between pre-history and history, then commercial biotechnology companies pushed the events into the pre-historical era. Obviously, such companies are reluctant to publish the details of their experimental work at least till after the respective patents are submitted and/or the products reached the market. A case that comes to mind is the long shelf life tomato. Biotechnology companies engaged in crop-improvement had an aim to produce transgenic tomatoes with a long shelf life. We should note that tomato fruits with long shelf life are of no benefit to both the tomato plant and the consumer, but they are of great interest to the grower and to the marketer — the fruits can be kept longer and do not rot on the way to the market or on the shelf. The seeds of transgenic tomatoes are sold to the grower, thus the companies that were engaged in tomato transformation had the grower in mind. For the benefit of the consumer an attractive brand name could be invented (e.g. *Flavr Savr*). The laboratory of Don Grierson (in Nottingham, UK) focused on the enzyme polygalacturonase (PG) that causes softening of the tomato fruit. This interest was also shared by the investigators at "Calgene" (Davis, California). The Grierson Laboratory and "Calgene" scientists, both introduced the antisense of the PG gene into transgenic tomato plants, in order to extend the shelf life of the fruits. The "Calgene" company reached the patent office a little earlier and received the patent-rights for slow maturing tomato fruits (e.g. *Flavr Savr*) induced by the antisense of the PG gene. The Grierson Laboratory did not give up. They found that the introduction of a sense transgene for PG can, in some transgenic lines, suppress the endogeneous PG expression, resulting in delayed maturation of the tomato fruits. Thus, the "Zeneca" company of the UK developed the latter type of transgenic tomatoes. Details of these events were provided in my previous book (Galun and Breiman, 1997). There is a final twist in this story. Although intended to benefit the growers, the latter were not too excited about the "Flavr Savr" and "Calgene" did not make a big profit from this cultivar. As for the "Zeneca" tomato — for many years, transgenic tomatoes could not be grown in the UK, so tomatoes were grown outside the UK, mashed, and only the concentrate was sold in the UK with a clear label indicating that the product was derived from GM plants.... Returning to our subject, the "Zeneca" tomatoes with the "sense" transgene for PG were assumed to show "co-suppression" of the PG gene but what was the molecular mechanism of this silencing was enigmatic and obviously the possible involvement of dsRNA was not recorded. Co-suppression of genes was not confined to plants; it was then amply observed in other eukaryotic organisms; also in the latter organisms the elucidation of the molecular basis of this phenomenon had to wait till later years.

Notes on the silencing history

These notes shall cover research on gene silencing up to 1998. The latter year was decisive for gene silencing due to the study by Mello and Fire (Fire *et al.*, 1998). These investigators showed, for the first time, that dsRNA is involved in gene silencing. A special section will be devoted to this study (i.e. Fire *et al.*, 1998 and later publications on RNA interference in the nematode *C. elegans*). The historical notes shall handle *co-suppression, viral-induced resistance* to the spread of viruses in plants and *changes in chromatin* that cause the silencing of TEs.

Co-suppression in transgenic plants

Co-suppression is the term given to the phenomenon that occurred when a transgene was introduced into an organism, causing a reduced expression of both the transgene and the homologous (or similar) — endogenous gene. For a period, this phenomenon was in the realm of *pre-history* (as indicated above), but then it was reported in reviewed publications.

Early records on co-suppression were reported in two papers, published in tandem in the Plant Cell journal (Napoli *et al.*, 1990 and van der Krol *et al.*, 1990) that came respectively from the laboratories of Jorgensen (Oakland, California) and Mol (Amsterdam, the Netherlands). Both publications touch on the pigmentation of petunia petals and both laboratories had the initial intention to enhance the pigmentation, but both were surprised by the results and found co-suppression. To better understand what was done and what happened, here are a few notes on petal pigmentation. An early key compound in this pigmentation is chalcone. It is the condensation product of one 4-coumaryol-CoA molecule and three molecules of malonyl-CoA that is performed by the enzyme chalcone synthase (CHS). In further metabolic steps the colorless dihydroflavonol is produced. The latter is processed further by the enzyme dihydroflavonol reductase (DFR), leading finally to the pigmented anthocyanin glycosides. At least in petunia CHS is rate-limiting for pigmentation. Thus, the Jorgensen Laboratory (Napoli *et al.*, 1990) focused on CHS. They attempted to overexpress this enzyme. A chimeric gene was constructed with the 35S (CaMV) promoter and the coding sequence for CHS. The chimeric gene was introduced into the genomes of several petunia varieties by *Agrobacterium*-mediated transformation. This resulted in many transgenic petunia plants with white flowers or with flowers that had patterns of white and pigmented petals (Fig. 1). The white flowers had a 50-fold lower mRNA for CHS than wild flowers. Some petals reverted to full pigmentation and the reverted petals had the normal level of CHS mRNA from both the endogenous gene and the transgene. Hence, the introduction of a transgene caused the co-suppression of the endogenous gene. The mechanism of this co-suppression was "unclear" to the authors but they suggested the possible involvement of methylation. The possibility of an involvement of dsRNA was not yet raised. The study of the Mol laboratory (van der Krol *et al.*, 1990) was rather similar to that of Napoli *et al.* (1990). The former investigators had previously shown that introducing the antisense coding-sequence for CHS into petunia plants reduced the expression of this enzyme and consequently reduced petal pigmentation, while the introduction of the code for DFR from maize caused novel petal pigmentation in the transgenic petunia petals. They therefore inferred that introducing the sense codes for CHS and/or DFR will enhance floral pigmentation. They were surprised to find that frequently a reduction of pigmentation was the result of such transformations. The reduction in pigmentation was correlated with a reduction of the steady states of mRNAs for CHS and/or DFR, respectively. The authors did not reveal the exact

Fig. 1. A petunia flower from the transgenic plant into which a chalcone synthase (*CHS*) transgene was added to supplement the endogenous *CHS* gene. The additional *CHS* gene suppressed pigmentation rather than enhancing it. (From Napoli *et al.*, 1990.)

mechanism of these gene silencings but suggested that it "may operate in naturally occurring regulatory circuits".

While Napoli *et al.* (1990) and van der Krol *et al.* (1990) furnished a colorful (or colorless) demonstration of co-suppression, they were not the first to report gene-suppression by a transgene. These authors were preceded by the Austrian team of M.A. Matzke and A.J. Matzke (Matzke *et al.*, 1989) working in Salzburg. The latter authors transformed tobacco plants. They tested the effects of two sequential genetic transformations with T-DNAs that shared some DNA sequences but differed in the coding sequences of the genes. One plasmid (T-DNA-I) contained the code for the kanamycin-resistance gene (for the selection of transformants) and the code for nopaline synthase as reporter gene. The other plasmid (T-DNA-II) contained the code for hygromycin-resistance and the code for octopine synthase as reporter gene. When tobacco plants were sequentially transformed by T-DNA-I and then by T-DNA-II, the T-DNA-I was silenced in many cases. In other words, when both T-DNA-I and T-DNA-II were introduced into the tobacco genome there was a frequent silencing of the expression of the nopaline synthase gene so that octopine but not nopaline was produced in these transgenic plants. The authors did not perform quantitative analyses of the mRNAs of the respective transgenes. On the other hand, they did analyze the methylation of the promoters. They found a correlation between the silencing of the T-DNA-I genes by the T-DNA-II and the methylation of the T-DNA-I promoters. They were surprised to reveal that when, after selfing or outcrossing, the T-DNA-II was removed from the sexual progeny, the suppression of the T-DNA-I genes as well as the methylation of their promoters — was partially or fully removed.

Rearrangement induced premeiotically in hyphal fungi

In parallel to the reports of gene silencing in plants by transgenes, a disturbance of stable genetic inheritance was reported in hyphal fungi (Selker *et al.*, 1987; Selker and Garret, 1988; Cambareri *et al.*, 1989; Goyon and Faugeron, 1989). The authors termed this process as RIP for <u>r</u>earrangement <u>i</u>nduced <u>p</u>remeiotically. It was revealed in *Neurospora crassa* when a sequence of DNA was inserted into a fungal genome that already contained this sequence. This caused a frequent rearrangement in the DNA sequence that was endogeneous in the host fungus as well as in the inserted DNA. The rearrangement happened during the dinucleotide stage that preceded meiosis, between fertilization and nuclear fusion. The cytosine residues in the rearranged DNA fragment became methylated. RIP can cause gene inactivation in the affected DNA sequences, and thus lead to frequent mutations. While apparently related to gene silencing, we shall not detail the RIP process in this book.

Reduction of viral pathogenicity in plants

I mentioned that in the *pre-history*, it was observed that infection of plants with a mild pathogen (especially a mild virus) appeared to cause resistance to a subsequent infection. These observations tempted several investigators to perform experiments in order to provide more information on such induced resistances. During the 1980s plant genetic transformation by engineered *Agrobacterium* plasmids became an efficient procedure (see: Breiman and Galun, 1997). Hence, R.N. Beachy and others performed genetic transformations in which the *Agrobacterium* plasmids contained the cDNA of genes from viral pathogens and evaluated the resistance of the transformed plants to the respective viruses (see: Beachy *et al.*, 1990, for review). While such transformations by Beachy and others were successful horticulturally, the mechanism of the induced resistance was enigmatic. Moreover, during these and subsequent years, the induced viral resistance in plants was the subject of many investigations. Among these, studies at opposite locations of the globe are especially noteworthy: studies by David C. Baulcombe and associates in Norwich, England, and by Peter M. Waterhouse and associates in Canberra, Australia. The early work of Baulcombe and associates (e.g. Baulcombe *et al.*, 1986; Harrison *et al.*, 1987) was on the reduction of viral symptoms by the introduction (by genetic transformation) of viral components (e.g. satellite of Cucumber Mosaic Virus) into the plant genomes. Indeed, Baulcombe *et al.* (1986) concluded that: "... plants infected with CMV cultures containing a benign satellite RNA showed milder symptoms than plants infected with satellite-free cultures (which) suggest that satellite RNA might be used to protect crop plants against effects of CMV infection". While such a procedure could be risky, the introduction of the code for a satellite CMV into the plants genome (by transformation) may furnish a practical solution. A subsequent study (Harrison *et al.*, 1987) indicated that the protection by satellite codes, in a given plant, against viral symptoms is not restricted to CMV but can be shown to operate also with tomato aspermy virus. By retrospect we should note that in the plasmid used for the *Agrobacterium* mediated transformation, the satellite codes were repeated and the resulting transcripts could form dsRNA sequences. As we shall repeatedly indicate below, dsRNA is a common start of the RNA silencing process.

Further studies on induced resistance against plants pathogens (mainly viral pathogens) led to the awareness of RNA involvement in this resistance and related it to RNA silencing. Thus, a special chapter shall be devoted in this book to induced viral resistance.

Early studies on silencing of transposable elements in plants

Transposable elements of eukaryotes transpose either directly or are retrotransposons that are first transcribed into RNA species, and the latter are retro-transcribed and inserted into a new location of the genome. Each of these two major groups of TEs can be sub-divided further as detailed in Galun (2003). The various kinds of TEs may comprise a major component of eukaryotic genomes. In the plant species *Arabidopsis thaliana*, which has a very small ("dense") genome, about 14 per cent of it are TEs and in the large genome of maize up to 80 per cent is composed of TEs. Also, in many animals the TEs occupy a major part of the genomes; about 40 per cent of the human genome is estimated to be composed of TEs. Active (mobile) TEs can be very harmful. Insertion of TEs into vital genes will destroy these genes, leading to lethality. On the other hand, rare transpositions of TEs could have selective advantages. Investigators therefore looked into mechanisms that will keep TEs at bay. Actually, the term TEs for this major component of eukaryotic genomes is misleading: almost all of these TEs are not mobile. Transpositions of TEs are commonly known rare events. The lack of mobility has three major reasons. Some TEs are only remnants of full size and functional TEs. They were included in the definition of TEs because they have nucleotide sequences that are identical or very similar to active TEs but otherwise are highly mutilated and indefinitely lost the transposition capability. Another group of TEs is deficient in the codes for enzymes that are required for transposition. If the enzymes are furnished in *trans*, in the same genomes, they may regain the transposition capability. These are non-autonomous TEs. A third group of TEs is potentially autonomous with respect to transposition but their transposition capability is *silenced*.

There are various possible mechanisms for the silencing of TEs. One obvious possibility is the insertion into a heterochromatic region of the genome as close to the centromere. Even when the TEs are inserted outside of the heterochromatin the latter can spread and "cover" the TEs. Soon after the elucidation of the molecular structures of some maize TEs (e.g. *Spm* and *MuDR*) it became evident that changes in the activities of these TEs were correlated with DNA-methylation of specific sequences in these TEs (e.g. Chandler and Walbot, 1986; Fedoroff, 1995). Indirect evidence for the role of DNA methylation in TEs silencing came from studies with mutated genes that normally code for enzymes that have a major role in such methylations (e.g. the *DDM1* gene of *Arabidopsis*). Under *ddm1* silenced TEs became hypomethylated and mobile. This subject was reviewed by Okamoto and Hirochika (2001). There is an accumulation of evidence for a linkage between DNA methylation and chromatin remodeling (e.g. methylation of lysine 9 in the tail of histone 3). The methylation of H3K9 is merely one of several other multistep processes that involve several modulating proteins, causing the inhibition of transcription from the affected chromatin (see Appendix II in Galun, 2003). Now comes the role of RNA. A relatively early study (Wassenegger *et al.*, 1994) indicated that there can be RNA-directed *de novo* methylation of genomic DNA in plants. When the cDNA of potato spindle tuber viroid (PSTVd) was introduced into transgenic tobacco plants, the PSTVd-specific DNA sequences were methylated. The methylation was site-specific and dependent on viroid RNA-RNA replication. It is plausible that specific RNA fragments serve as mediators to guide chromatin-modulations to specific locations by homologous binding to DNA sequences in the respective chromatin region.

Evidence for a more defined role of RNA in silencing TEs came from the laboratory of H.A. Plasterk in Amsterdam (Ketting *et al.*, 1999). These investigators looked into the situation of the nematode *C. elegans* that harbors the TEs *Tc1*. Typically, *Tc1* is active in the soma of *C. elegans* isolates but is silenced in the germline. An induced mutation (*mut-7*) also caused the activation of *Tc1* in the germline and worms with this mutation were also impaired in RNA interference (RNAi). The authors suggested that the wild-type gene (*MUT-7*) guides the *Tc-1* derived dsRNA to destroy the mRNA of the transposase of *Tc1*, thus silencing this TE.

A reader of this book may sense that it is skewed in favor of plants (e.g. angiosperms) over animals. This reader is probably right. The apparent inbalance has two reasons (one may claim: this is one reason too much). The first is that my research career was in plant biology.

The second reason is that indeed studies with plants led, historically, to the first clues of RNA mediated gene silencing. Presently, there is far more knowledge on RNA silencing in animals than in plants. No wonder, research activity in various animal systems is by far greater than in plants.

 Defence against Plant Pathogens: Viral Resistance and RNA Silencing in Plants

The "birth" of vaccination for protection against viral pathogens of man was narrated in a previous book (Galun and Galun, 2001). In the second half of the 18th century, milkmaids informed Edward Jenner that after being infected by the mild cowpox disease, they got protection against the deadly smallpox disease. Systematic vaccination to prevent smallpox started at the very end of the 18th century, and during the 19th century vaccinations against additional diseases were put to practice. The scientific elucidation of the immune system in mammals occurred during the 20th century. The phenomenon of induced "immunity" against plant viruses was recognized much later but its elucidation, while still not complete, took much lesser time. The agricultural/horticultural practice of inoculating crop plants with mild strains of viruses or viroids to prevent more virulent strains from causing severe plant diseases began many years ago; as indicated above; it was already practised in "pre-historic" times, namely, before this practice was orderly recorded. For more details, see the article by Abel *et al.* (1986).

The silencing of genes by homologous or homeologous nucleotide sequences can be regarded from three viewpoints. First, there is the silencing of invading pathogens (i.e. viruses). This process is considered favorable to those interested in eliminating virus diseases from crop plants. Then, there is the co-suppression that frequently follows the insertion of a transgene into transgenic plants. This phenomenon is obviously annoying to those interested to express a specific transgene

17

in these plants. The third viewpoint regards homologous silencing as neither good nor bad but rather interesting. The latter investigators are merely interested in understanding this phenomenon with the hope that understanding the reasons of this silencing will lead to additional biological knowledge. It is mainly with this last point of view that Matzke and Matzke (1995) reviewed the "homology-dependent gene silencing" in plants — until the end of 1994. We shall not summarize this review because it is full of hypotheses that were reasonable at that time but not all of them were substantiated by later studies.

In this chapter the emphasis will be on the defence of plants against viruses and on the involvement of small RNA sequences in this defence. Because historically many studies on this topic were performed before the awareness of RNA silencing in animals, this chapter precedes chapters on animals. We shall return to plants and handle additional aspects of RNA silencing in angiosperm plants in Chaps. 11 and 12.

Soon after methods for genetic transformation of plants became available in the 1980s, investigators started to introduce viral components into transgenic plants in order to reduce the sensitivity of these plants to viruses. We shall note the early work of Robert Fraley and Roger Beachy in Missouri, USA and of Baulcombe and associates in the UK. As for Baulcombe's research, it took place when he *changed lanes*, from Maris *Lane*, Trumpington (near Cambridge) to Colney *Lane*, near Norwich (e.g. Abel *et al.*, 1986 and Harrison *et al.*, 1987, respectively). The investigators on both sides of the Atlantic ocean (i.e. Beachy and associates and Baulcombe and associates, respectively) continued for many years to refine their procedures, yielding impressive results with resistance to viral infection in plants. The studies of Baulcombe started with the suppression of plant viruses and then proceeded into post transcriptional gene silencing (PTGS) and RNA silencing. We shall describe these studies in some detail, although there were contributions by other investigators who also showed that the introduction of virus-genome components into plants can cause resistance to the respective virus. One of these investigators was Milton Zaitlin of the Cornell University, NY. Zaitlin and associates (Golemboski *et al.*, 1990) who showed that when a cDNA of a part of the genome of tobacco mosaic virus (TMV) strain U1 was introduced, by *Agrobacterium*-mediated genetic-transformation, into tobacco plants, these plants became resistant to this TMV U1 strain. The resistance was manifested against either virions or the respective RNA.

Before going into details about PTGS and resistance to viral infection we have to consider a major issue concerning two different modes of gene silencing. One is PTGS, in which transcription proceeds normally (or almost normally) but the protein level that should result from a given transcript is strongly reduced or completely prevented. In such cases nuclear run-on reveals normal transcript production but a strong reduction (or elimination) of mRNA accumulation and the respective protein is lacking. The other phenomenon is termed transcriptional gene silencing (TGS). In the latter case there is a strong reduction (or elimination) of mRNA production. We shall not deal in detail with TGS but note that TGS was related, in many cases, with methylation of the promoter of the affected gene and/or remodeling of the chromatin region in which the affected gene resides. A relatively early example for an RNA-directed methylation of a viroid (PSTV) gene that was integrated in the tobacco genome, was presented by the Heinz Säger Laboratory of the MPI, Martinsreid (Wassenegger *et al.*, 1994).

Studies in Colney Lane: From Resistance to Viruses to Post Transcriptional Gene Silencing (PTGS) in Plants

After it became evident that infection with viral genomes as well as the integration of transgenes encoding structural components of plant viruses can impose resistance to viruses, the laboratory of Baulcombe focused on a viral gene that encodes an essential enzyme: the replicase of potato virus X (PVX). They produced several mutations of this replicase and tested their capability to induce resistance to PVX. Viruses with such mutations are non-infectious in *Nicotiana* plants or protoplasts. But when the respective mutated codes were introduced (by genetic transformation) into *Nicotiana* plants, some of these codes caused high resistance against the wild-type virus. Infection of the latter transgenic plants with PVX resulted in a very low level of viral RNA in the infected leaf cells and no viral RNA in the systemic leaves (Longstaff *et al.*, 1993). Similar results were reported by de Haan *et al.* (1992) as well as by William Dougherty and associates of Corvallis,

Oregon for the tobacco etch virus (Lindbo *et al.*, 1993) and potato virus Y (Smith *et al.*, 1994). In these cases it was suggested that the resistance to virus was correlated with posttranscriptional suppression of transgene mRNA accumulation.

In a further study by the Baulcombe Laboratory (Mueller *et al.*, 1995) the investigators crossed two transgenic tobacco plants; both contained the coding sequence of the viral RNA polymerase (RdRP) gene. But one plant was resistant and the other was susceptible to PVX. The two parents were either very low or normal, respectively, in the content of the RdRP mRNA. The hybrid plants were resistant to PVX and had low RdRP mRNA. This strongly suggested that gene silencing and virus resistance involve the same mechanism.

In a review of additional studies of their own published studies (English *et al.*, 1996) and their work which was then not published yet (i.e. studies reported later in Angell and Baulcombe, 1997; and in Ratcliff *et al.*, 1997) as well as of studies by others on resistance to plant viruses and gene silencing, Baulcombe and English (1996) came up with several conclusions and suggestions that can now be regarded as prophetic. Take, for example, they concluded that the virus resistance imposed by homologous transgenes should be termed "homology-dependent resistance" to emphasize the similarity of this process to homology-dependent (other) gene silencing. They also concluded that in both these similar processes there should be mRNA degradation, and that degradation should be sequence-specific. The authors came even closer to what was later found about PTGS and RNAi, by identifying small RNA fragments that accumulated after gene silencing. On the other hand, some of the hypotheses had less rigid foundations. The latter hypotheses concern the idea of mRNA threshold beyond which a silencing mechanism is initiated. This "threshold hypothesis" was refuted by Baulcombe and English on the basis of logical arguments. There was also a notion of a role for transcripts that are abberant and that these abberant RNAs may move to the cytoplasm where they lead to sequence specific mRNA degradation. Experimental evidence for a role of an abberant RNA (aRNA) in gene silencing and viral resistance was lacking in plants but clues for its existence came from the fungus *Ascobulus immersus* where transgene methylation can influence the formation of aRNA. There is another aspect that was emphasized by the Baulcombe reviews (e.g. Baulcombe, 1999; Baulcombe and English, 1996): viruses might gradually trigger the activation of gene silencing. In tobacco plants that contain transgenes based on the cDNA of some potyviruses there is a form of potential homology-dependent resistance. This resistance (and only in the upper leaves) was activated several weeks after the plants were infected with the virus. The plants first showed symptoms of viral infection, but in the upper leaves that develop after infection, the symptoms were fading and then disappeared. These symptomless leaves were resistant to secondary infection. The resistance was always limited to viruses with homology to the transgene.

Marano and Baulombe (1998) explored further the "homology-dependent resistance" to plant viruses. They intended to verify the requirement of high homology between the transgene and the virus as well as the role of RNA, rather than protein, in the induced resistance. They found that even a short fragment (383 nt) of the code for the 54 kDa of the TMV-U1 virus, when introduced (by genetic transformation) into tobacco plants, gave protection against infection with this virus. But the plants were not resistant against another strain (a crucifer strain) of TMV. Obviously, the 383 nt transgene had no protein product. These results therefore did not support the protein-mediated virus-resistance hypothesis that was suggested by some investigators, but they also did not refute this hypothesis completely. Moreover, the target of the transgene-induced resistance was the very same 54K RNA from which the short transgene was derived. This led, already in June 1997 (when the Marano and Baulcombe paper was submitted), to a suggestion that dsRNA was an intermediate of the gene silencing. Such dsRNA could be formed, according to these authors, when a 54K sense RNA (or a fraction of it) will anneal to complementary RNA in the viral replication intermediate. Because the virus resistance by the transgene is orientation-dependent such a possibility was plausible. Hence, the role of dsRNA in gene silencing was suggested (and submitted for publication) by the Baulcombe team on the basis of rational thinking, a few months before experimental evidence for this role became available (Fire *et al.*, 1998).

Further Studies on Homology-Dependent Gene Silencing in Plants

In parallel to the Baulcombe team that initially intended to understand the induced virus resistance in plants, there was, in the same location (the John Innes Centre, Colney Lane, Norwich) the team of Richard B. Flavell. The latter team was engaged to solve other riddles of gene silencing in plants. In one of their studies they focused on silencing the expression of the chalcone gene in petunia by a transgene (Metzlaff *et al.*, 1997). The expression of this gene was mentioned in the Introduction as an early case of co-suppression. The Flavell team arrived at the conclusion that PTGS goes through a step of RNA-RNA (i.e. dsRNA) hybridization and that this dsRNA is a key component in the specific cleavage of the chalcone mRNA by an endonuclease of the type of the *E. coli* dsRNA-specific RNase III. Notably, neither the Flavell team nor the Baulcombe team put their dsRNA assumption to test by introducing into plant cells, specific dsRNA fragments (i.e. fragments with homology to sequences in the chalcone mRNA). Most probably, this would have been involved with technical hurdles. Here, the advantages of the tiny worm *C. elegans* over flowering plants become eminent.

By the end of 1997 and early in 1998 several research groups concluded that virus-induced gene-silencing (VIGS) and other cases of homology-dependent silencings in plants are inter-connected and that VIGS has evolved in plants as a mean of defence against pathogenic viruses. This assumption emerged from the investigations of the Baulcombe team as well as from studies of other investigators (see: Depicker and Van Montagu, 1997). Hence, even before the mechanism of VIGS was fully understood it was asked how the VIGS could be utilized for two purposes: to protect plants from viral infection and to increase knowledge on molecular genetics of plants. To further explore the VIGS/homology-dependent gene silencing the team of Ruis, Voinnet and Baulcombe (1998b) devised a somewhat complicated experimental system. This team utilized the possibility to engineer the PVX genome so that it will also contain a transgene. Two such transgenes were chosen. In one case the cDNA for phytoene desaturase (PDS) was included in the PVX vector. PDS is essential for carotenoid biosynthesis and a plant in which PDS is silenced will be susceptible to photobleaching. Actually, a central sequence of the PDS cDNA of 415 nt was inserted into the PVX vector in either the sense (PDS) or the antisense (SDP) direction. In addition, the cDNA of the green-fluorescent-protein (GFP) reporter gene was also inserted into another PVX vector. The resulting DNA sequences were transcribed (*in vitro*) into the respective RNAs and these RNAs were used to infect *Nicotiana benthamiana* plants. Such transcribed RNAs were commonly used to infect *N. benthamiana* plants with PVX virus. It was found that the PVX vector with the PDS code caused the silencing of the endogenous PDS, rendering the respective plants sensitive to bleaching. The bleaching sensitivity spread into the systemic leaves (i.e. leaves that were not directly infected) after about 12 days and bleaching sensitivity extended subsequently throughout the plants. There was no difference whether the PDS sequence was engineered into the PVX in the sense (PDS) or in the antisense (SDP) direction. Infection of the plants with a GFP containing PVX vector did not cause silencing of the endogenous PDS. Further analyses showed that only exon sequences (but not intron sequences) of the PDS caused silencing, indicating that the silencing of the PDS mRNA took place in the cytoplasm. On the other hand, several exon regions were effective in silencing. Bleaching sensitivity was correlated with the reduction of endogenous PDS mRNA. Interestingly, the PDS containing PVX vectors did not cause silencing of itself: the mRNAs levels of this vectors stayed high in the infected plants. The investigators had also other *N. benthamiana* plants that expressed the GFP transgene. Such GFP plants were infected with either of two modified PVX RNAs. In one there was a full-length GFP code and in the other a truncated (GF) code. When the vector with the full length code was used, the "endogenous" fluorescent was superimposed by the GFP of the infecting vector. But when the GF containing vector was used the infected plants initially showed only the "endogenous" fluorescence (as expected). But at later stages the fluorescence was silenced. Both GFP and GF containing vectors caused this later silencing. After some time there were additional changes. Silencing was not apparent in floral and vegetative apexes. Much later in plant development (i.e. 41 days after infection) the silencing subsided in the growing points. It appears as if a "war" was going on between the silencing and the re-expression of the GFP gene in the plants which had the GFP transgene. Further experimentation showed that the VIGS of the GFP was targeted against the PVX-GFP: initially, the VIGS of the GFP was initiated in all the green tissues of the infected plants, but after about 30 days, the GFP was no longer expressed in newly emerging leaves but was maintained in the older leaves where it was initiated previously. This indicated to the authors that the *initiation* of VIGS requires the presence of the virus but the maintenance of it is virus independent. Still, how exactly this happens was not clear.

In another publication (Voinnet and Baulcombe, 1997) the authors reported on one experiment in which the virus was left out of the picture. The authors used a *N. benthamiana* plant that contained a GFP transgene and infiltrated its leaves with an *Agrobacterium* that carried the GFP reporter gene. The superimposed GFP caused a strong fluorescence (in addition to the weaker fluorescence expressed in the plants before infiltration). But then, a zone appeared in the leaves in which the GFP gene was silenced. Subsequently, the silencing spread to the upper leaves. This was already a clear indication that silencing can spread in the plant without the involvement of a virus. As it became clear that the silencing signal can move in the *Nicotiana* plants, the investigators (Voinnet and Baulcombe, 1997) concluded that this signal should be nucleic acid but the moving signal was not yet characterized.

More evidence on the ability of the PTGS signal to move in plants came from studies by the team of H. Vaucheret of INRA, Versaille, France, who found that this signal can cross a graft-connection between a stock and a scions (Palauqui *et al.*, 1997). Another study by the Norwich team (Voinnet *et al.*, 1998) provided additional support for the specificity of the silencing and for the movement of the PTGS signal, showing long-range movement of the PTGS signal and indicating that the long-range movement is through the phloem and that there is intercellular movement through plasmodesmata. The signal production can be amplified during movement. The speed of the signal movement was estimated to be similar to the speed of viral movement. Moreover, the pattern of signal movement (i.e. first to leaves on the same side of the stem as the infected leaves) was also similar to virus movement patterns. There was additional similarity between the

PTGS signal and viral movement: both usually do not enter actively dividing cells such as in apical meristems. On the other hand, these authors (Voinnet *et al.*, 1998) could not yet decide if the silencing required homologous DNA/RNA or RNA/RNA interaction. Actually, they preferred the DNA/RNA interaction. The study by Fire *et al.* (1998) that was published in 1998 and that pointed to dsRNA as the initial step of silencing did not change the tendency of Voinnet *et al.* (1998) to adopt the DNA/RNA model. However, the latter authors repeated and detailed the idea that PTGS is a mechanism that evolved in plants in defence against viral infection. The PTGS may move ahead of the virus systemic movement and repress the virus in the upper leaves. The authors even compared the PTGS-signal movement to the movement of the flowering signal (Florigen) that would move across grafts. The problem is, there is no rigorous proof for the existence of Florigen — only its effect was observed while its chemical identity was never defined. We are faced with the persistence of the grin of the Cheshire cat in the absence of a cat.

Some plant viruses developed means to fight back to overcome the PTGS of plants. This was documented by a Norwich-Singapore collaboration (Brigneti *et al.*, 1998). These investigators provided evidence that the potato virus Y (PVY, a potyvirus) and the cucumber mosaic virus (CMV) encode virus-suppressor proteins HCPro (or HC-Pro) and 2b, respectively. The HCPro blocks the maintenance of PTGS in tissues where silencing had already been set. The 2b protein prevents the initiation of silencing at the growing point of the infected plants. The PVX does not encode such suppressor proteins — this can be considered lucky for the Norwich investigators who made ample use of PVX in their earlier studies. Had they used PVY as vector their results would have been very different. An indication that some plant viruses (e.g. the potyvirus PVY) developed a means to overcome the natural resistance of plants against viruses was already reported by Vicki Vance of Columbia, South Carolina (1991). Vance and collaborators subsequently elaborated the studies of a specific suppressor encoded in potyviruses (HC-Pro). Thus, for example, it was found (*Pruss et al.*, 1997) that coinfection of plants with PVX and a potyvirus evolves a synergism: the pathogenicity of the PVX is then much more severe than if PVX alone is infecting the plant. This increased pathogenicity was attributed to the HC-Pro that is encoded in the potyvirus. The story actually got more complicated. The Vance Laboratory identified a calmodulin-related protein (rgs-CaM) that interacts with HC-Pro but can also suppress PTGS by itself (Anandalakshmi *et al.*, 2000). The rgs-CaM protein is encoded by the plant's genome. Thus, here we have a *Quisling* situation: a plant gene that joins the virus-enemy to break the plants defence against viral infection. Tirans have no inhibitions while invading foreign territories. With this respect it does not matter whether the tiran is a highly educated human being (as Alexander the Great, who was educated by Aristotle), or the simplest reproducing entity — a virus. Additional studies by the Baulcombe Laboratory with several plant viruses provided evidence that PTGS of nuclear genes is a manifestation of a natural defence mechanism that is induced by a wide range of plant viruses (Ratcliff *et al.*, 1999). Furthermore, a survey of several RNA and DNA plant viruses (Voinnet *et al.*, 1999) revealed several PTGS suppressor-proteins that are encoded in these viruses. The various suppressors are chemically and functionally different, indicating that these counter-defense proteins evolved independently during the virus-plant wars. In a further study, the Norwich Laboratory (Hamilton and Baulcombe, 1999) looked at the nucleotides that are produced during transgene-induced PTGS and virus-induced PTGS. In both of these systems short antisense RNAs were found with a length of about 25 nt. The experimental results suggested that an RNA-dependent RNA polymerase is involved in these PTGS phenomena. In further studies on PTGS and on the requirement of an RNA-dependent RNA polymerase (RdRP), the Norwich investigators (Dalmay *et al.*, 2000a, 2000b) found that RdRP is required for both transgene PTGS and virus-induced PTGS. In both cases the RdRP is required for the synthesis of a dsRNA initiator of PTGS. In VIGS the virus supplies this enzymatic activity, while in transgene induced PTGS a plant gene (in *Arabidopsis* — *SDE1*) encodes the respective enzyme. The *SDE1* is not only the plant-encoded gene that is essential for transgene-induced PTGS but also for virus-induced silencing. There is also an RNA helicase that is encoded in the *SDE3* gene (found also in *Arabidopsis*) that is redundant in virus-induced silencing but essential for transgene-induced PTGS (Dalmay *et al.*, 2001). The laboratory of H. Vaucheret (Mourrain *et al.*, 2000) also contributed information on this topic. They revealed two *Arabidopsis* genes that are essential for PTGS: *SGS2* and *SGS3*. Of these *SGS2* is probably a RdRP. The *Arabidopsis* mutants *sgs2* and *sgs3* are both defective in PTGS. The "war" between viruses and plants is even more complicated than is described till now. From the above information a race between the virus and the PTGS takes place in virus-infected plants. If the PTGS signal is quickly replicated (e.g. by the *SDE1* gene) and moves ahead of the movement of the virus, the upper (systemic) leaves will be protected against the later-arriving virus. But if the replication/movement of the PTGS is inhibited — the virus will have the upper hand. Indeed, Voinnet, Lederer and Baulcombe (2000) found that the virus-encoded movement protein (that helps the virus to move into the plasmodesmata of the plant cells) interferes with the spread of the PTGS signal in *N. benthamiana*. Additional information on the PTGS was revealed by Vaistij, Jones and Baulcombe (2002). First, when a transgene that contained only a part of a target gene was used, the silencing also affected 3′ and 5′ regions, downstream and upstream of the target, respectively — nucleotide sequences that did not exist in the transgene. Also, there was a DNA — methylation of the target gene. These effects were dependent on the transcription of the transgene. A recent review from the laboratory of Vicki Vance at the University of South Carolina (Roth *et al.*, 2004) surveyed the plant viral suppressors of RNA silencing and lists 16 such suppressors.

Three reviews from the Norwich (John Innes Centre) Laboratory described a gradually evolving picture of PTGS, virus-induced gene silencing (VIGS) and the possible utilization of these mechanisms for plant protection and genetic investigations: (1) Baulcombe (1999); (2) Voinnet (2001); and (3) Lu *et al.* (2003). As we shall see in subsequent chapters, RNA silencing was revealed in fungi in parallel to the silencing in plants and was also found later in animals. The review of Voinnet (2001) tells the story of the coevolution of knowledge on RNA silencing in plants and animals. The review of Lu *et al.* (2003) focuses on practical and theoretical issues of utilizing the VIGS in studies intended to identify specific genes as those required for disease resistance in plants. This latter review details five specific protocols. Their approach is illustrated in Figs. 2 and 3.

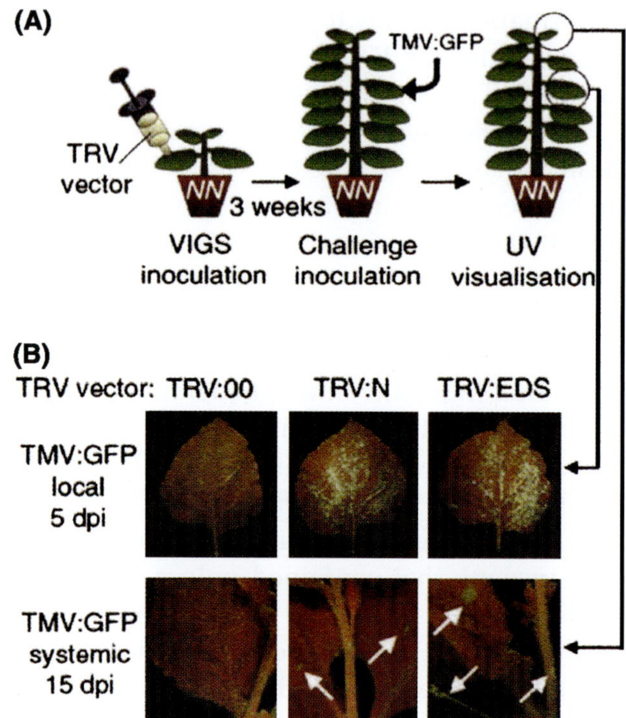

Fig. 2. Virus induced gene silencing (VIGS) of the *N. benthamiana* gene *EDS1* (*NbEDSI*) compromises *N*-mediated resistance to tobacco mosaic virus (TMV). (A) Schematic representation of the VIGS procedure to test the requirement of *NbEDSI* for *N*-mediated resistance. (B) TRV:00, TRV:N or TRV:EDS plants were challenged — inoculated with TMV:GFP sap and TMV:GFP was monitored by GFP fluorescence; white arrows indicate foci of GFP. GFP-Green Fluorescent Protein. (From Lu *et al.*, 2003.)

A Widespread Interest in Virus-Defence and Gene Silencing in Plants

The interest in virus defence and gene silencing in plants was not confined to the Norwich Laboratory. This interest resulted in ample publications, especially since 1997. The publications came from various locations such as France, Germany, Austria, Australia, Singapore and the United States.

A team from various laboratories in North Carolina (Tanzer *et al.*, 1997) returned in a way to the approach taken previously by Beachy and associates (see: Beachy *et al.*, 1990). The investigators used

Fig. 3.   Selected VIGS (virus induced gene silencing) phenotypes. From a VIGS survey of 5000 different *N. benthamiana* cDNAs approximately 15 per cent produced pronounced symptoms resulting in suppression of plant growth or development. These images illustrate three of the symptom types due to silencing of (A) ubiquitin, (B) magnesium chelate or (C) an unknown gene. (From Lu *et al.*, 2003.)

transgenic tobacco plants that expressed the tobacco etch virus (TEV) coat protein (CP). Such plants had resistance to TEV and after TEV infection the level of the CP mRNA was reduced. Two types of reactions were revealed. Some of the plants were immune to TEV while in other plants there were initially TEV symptoms but then there was a gradual recovery. An exact and correct mechanism of PTGS and/or VIGS was not derived from this study.

Wassenegger and Pelissier (1998) of the Max Planck Institute in Martinsried, Germany, reviewed the literature on TGS, PTGS and VIGS, and suggested a general model for these mechanisms. This model had some original and interesting aspects. Take, for example, they insisted that methylation of specific DNA sequences that leads to TGS is not necessarily a result of DNA-DNA interaction (as suggested previously) but may well be the result of RNA-guided DNA methylation. This specific mechanism was a plausible suggestion that was supported by later investigations. These authors also rightfully indicated that RdRP is involved in the silencing processes. Other components of the silencing model attributed a major role to antisense mRNA and to the formation of DNA/RNA hybrids. These latter components were not verified in later studies as the main features of silencing. The review of Wassenegger and Pelissier was published in 1998 although submitted in September 1997 (and resubmitted after revision in January 1998). This was just a few weeks before the publication of the Fire *et al.*'s (1998) paper (on silencing in *C. elegans*) that led to a correct model of silencing. Dates of submission and publication can be interesting especially in fields of very active and competitive research. We shall meet this issue frequently in future chapters. In May 1998 Waterhouse and associates (Waterhouse *et al.*, 1998) submitted their manuscript on virus resistance, and gene silencing by sense and antisense RNA to the *Proceedings of the National Academy of Science* (*USA*). This happened only a few months after the publication of the Fire *et al.*'s (1998) in *Nature*, where it was shown that the dsRNA, composed of the sense and antisense of a mRNA, is the actual mediator of silencing (RNAi). It is fair to assume that Waterhouse and associates started with their research much before they knew about Fire *et al.*'s publication. Thus, it is interesting that Waterhouse and associates followed the same general approach. They asked whether dsRNA made of sense and antisense sequences will be more efficient in silencing than either the sense or the antisense ssRNAs. Their experimental systems were transgenic tobacco or rice plants that were obtained by transformation with vectors that contained the coding sequence of the PVY gene for a protease (Pro). The vectors were constructed in either of three ways: (1) containing only the sense sequence; (2) containing only antisense sequence; and (3) containing both the sense and antisense so that the transcripts would hybridize in the plant cells forming dsRNA or a hairpin structure in which the sense and antisense RNA would form dsRNA. They also obtained dsRNA by first expressing either the sense or the antisense in individual plants and then crossed these plants. The transgenic plants that contained the dsRNAs were by far the most resistant to virus and they transmitted this resistance to their sexual progeny. The authors thus obtained clear evidence (similar to the evidence of Fire *et al.*, 1998) for the decisive role of dsRNA in PTGS and VIGS. While Whitehouse and associates also suggested that dsRNA is instrumental in recruiting an RNase to cut the target mRNA, the details of their suggestion were not verified.

Even experienced investigators are frequently surprised by the many "layers" of sophistication that exist in "brainless" organisms. They should not be surprised because these are wisdom that accumulated during billions of years. The "wars" between hosts and parasites are good examples for such a sophistication. Seven investigators in Singapore (Li *et al.*, 1999) analyzed such a "war". They called it: "defence, counter-defence and counter counter-defence" (the British spelling of <u>defence</u> is the norm in Singapore). In order to understand the experiments of Li *et al.* (1999) we should look at the background. I had already mentioned that at this time, two genes were known in viruses that counteracted the VIGS of plants: HCPro and 2b. These genes were found in PVY and CMV, respectively. When these genes are expressed and the respective proteins synthesized, the defence mechanism of plants against viral infection is suppressed. There is a protein that is similar to the 2b protein, termed tomato aspermy cucumovirus 2b (Tav2b). The code for Tav2b can be introduced into the TMV genome. When such a chimeric TMV (with the Tav2b code) is used to infect tobacco plants it activates a very high resistance in the plants. This resistance is similar to the well known gene-for-gene interaction that is common in many host-pathogen systems: a hypersensitive response (HR) which results when a host that carries an *R* resistance gene is infected by a pathogen with a matching *avirulence* (*Avr*) gene. This nomenclature can be confusing because in this case the *Avr* causes uninhibited infection. Anyway, the HR is a severe attack by the pathogen that quickly kills the plant cells at the infected site, causing a necrotic lesion. Because of the dead host cells, the pathogen does not move away from the infection site and the end result is actually resistance to the *R-Avr* infection.

While Tav2b is similar to the PTGS-suppressor 2b of CMV, the former's activity is functionally very different from 2b; it causes HR in the host. Li *et al.* (1998) introduced the code for *Tav2b* into the genome of TMV and used the chimeric TMV to infect tobacco plants. This resulted in a HR, meaning that actually the infected tobacco plants became resistant because the TMV did not spread in the infected plants. Furthermore, Li and associates found that *Tav2b* differs very little in its sequence from the 2*b* of CMV; there were two coding changes: lysine 21 was converted to valine and arginine 28 was converted to serine. Also, while in tobacco (*N. tabacum*) the Tav2b caused HR, it caused suppression of PTGS in *N. benthamiana*.

After the VIGS phenomenon was clarified, at least at the level of functionality, it could be recruited for plant protection. This was performed by Wang, Abbot and Whitehouse (2000a) of the CSIRO Plant Industry in Canberra, Australia. They planned to render barley resistant against the most serious and widespread virus of cereals: barley yellow dwarf virus-PAV (BYDV-PAV). Their strategy was based on the then available knowledge that introducing into the plant, a coding sequence that will express dsRNA that is homologous to a part of the viral genome will cause VIGS. In order to obtain in the barley cells such a dsRNA, a liner sequence of RNA that will fold back into a hairpin structure could be utilized. The "stem" of such a hairpin will produce after cutting ("dicing") dsRNA. Wang *et al.* (2000a) engineered such a sequence (a potential hairpin-forming sequence from the polymerase coding sequence of BYDV-PAV) into an *Agrobacterium* vector that also contained a hygromycine-resistance gene. *In vitro* cultured barley scutella from immature embryos were infected with a suspension of vector-containing agrobacterial suspension and the regenerating barley shoots were selected on the appropriate culture medium. Transgenic plants were regenerated and could be propagated sexually. Nine out of 25 such transgenic plants that expressed the potential hairpin RNA from BYDV-PAV were resistant to BYDV-PAV. Two lines were then propagated (probably by self-pollination) and maintained this resistance in the progenies. This study also furnished additional information. There is a related virus termed cereal yellow dwarf — PAV (CYDV-PAV). When the BYDV-PAV resistant barley plants were infected with both BYDV *and* CYDV the plants showed immunity to BYDV but were susceptible to CYDV. This suggested either or both possibilities: (1) The CYDV does not contain a VIGS suppressor (i.e. of the HC-Pro kind) and (2) the CYDV has no product that protects the BYDV-PAV from the VIGS reaction of barley.

Here I would like to introduce a future issue of this book. The very same strategy used by Wang *et al.* (2000) to obtain a specific dsRNA (representing a segment of the polymerase of BYDV-PAV, namely to put two sequences in tandem but in opposite directions), was "invented" hundreds of million years ago, already by eukaryotic organisms. This is the structure of microRNA (miRNA) genes that shall be discussed later. But the miRNA genes were not known yet when Wang *et al.* (2000) were engaged in their experiments.

Another case of "war" that goes on between the viral pathogen and the plant host was described by a team of Scandinavian investigators: Eugene Savenkov from Uppsala, Sweden and Jari Valkonen of Helsinki, Finland (Savenkov and Valkonen, 2002). These investigators reviewed the existing information on VIGS in plants and had already mentioned the "movement" of a silencing complex that included an RNase guided by a unique short sequence of RNA to degrade the mRNA of the invader. The experimental work of these investigators concerned the HC-Pro of potyviruses (their designation is: HC[pro]) such as potato virus A (PVA). They used transgenic *N. benthamiana* plants that expressed HC[pro] and infected them with PVA. In several cases they found a peculiar *lanceolate leaftip* (LLT). At a certain lapse of time (ca 3 weeks) after infection in systemic leaves there was a leaftip (the older part of the blade) which showed strong PVA symptoms but the rest of the leaf had no symptoms. Later, developing leaves (e.g. leaves positioned 8 nodes and higher above the infected leaf) showed resistance (recovery) to the PVA infection. It seems that the plant found a means to "fight back". The investigators found that the HC[pro] transgene that is considered to suppress the VIGS of the infected plants may undergo DNA methylation. There was a correlation between this methylation and the ability of the leaves to recover from PVA infection and become resistant to further PVA infection.

The summer of 2001 was a good season for reviews on gene silencing and defence against viruses in plants. Three reviews on this subject appeared in June 2001: Dangl and Jones (2001), Waterhouse *et al.* (2001), and Vance and Vaucheret (2001). Two additional reviews were also published in 2001: Vaucheret *et al.* (2001), and Wang and Waterhouse (2001). Because much more information on PTGS in plants was added since these reviews I shall not summarize these five reviews but note that they included two important conclusions: the key to silencing are short sequences of dsRNA and that silencing can be mediated by two processes; the specific cutting of target mRNA and methylation of DNA that encodes the target mRNA. The latter process is actually TGS rather than PTGS and will not be detailed here. The two reviews in which Vaucheret participated in (Vance and Vaucheret, 2001 and Vaucheret *et al.*, 2001) compared the plant-PTGS and viral-defense to gene silencing in other organisms, indicating the common and unique features of this silencing in various eukaryotic organisms. This tendency to have a broader view on PTGS by Vaucheret and associates resulted in a publication that went beyond the VIGS of plants (Fagard *et al.*, 2000). These authors pointed out that several similar proteins that are involved in RdRP activity exist in very different eukaryotes: QDE-1 in *Neurospora*, SGS2 in *Arabidopsis* and EGO-1 in *C. elegans*. In all these very different organisms these similar proteins are essential for PTGS. Thus, the PTGS appears as a very early-developed system in the evolution of eukaryotes. The authors focused on another group of similar proteins that are essential for PTGS: QDE-2 in *Neurospora*, RDE-1 in *C. elegans* and AGO1 in *Arabidopsis*. The authors actually isolated AGO1 mutants and found that the sequence of the mutant differed in one amino acid from the wild type. This change was in a very conserved sequence that exists in the QDE-2 (*Neurospora*) and the RDE-1 (*C. elegans*). Moreover, the authors found that *Arabidopsis* with the AGO1 mutation displayed several developmental abnormalities. Thus, the wild-type protein is essential not only for PTGS in *Arabidopsis* but also for normal differentiation. The reason for this phenomenon will be clarified in Chap. 12 that deals with miRNAs in plants as regulators of biochemical and structural differentiation.

Because the control of viral diseases in crop plants is an important biotechnological issue and of great economical significance, the application of RNA interference for this control is an ongoing activity. Information on this activity can be obtained from two publications that were submitted by veterans in this field. One of these is from the team of Diaz-Ruiz of the CSIC in Madrid, Spain (Tenllado *et al.*, 2004) and the other from the laboratory of Vicki Vance of the University of South Carolina, USA (Roth *et al.*, 2004). The latter publication stresses the issue of viral inhibitors of RNA interference.

Chapter 3

Gene Silencing in Fungal Organisms

RIP, MIP and Quelling

In Chap. 1, I mentioned that the *rearrangement induced premeiotically* or the *repeat-induced point mutation* (RIP) was an acronym given by Eric Selker and associates of the University of Oregon in Eugene, Oregon, in 1986. It is not clear why the same acronym was given to different full names. Moreover, why use the acronym RIP at all? This acronym commonly stands for the Latin *Requiescat In Pace* and as such it is inscribed on Christian tombstones. The RIP of Selker means the opposite of "Rest in Peace". Now, back to the investigations of Selker and associates. Briefly, a duplicated nuclear sequence was observed to cause an extreme genetic instability in *Neurospora crassa* that was displayed only in the sexual life cycle of this fungus. Linked duplications, and to a lesser rate, unlinked duplications, caused DNA modifications in the duplicated sequences. These modification were *de novo* methylation and base-pair alterations (from G:C to A:T). Every duplication that is longer than 1000 base pair is potentially subject to RIP. Because RIP destroys both copies of duplicated genes it was used by *Neurospora* geneticists to inactivate endogeneous genes. This type of gene silencing is restricted to *Neurospora* and was reviewed by Selker (2002) who revealed it. The RIP system will not be detailed in this book.

Investigators looked for phenomena that are similar to the RIP of *Neurospora* in other fungal organisms and found MIP (methylation-induced premeiotically). Like RIP, MIP detects linked and unlinked sequence duplications during the period between fertilization and karyogamy (nuclear fusion) in the sexual phase of hyphal fungi. The MIP causes inactivation in these sequences in a pair-wise manner. But unlike RIP, in MIP there is only DNA methylation that causes the inactivation. MIP does not cause mutations by alteration of bases in the affected sequences. The details of MIP will also not be given in this book and the interested reader is referred to the review of Selker (2002) for details and literature.

Contrary to RIP and MIP, *quelling* was discovered (Romano and Macino, 1992) as a posttranscriptional mechanism. The discovery was made in *Neurospora crassa* that was transformed with a portion of a gene (involved in carotene synthesis) that already existed in the transformed fungus. The transformation caused the opposite of the expected: rather than enhancing the expression, the transgene caused the silencing of the endogenous gene (albino fungi were produced). Further, studies showed that the transgene reduced the level of protein that was translated from the respective transcript. The rate of transcription was not reduced but the level of mRNA was reduced posttranscriptionally. Hence, quelling is actually the *Neurospora* version of PTGS that was found in plants and described in Chap. 2. It is also the same phenomenon that was termed RNAi (RNA interference) in protozoa and animals. The latter phenomenon will be detailed in subsequent chapters.

Quelling in *Neurospora*

Basic features of quelling

The RNA silencing in *Neurospora* was discovered by Romano and Macino (1992) in a study that provided unexpected results. Actually, the early studies on RNA silencing were paved with unexpected results. I have mentioned that the co-suppression of petunia pigmentation and the first experiments on RNA silencing (RNAi) in the nematode *C. elegans* also started with unexpected results. Probably, Galileo Galilee was also surprised when he recorded that light and heavy articles are falling with the same velocity.

The unexpected results of Romano and Macino (1992) were obtained with the transformation of *N. crassa*. The hyphae and conidia (vegetative spores) of *N. crassa* have a bright yellow pigmentation due to carotenoids. Albino mutants can be obtained in this fungus and they are caused by mutations in one of three genes: *al-3* (defect in geranylgeranyl phosphate synthase), *al-2* (defect in phytoene synthase) or *al-1* (defect in phytoene dehydrogenase). Plating a mutated conidium results in an albino colony. Thus, the *al* mutants are convenient for genetic studies because screening is quick and simple. One can also evaluate the *rate* of pigmentation. Rather than enhancing pigmentation, the transformation with *al-1* caused 40 per cent of the progeny to show a *range* of phenotypes from albino to dark yellow. Different results were obtained after transformation with a construct of *al-3*. In the latter case only 0.5 per cent of the progeny was silenced. The authors had an explanation for the difference. The *al-3* gene codes for an enzyme that is essential for the vitality of the fungus (i.e. geranylgeranyl phosphate synthase) not only for the pigmentation. Therefore, many of the silenced progeny were not vital. In a further study (Cogoni and Macino, 1997) they used a fraction of the *al-2* gene for transformation. In this case the frequency of silencing was 10 per cent.

Quelling in *N. crassa* was reviewed in detail by Pickford *et al.* (2002) and these authors listed the quelling with other *Neurospora* genes by the Roman team of Macino, Cogoni and associates. Moreover, other research teams detected additional cases of quelling in *Neurospora* when the fungus was transformed with transgenes that are also endogenous.

Further studies provided additional information. First, contrary to what was observed with RIP, the inactivation of the (*albino*) genes did not involve rearrangements of the endogenous genes. Rather, a strong reduction of steady-state levels of the respective mRNAs was recorded. There was an additional observation: the silencing of the *al* genes was reversible. When the pigmentation was restored, the level of mRNA was also increased. Moreover, the reversion was correlated with the loss of the transgene (probably due to homologous recombination in the vegetative hyphae). This caused the Roman team to suggest that the presence of the transgene is not only required for the initiation of quelling but also for its maintenance.

The question of the involvement of DNA methylation in the quelling of *Neurospora* was also a theme of the studies by Cogoni and Macino. The final conclusion, based on the studies of Cogoni and Macino (1997), was that methylation should be ruled out as a cause for quelling.

The hyphae of *Neurospora* contain many nuclei between two septa (they are coenocysts). Two hyphae from different colonies, each with a different nuclear composition, can be fused. The fused hyphae will then be heterokaryotic. When the heterokaryotic colony produces conidia, the colonies developing from each conidium can be homokaryotic and thus the nuclear composition of the heterokaryon can be revealed. The heterokaryons of *Neurospora* are thus handy to test whether quelling is dominant or recessive. When nuclei with a recessive allele are combined in a heterokaryon that also contains nuclei with a dominant allele, the heterokaryon will always show the dominant phenotype. But when nuclei with the wild-type nuclei were combined in the same heterokaryon with quelled nuclei, the phenotype of the heterokaryon was quelled. This indicates that quelling is dominant. A component from the quelled nuclei must have caused quelling in the wild-type nuclei (Cogoni *et al.*, 1996). Because only the spliced (mature) mRNA was reduced in the quelled hyphae, it was obvious that the process takes place in the cytoplasm rather than in the nuclei.

Quelling-defective (qde) mutants

Cogani, Macino and associates attempted to isolate mutants that are defective in quelling (*qde*). For that they used a quelled strain that was stable and mutagenized it (by UV irradiation). Out of 100 000 colonies, 19 lost the quelling. Four of these had lost the transgene while 15 which did not lose it neither were in them a rearrangement of the transgene. Genetic studies by complementation between the 15 *qde* lines indicated that there are three different *qde* mutants. Even additional mutagenesis of quelled strain did not add more *qde* mutants.

The three *qde* genes were then isolated by insertional mutagenesis, by means of additional transformation. The genes were subsequently sequenced to reveal the encoded sequence of amino acids.

The *qde-1* was found to encode 1402 amino acids with significant homology to an RdRP of tomato. The homology was confined to the carboxy-terminal portion of the tomato enzyme. This furnished support the notion that indeed RdRP is involved in PTGS (quelling) in both tomato and *Neurospora*. We shall see in subsequent chapters that similar RdRPs are involved in RNA silencing in additional organisms such as *Arabidopsis* (e.g. SDE1/SGS2), *Schizosaccharomyces pombe* and *C. elegans (ego1)*. The role of the RdRP gene of *Neurospora (qde-1; QDE-1)* is not clear yet but it may be involved in propagating the small dsRNAs (of ca 22 nt) that are produced during RNA silencing.

The *qde-3* gene encodes 1955 amino acids with homology to a DNA helicase (RecQ-like DNA helicase). The RecQ enzymes are widespread among diverse organisms from bacteria to humans. Actually, some organisms have several RecQ proteins (*Arabidopsis* has six!). The RecQ enzymes in the diverse organisms have various functions in the genome such as DNA repair and processing of DNA at the replication fork. A role for RecQ in quelling of fungi is thus an interesting additional role. But how exactly the RecQ-like QDE-3 of *Neurospora* is involved in quelling and whether it interacts with a topoisomerase, is still an open question.

Relatively, little is known about *qde-2* (QDE-2), its chemistry and functionality. Interestingly, a *qde-2* homolog was identified in *S. pombe* but not in *S. cerevisiae*. This is significant because there is PTGS in *S. pombe* but not in *S. cerevisiae*. The latter yeast is rather unique with this respect — it is possible that during evolution the PTGS genes were eliminated from *S. cerevisiae*. The *Arabidopsis* genome contains several homologs to QDE-2. One of these *AGO-1* is involved in PTGS of this plant and also in leaf and flower differentiation. The other homolog, ZWILLE/PINHEAD, is involved in shoot apical meristems and axillary meristem activities but unknown to be involved in PTGS of this plant. In *Drosophila* there is a PIWI protein that has structural similarities with AGO1 and ZWILLE but with apparent different functions. In the nematode *C. elegans* there is a family of at least 23 genes that have a similarity to QDE-2. One of the nematode genes, *rde-1*, is implicated in dsRNA interference. Nematode mutants defective in *rde-1* are completely lacking in RNA interference (RNAi). Another complication comes along. While *rde-1* mutants of the nematode are defective in RNAi, such mutants still suppress transposon's mobility in these worms. It could be that the two types of mRNA degradations (the one involved in RNAi and the one involved in transposon silencing) are activated by different pathways. Recent results of the Roman team (Catalanotto *et al.*, 2002) indicate that there is a *QDE-2* complex that contains small RNA molecules. This suggested that QDE-2 of *Neurospora* could be part of a small RNA-directed ribonuclease-complex that is involved in sequence-specific mRNA degradation (quelling).

Gene silencing in Phytophthora

The plant-pathogen genus *Phytophthora* is an oomycete. It is thus not a "kosher" fungus. DNA analyses of oomycetes suggest that these hyphal organisms are closer to golden-brown algae than to ascomycetes and basidiomycetes. On the other hand, the hyphae of *Phytophthora* have similarities to the hyphae of *Neurospora* (note: *Phytophthora*, as other oomycetes and all Phycomycetes, does not have chitin in its hyphal walls). As in *Neurospora* hyphae, the hyphae of *Phytophthora* are also coenocytic. Thus, heterokaryons can also be established in the latter hyphae. In a relatively "old" study, van West and associates of Wageningen, the Netherlanders (van West *et al.*, 1999) investigated gene silencing (quelling) in *Phytophthora infestans*. The Wageningen investigators transformed diploid *P. infestans* hyphae with the sense, the antisense and the promoter-less construct of the coding sequence of the elictin gene *inf1*. This resulted in the silencing of both the transgene and the endogenous gene. This could already indicate that the silencing is dominant because transformation will never affect *all* the nuclei of a given hyphae. Nevertheless, the investigators produced heterokaryons between a silenced strain and a wild-type strain. The heterokaryons were silenced. This substantiated the dominance of the silencing. The investigators then analyzed progenies of the heterokaryotic hyphae by establishing homokaryotic hyphae of which those that did not contain the transformed nuclei were *not* silenced. This is different from results obtained with similar hetero-karyotes of *Neurospora* as noted above. This difference between RNA silencing of *Neurospora* and of *P. infestans* could result from the great phylogenetic distance between these two organisms: the details of quelling in these two organisms could be significantly different.

Transposon silencing in Magnaporthe grisea

A research team at the University of Kobe, Japan (Nakayashiki *et al.*, 2001) looked at gene silencing in fungi from a completely different angle. They investigated the taming of a TEs that belongs to the Long-Terminal-Repeat retrotransposons (LTR-retrotransposons), MAGGY (see: Galun, 2003 for a detailed description of these LTR-retrotransposons). Briefly, these TEs are very common in eukaryotic organisms. They are usually "sleeping" and only transpose occasionally. In some cases, transposition is frequent and when they transpose into new sites in the genome, they may cause mutations. Their transposition involves the synthesis of an RNA copy that goes through a retrotranscription into a cDNA that is converted into a dsDNA and then re-inserted into a new site in the nuclear genome. Thus, each transposition doubles the number of the transposed LTR-retrotransposon in the nuclear genome.

In *M. grisea* some strains harbor MAGGY elements while in other stains (naïve strains) there are no MAGGY elements. The Kobe investigators introduced a MAGGY element into a naïve *M. grisea* strain by transformation with a plasmid that contains this element. They intended to follow the transposition of MAGGY in the naïve strain. It was found that soon after the transformation the number of MAGGY elements increased quickly (due to retrotransposition) until there were about 20–30 elements per nuclear genome. But then when the increase of MAGGY stopped, it was silenced. Was this silencing caused by methylation? Such methylations of transposon promoters were reported in the past to be correlated with transposon silencing. The investigators found that *de novo* methylation of MAGGY indeed occurred immediately after its introduction into the fungi, but then the methylation was constant and did not increase. Furthermore, 5-azacytidine treatment caused demethylation but did not affect the transposition of MAGGY. Further studies such as searching for the correlation between the methylation in MAGGY-containing strains and the MAGGY transposition clearly indicated that the methylation status did not correlate with MAGGY copy number and transposition activity. Consequently, the investigators suggested that the silencing of MAGGY activity is caused mainly by posttranscriptional suppression.

A Spanish team at the University of Murcia (Nicolas *et al.*, 2003) studied RNA silencing in *Mucur circinelloides.* Throughout their publication these authors considered *Mucur* a "filamentous fungus". *Mucur* is a zygomycete and as such it belongs to the Phycomycetes group, as the aforementioned *Phytophthora*. The Phycomycetes are close to golden-brown algae, rather than to the "real" fungi as ascomycetes and basidiomycetes. I raised the taxonomic issue to avoid a consideration that findings in the RNA silencing of *Mucur* are necessarily relevant to RNA silencing in real fungal organisms — "it ain't necessarily so..." (Porgy and Bess).

M. circinelloides is also able to synthesize carotenoides in its hyphae, but it differs from *Neurospora crassa* with respect to this synthesis that while in *N. crassa* these carotenoides are produced constitutely, the production of carotenoids in *M. circinelloides* is light-dependent. When the hyphae of the latter organism are grown in the dark no carotene is produced. But right after blue light illumination there is an accumulation of β-carotene. Blue light activates the gene that encodes phytoene dehydrogenase (*car* β). When *car* β is inactivated the hyphae stay albino in spite of blue light. In this case the colorless precursor, phytoene, is accumulating. The investigators transformed a wild-type *M. circinelloides* with plasmids that contained either the full length of the *car* β coding region or with fragments of this gene. Several markers and selective-genes were added to the plasmid to facilitate selection and analysis of the results. When the organism was transformed with the full-length *car* β gene 3 per cent of the transformants stayed albino after illumination. Even a higher proportion of the transformants stayed albino, after illumination, when the transforming plasmid contained only a small fraction of the whole *car* β (a total of 677 bp that contained the promoter and 166 bp of the coding region; the total length of the gene is about 2.5 kbp). It was found that in the transformants in which the *car* β was silenced there was a strong (up to 18-fold) reduction in mRNA of *car* β. This reduction was not caused by reduced transcription. This indicated that the system that was analyzed by transgene-induced silencing in *M. circinelloides* is PTGS. Experiments clearly indicated that the endogenous *car* β remained intact in the silenced transformants. In addition, the transformed mycelium had copies of the intact transgene. Furthermore, methylation of the endogenous gene and the transgene was not changed in the silenced transformants. This indicates the silencing in this system is not mediated by methylation.

When Nicolas and associates performed their experiments on silencing in *Mucur*, small RNA species were already known to be implicated with PTGS. These authors looked for such small RNA species in wild-type and silenced mycelia. A predominant antisense RNA (for *car β*) of about 21 nt was found in silenced mycelia but not in wild-type mycelia. But there was also a predominant sense RNA; this had a length of about 25 nt. Further, more detailed analyses of the antisense RNA species indicated that at first there were 25 nt antisense RNAs but after 48 hours these were reduced in amount and the 21 nt antisense RNAs predominated. There was no such shift in size of the sense RNA species — they retained the 25 nt size. The authors' hypothesis was that the short antisense RNAs (21 nt) are involved in degradation of the mRNA while the longer antisense RNAs (25 nt) are involved in the propagation and movement of the silencing signal along the hyphae.

Finally, a "spreading" of the silencing was indicated in the *Mucur* system — as was also found previously in other PTGS systems. When the truncated *car β* gene was used as a transgene, the investigators found in silenced mycelia, small RNA sequences that were homologous *car β* sequences that were *not* represented in the transgene.

Degradation of pre-mRNAs in S. cerevisiae

This degradation of pre-mRNA in the budding yeast (*Saccharomyces cerevisiae*) was revealed recently by Guillaume Chanfreau and associate of the University of California, Los Angeles (Danin-Kreiselman *et al.*, 2003). Before we go into the details of this degradation it should be noted that while budding yeast and fission yeast (*Schizosaccharomyces pombe*) share the term "yeast" and even though their genus name is rather similar, these two yeasts are phylogenetically very far apart. They were separated about a billion years ago. Thus, for example, the transcriptional silencing in fission yeast (e.g. Grewal, 2000; Volpe *et al.*, 2002) has nothing in common with silencing in budding yeast. Danin-Kreiselman *et al.* (2003) studied the RNase III-like proteins of a family of dsRNA endonucleases that are involved in the cleavage of cellular RNAs in the budding yeast (*S. cerevisiae*). The RNase III was studied in the past in bacteria (e.g. *E. coli*). In eukaryotes the RNase

III-like proteins are involved in several kinds of RNA degradations such as the maturation of ribosomal RNA (rRNA) and the cleavage of small nuclear and small nucleolar RNAs, as well as other non-coding small RNAs. We shall see that the *Dicer* that cleaves dsRNA in RNAi and also cleaves microRNA genes, contains RNase III-like motifs. The *S. cerevisiae* RNase III, Rnt1p, specifically cleaves dsRNA that is capped by a tetraloop sequence with the (weak) consensus nucleotide sequence AGNN. The cleavage site is then 14–16 bases away from the tetra-loop. For molecular biologists this means that the yeast Rnt1p acts as an "RNA helical ruler". The budding yeast has a few intron-containing transcripts. These introns may form stem-loop structures. After these intron sequences are cleaved, they are subject to further degradation. The study of Danin-Kreiselman *et al.* (2003) was focused on the cleavage by Rnt1p of intron-containing transcripts in the budding yeast. The investigators choose the transcript of the ribosomal protein gene *RPS22B* that has two introns: intron 1 and intron 2. The splicing is sequential, a partial splicing of the precursor and then a full splicing. The splicing also required, in addition to Rnt1p, another enzyme encoded in the *DBR1* gene. Only in the presence of the two proteins is there a full degradation of the lariat introns. In summary although the *Rnt1p* is an ortholog of RNase III, the function of Rnt1p is not in RNA silencing but rather in maturation of RNA transcripts.

Unpaired DNA and RNA-mediated silencing

Two Australian investigators (Hynes and Todd, 2003) summarized two earlier studies by Shiu *et al.* (2001) and Shiu and Metzenberg (2002). Hynes and Todd attributed strong emotions to *Neurospora crassa* by noting that this fungus "... has an intense dislike for aberrant or duplicated nucleic acids..." If dislike is causally related to the destruction of the disliked, these investigators are probably right. RNA-mediated destruction of unpaired DNA has its own acronym: MSUD for meiotic silencing by unpaired DNA. This silencing occurs during meiosis in *N. crassa*. This process in *N. crassa*, as in other ascomycete fungi, is very different from meiosis in plants and mammals. Vegetative hyphae of *N. crassa* are either of mating type A or mating type B. Two different mating types may fuse (fertilization) and ascogenous hyphe are produced. The latter have two kinds of nuclei of type A and of type B, in each of the hyphal cells. At a further stage there is a pair-wise fusion between A and B nuclei that is termed karyogamy. Diploid nuclei (zygotes) are thus formed. The diploid nuclei go immediately through two meioses and one mitosis. This leads to the formation of asci. Each ascus contains 8 (haploid) ascospores. The ascospore eventually germinates into hyphae that are, again, either A or B type. The MSUD happens right *after* karyogamy (RIP is one step earlier, during the phase of ascogenous hyphae). Shiu *et al.* (2001) of the Stanford University, California studied an MSUD case in great detail. The study concerned a gene *Asm-1* that is required for the maturation and melanization of ascospores. Unmelanized acrospores are not viable as they fail to mature. It was found that the inability to form melanin is a consequence of a specific situation: when the two *Ams-1* genes are *not paired* in the zygote. When this happens all ascospores in the resulting asci are immature. This appears to be a new general silencing system that takes place in the zygote: unpaired segments are silenced as are genes present in any odd number of copies or homologous genes that are single-copy in each parent (of the zygote), but that occupy non-homologous positions. The authors suggested that DNA that is unpaired in the early stages of meiosis in *Neurospora* causes both self-silencing and transsilencing of all DNA homologous to it, whether paired or not. Moreover, the silencing was suggested to be posttranscriptional. Evidence for PTGS came from the role of another gene in MSUD: *Sad-1* which is required for RdRP (and thus for PTGS). In the absence of functional *Sad-1* there was no MSUD in *Neurospora*. The same team (Shiu and Metzenberg, 2002) further analyzed the MSUD system genetically. Several mutants of the *Sad-1* gene were produced and their impacts on MSUD and meiosis was investigated. Interestingly, the wild-type *Sad-1* is also fully expressed during the meiotic phase in the normal cases where all the pairings are homologous, but its role during this phase was not established yet. It is also noteworthy that the MSUD is always confined to the specific ascus where there were unpaired DNAs. The silencing did not spread to the neighboring asci in the fruiting body (perithecium) having a mixed genetic constitution. It is also worthwhile to note that methylation does not seem to be required for MSUD. This was suggested by the use of a methylation mutant *dim-2*. The wild-type *dim-2* gene mediates cytosine methylation. MSUD goes on even in the presence of a mutation of *dim-2*. This suggestion is still far from being confirmed. It is possible that a specific gene that is mediating methylation during meiosis exists in *Neurospora* so that *dim-2* is not essential for methylation during this phase.

The modest bread mold *N. crassa* made an impressive career. It came to fame at Stanford University and the California Institute of Technology (Caltech) during the first half of the 20th century. It led to the "one gene-one enzyme" concept (that was later amended) and made Beadle and Tatum Nobel Prize laureates. *N. crassa* was also studied by another Nobel Prize laureate, Barbara McClintock who analyzed its cytology (but received the prize for Transposable Elements). More recently and related to our theme it was found to harbor four silencing systems: quelling, RIP, MIP and MSUD.

Haynes and Todd (2003) predicted an interesting future for MSUD and *Neurospora*. Further insight may be provided for the mechanism of maintaining genome integrity, in cases of crosses between strains that contain one or more transposons that are lacking in the other strains. The MSUD has still several riddles that should be solved. One of them is that the normal genome of *N. crassa* has several duplicated sequences such as genes for ribosomal RNA. Why are these not causing MSUD?

Chapter

4

RNA Interference in the Nematode *C. elegans*

There is a maxim by Amiel Ben David Halevi* that says: "Short is Pleasing". Three *short* articles published in the journal *Nature* had immense impacts on their respective fields. The first article (Joliot and Curie, 1934) written by the couple, Joliot and Curie, was on alpha-particle bombardment of elements that opened a new field of investigation and won them the Nobel prize. The second on the three-dimensional structure (double helix) of DNA was written by Watson and Crick (1953) and was also a Nobel Prize number. The third was by Fire *et al.* (1998) on the role of short dsRNA in RNA interference in a nematode that had a major impact on molecular-genetics of RNA silencing. This third article is of great relevance to this book.

Studies by Fire, Mello and Associates

The article of Fire *et al.* (1998) was mentioned briefly in previous chapters. I shall provide more details on this study, but first present a few introductory remarks.

The interest of Andrew Fire in the nematode *C. elegans* was already indicated in the mid 1980s when he was working at the MRC Laboratory in Cambridge, England (the laboratory where Sydney Brenner developed *C. elegans* as a metazoan model-organism). In the MRC Laboratory, Fire established the genetic transformation of this nematode and subsequently used this transformation (Fire and Waterston, 1989) to obtain transgenic nematodes. After moving to the Carnegie Institution of Washington in Baltimore, Fire and associates published

---

*This is the pseudonym of (Amiel) Esra Galun.

the first paper on the inhibition of gene expression in *C. elegans* by anti-sense RNA (Fire *et al.*, 1991). It should be noted that by 1991, the co-suppression of genes by transgenes was already well documented in plants (see: *A short history of RNA silencing* in the Introduction of this book, e.g. Matzke *et al.*, 1989; Napoli *et al.*, 1990; van der Krol *et al.*, 1990). No references on antisense RNA that inhibits the expression of plant genes were given by Fire *et al.* (1991). The investigators of the latter study have used antisense strategy to disrupt the expression of two genes that encode myofilament proteins in the body-wall muscles of *C. elegans*: *unc-22* and *une-54.* Segments of these genes were placed in reverse orientation in appropriate vectors. These vectors (plasmids) were injected into oocytes. The progeny resulting from the injected oocytes had defects that were similar to those of worms with the null mutations of these genes. Analyses of the RNA transcripts of the *unc-22* gene showed that the transcript was normal and not modified but the level of *unc-22* protein was greatly reduced. The antisense RNA from the transgene was more abundant than the sense RNA from the endogenous gene. The authors suggested that there was "interference with a late step of gene expression, such as transport into the cytoplasm or translation". Conspicuously, the investigators found that a fraction of worms that were defected by the antisense transgene transmitted this defect to their progeny. This transmission into the next sexual generation was an unsolved puzzle because endogenous RNA transcripts are rapidly degraded in the worm's embryo. Already in 1991, Fire *et al.* observed that not only did the antisense sequence of *unc-22* but also the injection of the sense sequence cause a strong reduction in the protein encoded by *unc-22.* One of the authors' hypotheses that was refuted by later studies was "that double-strand specific RNase that mediated degradation of transcription does not play a role in the observed disruption of expression". Another unsolved puzzle came from another research team of the Cornell University. Guo and Kemphues (1995) studied a gene (*par-1*) involved in the asymmetric division in *C. elegans* eggs. They had a surprise. Both the sense and the antisense RNA transcripts for the *par-1* gene caused specifically embryo lethality when injected into the gonads of the parent worms.

The title of the article by Fire and associates that was published 7 years after their aforementioned publication already shows that there was a change in opinion. The article of Fire *et al.* (1998) carries the title "Potent and specific genetic interference by *double-stranded RNA* in *Caenorhabditis elegans*". The study reported in this latter article clearly indicated the advantages of *C. elegans* as a model organism. There is no doubt that the choice of Sydney Brenner (e.g. Brenner, 1974) was a very proper one. Moreover, the vast genetic and morphogenetic information that accumulated during over 25 years of *C. elegans* investigations paid off nicely. One of the features of *C. elegans* is that this worm is the most *sensitive* creature — one half of its 600 somatic cells are neurons. The total number of cells in the mature worm is 959; this number includes gonad cells; there are 302 nerve cells. The 1998 study of Fire, Mello and associates intended mainly to provide answers to two questions: (1) why is interference imposed by transcripts that have either the sense or the antisense orientation, and (2) how could transcript-induced interference be carried on to subsequent generations. The investigators found that the previously used sense and antisense RNAs were actually not pure. They contained traces of the opposite orientations and thus not only ssRNA but also dsRNA was included in the injections. Thus, they further purified the ssRNAs and also deliberately produced dsRNA from the sense and antisense RNAs. The study first focused (again) on the *unc-22* gene. One of the advantages of this gene is that it encodes an abundant but nonessential myofilament protein. Normally, several thousand copies of *unc-22* mRNA are present in each striated muscle cell of *C. elegans*. Moreover, there is a fair correlation between the abnormal phenotype and the decreased expression of *unc-22* until null expression correlates with the lack of motility of the worms.

When either pure sense or pure antisense RNAs covering 742 nt of the *unc-22* gene were injected into worms and the larvae hatched from the injected animals were analyzed, the authors found very little interference effects. In contrast, a sense-antisense mixture produced a strong interference with the activity of the endogenous *unc-22* gene. It was calculated that the 500 embryo cells received only a few molecules of the mixture per cell. This low number of mixture-molecules was sufficient to cause a strong interference impact. The mixture caused the production of dsRNA. When the sense and antisense constructs were injected separately with a 1 hour time-lapse, the effect was reduced dramatically. Similar results were obtained when transcripts of additional genes were injected as either ssRNA or as a mixture (e.g. causing the formation of dsRNA) of sense and antisense transcripts. The interference was observed only when coding sequences were injected; introns and promoters' sequences were not effective. It was also found that the injected RNA caused a strong reduction in the corresponding endogenous mRNA. An additional observation was that the injection-effect was *mobile*: an injection into one location crossed cellular boundaries. When the injection was made into specific worm organs as body cavity, head or tail, the interference was expressed in the sexual progeny, just as an injection directly into the gonads. The authors thus suggested that there should be an effective RNA-transport mechanism in *C. elegans*. The important lesson from Fire *et al.* (1998) was that a dsRNA that is homologous to part of the coding of a specific gene will silence this gene in a specific and effective manner.

Following the finding that the dsRNA effect can *spread* so that injection of dsRNA into the extracellular body cavity will affect a broad region of the worm, a technical question was asked: can the dsRNA be applied by a simpler method? *C. elegans* normally feeds on bacteria, ingesting and grinding them in the pharynx and subsequently absorbing the bacterial contents in the gut. Timmons and Fire (1998) engineered bacteria that expressed dsRNA with homologous sequences to either of two *C. elegans* genes: *unc-22* or *fem-1*. The null mutant of *fem-1* is defective in sperm production. The larvae of *C. elegans* were fed with either of these two kinds of engineered bacteria. The results were similar to those of injection: defective motility and defects in sperm production, respectively. When feeding was subsequently changed to wild-type bacteria, the interference was removed. In an additional experiment, transgenic *C. elegans* larvae that express GFP transgene were used. These worms were fed with bacteria that expressed dsRNA that corresponded to GFP. This resulted in a decrease in GFP fluorescence in about 12 per cent of the fed worms. In each of the three feeding experiments the interference of the dsRNA was gene-specific. Further, experiments showed that not only feeding with dsRNA producing bacteria but also soaking the worms in a solution containing dsRNA will cause interference. The bacterial-feeding method was further elaborated in a subsequent publication (Timmons *et al.*, 2001). Tabara *et al.* (1998) soaked worms for 24 hours in a solution that contained dsRNA from the unmutated *pos-1* gene which caused lethality in the derived embryos.

Once RNAi was revealed, the work on RNAi in *C. elegans* progressed quickly in the Fire Laboratory of the Carnegie Institution of Washington in Baltimore. Montgomery *et al.* (1998) injected dsRNA into the gonads of worms and followed the results in detail. Briefly, it was found that the dsRNA did not change the sequence of the respective nuclear DNA. The investigators also revealed that the dsRNA could interfere with the expression of sequences that are upstream of the gene from which the dsRNA was derived, but there was no downstream effect. A third finding showed that while there was a slight reduction of nascent transcript accumulation in the nucleus, after RNAi, the cytoplasmic accumulation of transcripts was completely eliminated. From their results the authors drew the conclusion that endogenous mRNA is the target of RNAi. Consequently, they suggested a model of how dsRNA may function in the mechanism to target homologous mRNA for degradation. In fact, they suggested that an "RNAi co-factor" (protein or riboprotein) is complexed with the dsRNA in the early sequence of events and that this complex is cleaving a specifically mRNA that contains a sequence that is homologous to the RNA of the complex. This suggestion was verified in later studies (see below) and the "RNAi co-factor" was then given the name *Dicer.*

In further efforts to understand the RNAi mechanism, Fire, Mello and associates (Tabara *et al.*, 1999), now residing in two locations (University of Massachusetts Cancer Center, Worcester, MA and the Carnegie Institution of Washington, Baltimore, MD) continued their collaboration in the study of RNAi. By September 1999, when their manuscript was accepted for publication (Tabara *et al.*, 1999) they were already aware of the dsRNA silencing of genes in several other organisms as plants, trypanosomes and *Drosophila.* To further understand RNAi and its possible role in *C. elegans* and other organisms they decided to investigate nematode mutants that are defective in RNAi. Such mutations were reported in *Neurospora* (Cogoni and Macino,

1997) and *Arabidopsis* (Elmayan *et al.*, 1998). The *Neurospora* mutant was defective in the gene *qde-1* that encoded RdPR and this suggested that RNA synthesis from an RNA template was essential for the dsRNA mediated silencing.

For screening new RNAi mutants, the authors used the *pos-1* gene that was mentioned above. When a dsRNA from this gene is applied to hermaphrodite worms, the worms themselves are not affected but they produce dead embryos. For easier screening they used a *C. elegans* strain that is defective in egg laying. Such worms attain at maturity the form of "a bag of eggs" from which viable embryos are released. These worms are very different from worms that contain non-viable eggs (with bags of dead embryos). Thus, the screening of worms that produce eggs with dead embryos is simplified. When hermaphrodite worms were fed with bacteria that express a dsRNA from the *pos-1* gene, the worms themselves were not affected but they produced eggs with dead embryos. In order to identify mutations in the RNAi mechanisms worms were muta-genized and the F2 generation was screened to reveal individ-ual worms that produce a viable (F3) progeny. The authors used chemical mutagenesis and also looked for spontaneous mutants. Seven RNA interference deficient-1 (*rde-1*) mutants were revealed and they were verified demonstrating their inability to cause RNAi by the injection of dsRNA. The mutations were mapped and found to comprise complementation groups as shown in Fig. 4. There were three alleles of *rde-1* (chromosome LGV), two alleles of *rde-4*

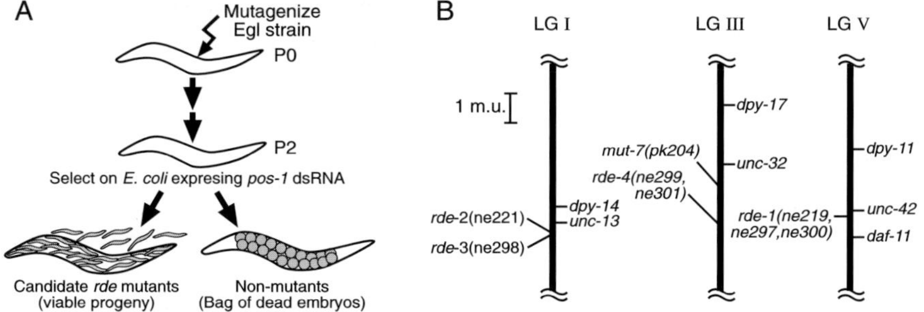

Fig. 4. Identification and linkage group analysis of RNAi-deficient mutants. (A) Genetic scheme for the identification of *rde* mutants. (B) Summary of genetic map-ping data. (From Tabara *et al.*, 1999.)

(LGIII) one allele of *rde-2* (LGI) and one allele of *rde-3* (LGI). In addition, two strains that have elevated transposition of TE were also found to be impaired in RNAi: they had the mutations *mut-2* and *mut-7*, respectively. The mutator activity (i.e. frequent transposition) and lack of RNAi were genetically linked traits. Of these mutations only *mut-2* was weakly dominant. The other mutations were recessive. The investigators then used various dsRNAs, derived from genes that were required for several functions of *C. elegans*. These dsRNAs were injected into the mutants and into control worms. All the mutants had resistance to RNAi but the level of resistance to dsRNA from genes that are somatically expressed (e.g. *unc-22*) varied among the mutants. The mutants *rde-2* and *mut-7* were only partially resistant to dsRNAs of the latter genes. In the presence of these two mutants, the transposition of TEs was not inhibited. Finally, the authors attempted to characterize the *rde-1* gene and its encoded protein (RDE-1). They found that part of the RDE-1 is a very conserved motif found in other animals, in plants and also in budding yeast but its function in these organisms is quite different. ZWILLE is essential for maintaining shoot apical meristems as undifferentiated tissue. There is a *Drosophila* homolog, *piwi*, that is required for germline maintenance in this fly. The homologous sequence of the *Drosophila piwi* was previously detected by the H. Lin Laboratory of Duke University, NC, in *C. elegans* and in *Arabidopsis*. The latter investigators thus used the term *piwi*-like genes and pointed out that a part of the *piwi* gene-sequence is well conserved since the early evolution of eukaryotes. Tabara *et al.* (1999) thoroughly discussed their results and suggested that transposon silencing is a natural target of the RNAi mechanism. Because later studies at least partially resolved questions discussed by the authors (as a possible link between transcription silencing and post transcriptional silencing) I shall not detail this discussion now.

In a further collaborative study by A. Fire and C. Mello, the requirements for efficient RNAi during the life-cycle of *C. elegans* were studied further (Parrish *et al.*, 2000). The investigators utilized the advantage of the RNAi system in this organism: a minute quantity of injected dsRNA can cause RNAi. One essay to provide answers for two questions was based on interference with GFP expression. The questions were whether dsRNA shorter than ~1000 nt will cause interference and whether interference is caused by different regions of the *gfp* coding sequence. The answers were that dsRNA of 62–242 nt were already effective and that various regions of the coding sequence from the *gfp* gene caused RNAi. Another experiment with synthetic dsRNA that had homology to *unc-22* indicated that while even 26 nt of dsRNA had some RNAi effect the effect of 81 bp dsRNA was at least 250-fold stronger. The purified ssRNA of either sense or antisense orientation had no effect on RNAi even when very high amounts were injected into worms. Furthermore, there must be an identity in sequence between the target gene and the dsRNA of about 96 per cent. When the identity was reduced to about 75 per cent, the RNAi was drastically reduced or eliminated. Certain modifications of the bases were tolerated in the sense sequence of the dsRNA but not in the antisense. Of great significance for future studies was that whatever the length of the dsRNA injected, the length of the dsRNA that was retained in the worms was always about 25 nt. Similar results were actually obtained in other organisms. Parrish and Fire (2001) further investigated the two proteins (RDE1 and RDE4) that were found previously to be essential for RNAi in *C. elegans*. The rationale for their investigation was that the RNAi/PTGS mechanisms were conserved across diverse phyla from all four eukaryotic kingdoms and were thus of ancient origin. These mechanisms may therefore represent a primitive nucleic-acid-based immune-response. Understanding the details of this mechanism in *C. elegans* should be instrumental for an overall acquaintance with this response in all eukaryotes. Indeed, from their own previous studies, as well as from studies on plant PTGS it was known that in the early stage of dsRNA-induced silencing a population of small interfering dsRNAs (siRNA) is produced in various organisms such as nematodes, flies and plants. Hence, a common denominator of a step in this silencing was already established. Moreover, the siRNAs have the same basic features in the various organisms; they had a 5′ phosphate and a 3′ hydroxyl and a two- or three-base 3′ overhang on each strand of the dsRNA. This structure strongly suggested that an RNase III-like enzyme is involved in the production of siRNA in all organisms that show PTGS or RNAi. Indeed, a nice name was given to the enzymatic complex that cuts dsRNA into siRNA: *Dicer.* All the investigators of RNA silencing accepted this name.

Previous studies by Fire and associates started to characterize two mutants that are not essential for the vitality of *C. elegans* but that suppress RNAi: *rde-1* and *rde-4*. Of these the mutation *rde-1* does not prevent accumulation of siRNA while *rde-4* strongly reduces this accumulation. To better assess the impacts of the mutations *rde-1* and *rde-4* the authors developed an essay that quantified the levels of ~25 nt dsRNA (siRNA) that were found in worms after the injection of a (longer) interfering dsRNA. The levels in the respective mutants were compared to the levels in wild-type nematodes. In *rde-1* mutants the levels of siRNA, after triggering with interfering dsRNA, was comparable to the levels found in wild-type worms. This indicated that the RDE-1 protein is not involved in the early stage of RNAi (formation of the 21–25 nt siRNA). A different picture emerged when the dsRNA-trigger was injected into *rde-4* mutants. In the latter case there was a dramatic lower level of siRNA in comparison to wild-type worms. The rate of siRNA-level reduction was temperature dependent. The decrease in siRNA levels was stronger in elevated temperatures. Actually at 16°C there was some accumulation of siRNA while no such accumulation was observed in *rde-4* mutants cultured at 20–23°C. This reaction to temperature was parallel to the suppression of RNAi in this mutant. The experiments with the *rde-4* mutant indicated that the RDE-4 protein is involved in the very early stage of RNAi process, meaning in the conversion of the trigger-dsRNA to siRNA. This suggestion was verified by the use of synthetic dsRNA of 22–26 nt that had the characteristics of siRNA. When such synthetic dsRNAs were injected into *rde-4* mutants, the prevention of RNAi by the mutant was bypassed: the RNAi did take place. On the other hand, the synthetic dsRNA could not bypass the inhibition of RNAi in *rde-1* mutants. The authors thus developed a model that proposed roles for RDE-1 and RDE-4 in the RNAi of *C. elegans* (Fig. 5).

By 2001, several aspects of RNA interference in *C. elegans* and RNA silencing in other organisms were clarified as reviewed by Chicas and Macino (2001), Matzke *et al.* (2001a), Moazed (2001) Ruvkun (2001), Sharp (2001), Sullivan *et al.* (2001) and Vaucheret and Fagard (2001).

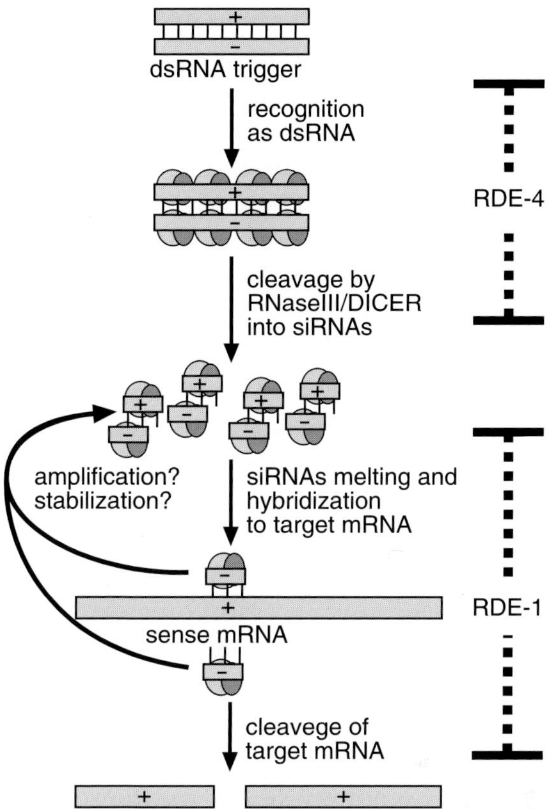

Fig. 5.   A working model showing the roles of RDE-1, RDE-4 and siRNAs in the interference reaction. (From Parish and Fire, 2001.)

It became evident that RNA-mediated gene silencing is an ancient eukaryotic mechanism although the details in this mechanism are specific to the various groups of organisms. Certain aspects of the silencing were at the level of established phenomena but the mechanism was still enigmatic. One of these phenomena triggered a transcontinental collaboration in which Donald Plasterk and associates of the Hubrecht Laboratory in Utrecht, The Netherlands and Andrew Fire and his associates, in Baltimore, MD, participated. Sijen *et al.* (2001) were intrigued by the catalytic aspect of RNA silencing: a few molecules of dsRNA were sufficient in *C. elegans* and in *Drosophila* to start the degradation of a much larger population of target mRNAs.

There were several ideas how this could happen. There was one "simple" possibility. If the trigger dsRNA is a relatively long sequence, let us assume that of 1000 nucleotides, this can be cut into about 40 different small RNAs (siRNAs) of 22–25 nt each. Each of these could contribute its ssRNA to a *Dicer* complex and this will create a multitude of *Dicers*. One cut per mRNA would be sufficient for the silencing of such a mRNA. Thus, one dsRNA trigger of 1000 nt could degrade 20- to 40-fold more mRNA. But the actual mRNA degradation is at least 10 or even 100-fold greater. Moreover, the mRNA degradation (as well as the degradation of alien dsRNAs, such as dsRNA of viruses) can spread to other tissues. This led to the assumption that an amplification process is required. Obviously, the involvement of RdRP seemed plausible. As seen above, one of the *C. elegans* genes (*rde-1*) that is required for dsRNA-mediated silencing was coded for RdRP activity. A similar enzymatic activity was found in tomato (see below) and is required for quelling. The paradox was that in *Drosophila* no measurable level of RdRP was found in embryonic extracts although these embryos do have dsRNA mediated silencing. Moreover, no obvious coding sequences for RdRP were detected in *Drosophila* and in mammalian genomes by 2001. Do these organisms use a different protein for RdR? An additional consideration was that there is a more stringent requirement for homology of the antisense RNA than for the sense RNA to guide the specific degradation of mRNA. If the siRNA that triggers a specific mRNA is replicated directly by an RdRP, the asymmetry of a siRNA that has an antisense that is not perfectly complimentary to its sense sequence will not be retained. An exponential amplification by RdRP would result in a loss of the "memory" of the difference between the two strands. There is a modified model for the amplification of siRNA. Only the antisense of a primary siRNA will anneal to the target mRNA. The antisense sequence will then be extended, complementarily to the mRNA. The RdRP will perform this extension. The extended dsRNA will then be cut into many secondary siRNA fragments. This will result in a net-amplification of the primary siRNA. Such a model was put to test by Sijen *et al.* (2001). If indeed this model provides the real events, then when short dsRNA trigger is applied, one should find a population of secondary siRNAs with homologies that are *upstream* of the sequence, in target mRNA, that is homologous to the trigger. By applying appropriate procedures such as elimination of interfering ssRNA and triggers from specific *C. elegans* genes (*unc-22* and *pos-1*) the authors did reveal secondary siRNA. Some of these had indeed nt sequences that were upstream of sequences (in the target mRNA) that had homologies to the trigger although the secondary siRNAs were less abundant than the trigger-coincident (primary) siRNAs. Moreover, the secondary siRNA species appeared to decrease (in abundance) as a function of distance (upstream) from the primary trigger. A further test by the authors was devised by a "transitive RNAi" assay. In this assay a chimera target was used. One part of the "target" was homologous to the trigger but an upstream part had no such homology. Using such a system it was found that the non-homologous (linked) transgene is also silenced. On the other hand, when a chimeric transgene with the homology upstream to the non-homologous sequence was used, only the homologous coding region was silenced. The authors proposed a model for RNA interference in *C. elegans* that was based on their own results and on experimental results in RNA silencing in other organisms. In their model the first step is the uptake of dsRNA (trigger) by the cells and an inefficient cleavage of the original trigger into short fragments (primary siRNA). The cleavage is performed by the RNase III-like complex, *Dicer*, that interacts with the protein RDE-4. One of the several possibilities is that the resulting complex binds to the target mRNA and the antisense RNA is extended by RdRP. Further steps such as establishing of secondary siRNA may involve the RISC component.

The Use of *C. elegans* and RNAi for Studies on Functional Genomics

After the RNAi phenomenon was well known in *C. elegans*, investigators started to contemplate how this phenomenon can serve to increase knowledge in molecular genetics. One obvious idea was to use it to reveal functions of specific coding sequences. The general idea was actually the same as the wisdom of King Solomon (mentioned in the motto of this book). King Solomon used a sword; he threatened to cut a baby into two in order to reveal who the real mother of the baby was. In the case of intending to reveal the function of a coding sequence in *C. elegans* the way was to cleave, specifically, the mRNA coded by a certain DNA sequence and then find out what went wrong. Once the DNA sequence of a chromosome (or an entire organism) is known and by their regulatory motifs, coding sequences could be predicted, it became possible to synthesize short dsRNAs sequences that are homologous to the coding sequences. Introducing such siRNAs into *C. elegans* should result in mutations that will indicate which function was destroyed by each siRNA. In *C. elegans* it is possible to introduce short dsRNAs by feeding. As mentioned above, when worms are fed with bacteria that produce a specific dsRNA, this dsRNA will initiate specific RNAi. Bacteria (i.e. *E. coli*) can be swiftly engineered to produce such specific dsRNAs so that each engineered strain of bacteria can cause the cleavage of a specific transcript of a given sequence of the *C. elegans* genome. This idea was implemented in a study reported by Fraser *et al.* (2000), a team of investigators from the Tennis Court Road (Wellcome/CRC) in Cambridge, UK. These investigators focused on chromosome I of *C. elegans*. There are 2769 predicted genes in this chromosome. Independent clones representing 2416 genes (about 87 per cent of the predicted genes) were cloned and bacteria with the respective dsRNA were engineered and fed to wild-type *C. elegans*. Mutated phenotypes were revealed by 14 per cent of the dsRNAs that were fed to the worms. Thus, about 300 more cases of mutated phenotypes could be correlated to specific coding sequences in chromosome I of *C. elegans*. The complete sequence of genomic DNA in all the chromosomes of *C. elegans* is now available. Hence, providing functional genomics to the total genome of this worm seems possible now. It would require substantial investment of effort and time, but as usual one has to pay for the satisfaction of curiosity. As for the fate of dsRNA introduced by feeding the nematodes with dsRNA-expressing bacteria, the situation is not plain and simple. The Plasterk Laboratory in the Netherlands (Tijsterman *et al.*, 2004) found that "RNAi spreading defective" genes, termed *rsd*, are required for systemic RNAi. Some *rsd* mutants are completely defective in cellular uptake of dsRNA while other *rsd* mutants are defective in the distribution of the dsRNA from somatic tissue into the germline. The Ruvkun Laboratory at the Harvard Medical School found a nematode mutant, *eri-1*, that had an enhanced response to RNAi. Worms with this mutation were not refractory anymore to RNAi in their nervous system (Kennedy *et al.*, 2004).

There is one warning. While *C. elegans* feasts on bacteria, some bacteria will kill this worm. A team from the Mass General Hospital of Harvard University, consisting of Gary Ruvkun, Frederick Ausubel and associates (Garsin *et al.*, 2003) noticed that several Gram-negative bacteria (e.g. *Pseudomonas aeruginosa, Salmonella enterica*) and Gram-positive bacteria (e.g. *Enterococcus feacalis, Staphylococcus aureus*) that are human pathogens are also killing *C. elegans*. Moreover, even some strains of *E. coli* (e.g. OP50) may be pathogenic to *C. elegans*. There is a remarkable overlap in pathogenicity of bacteria to humans and to *C. elegans*. While humans utilize their immune system and medicine to resist pathogenic bacteria, *C. elegans* developed resistant mutants, for example, worms with the *daf-2* mutation are resistant to *E. faecalis*, *S. aureus* and *P. aeruginosa*.

A similar study to this performed by Fraser *et al.* (2000) was performed by 21 investigators from Germany, UK and Canada (Gönczy *et al.*, 2000). Their report appeared in tandem, in the same *Nature* issue, as the article by Fraser *et al.* (2000). But, rather than feeding the worms with engineered bacteria, Gönczy *et al.* (2000) injected the dsRNA into the worms. The latter team analyzed chromosome III (rather than chromosome I) and focused on one specific process: cell division. Thus, dsRNA representing all the open reading frames (ORFs) of the DNA sequence in chromosome III were synthesized. There are about 2300 ORFs in this chromosome and the analysis, by dsRNA silencing, was performed on 96 per cent of these ORFs. By this screening the investigators found 133 genes that are required for proper cell division. About half of these genes seemed to be orthologues to genes in other eukaryotes. The authors therefore suggested that the *C. elegans* screening can provide putative gene functions for other eukaryotic organisms.

A soaking procedure was developed by Japanese investigators to perform a large-scale analysis of gene-function in *C. elegans* (Maeda *et al.*, 2001). The intent of these investigators was to perform a reverse-genetics analysis by disrupting the expression of as many as possible coding sequences. Briefly, these investigators prepared non-redundant cDNAs that represented about half of the total number of *C. elegans* genes (i.e. about 10 000 cDNAs). These cDNAs served to derive the corresponding dsRNAs. Four worms of the L4 larval stage (the stage before mature worms) were soaked in each of many microwells which contained different dsRNA sequences. The effect of soaking was observed in the mature worms that developed from the soaked larvae for morphological defects that occurred already in the P0 generation. Then the progeny (F1) was also analyzed for such defects (Fig. 6). This procedure was actually followed and about 2500 genes were screened. In about one quarter of them the dsRNA caused developmental aberrant phenotypes. The F1 generation survey identified abnormal germ line development that could be related to 24 genes. The most frequent abnormality was embryonic lethality. About 7 per cent of the dsRNAs caused sterility in the P0 generation (the mature worms derived directly from the treated L4 larvae). The authors indicated that the soaking can also be performed with younger L1 worms (rather than L4 worms). In such soaking additional aberrant phenotypes could be revealed already in the generation of soaked worms (P0).

More recently, a team of investigators, mostly from Cambridge, UK (Kamath *et al.*, 2003) returned to the large-scale functional analysis of the *C. elegans* genome. They conducted a study that was actually an extension of the work of Fraser *et al.* (2000) on chromosome I. This extension was based on two reasons. One reason was that in spite of previous efforts only about a third of the predicted genes of *C. elegans* were analyzed. The second reason was based on the accumulating evidence that about half of the *C. elegans* genes have homologs in humans. Thus, functional analysis in this worm should provide insight into human gene functions.

For their screening Kamath *et al.* (2003) constructed a library of 16 757 bacterial strains which corresponded to 86 per cent of the predicted number of genes in *C. elegans* (19 427 genes, but other investigators estimated 19 757 genes). This worm contains five autosomal chromosomal pairs and an X-chromosome. Hermaphrodite worms have two X chromosomes while male worms have a single X chromosome. The clones of the bacterial library were almost equally represented in the six chromosomes as follows: chr I: 2445, chr II: 2978, chr III: 2132, chr IV: 2643, chr V: 4152 and chr X: 2357. Wild-type hermaphrodite worms were screened to identify genes for

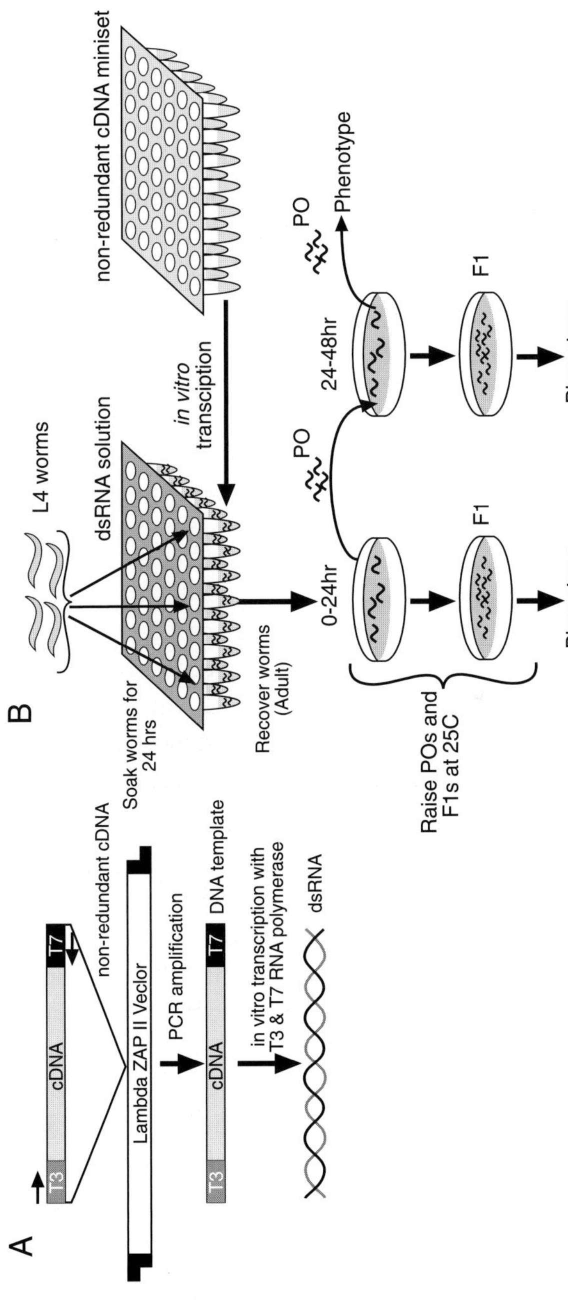

Fig. 6.   Strategy of the RNAi screening. (A) RNA preparation; (B) Diagram of "RNAi — by soaking". (From Maeda *et al.*, 2001.)

which specific dsRNA resulted in aberrated phenotypes as sterility, embryonic or larval lethality, slow post-embryonic growth or post-embryonic defects. Numerous acronyms were given to the defects as Ste and Stp for sterility of the worm and its progeny (respectively), Egl, for egg-laying defects, Lva for larval arrest, etc. The screening-identified phenotypes were assigned to 1722 genes. Most of these genes were not assigned to chromosomal sequences previously. They revealed an interesting phenomenon. Genes of similar functions were clustered in distinct, multi-megabase regions, of individual chromosomes. Such genes frequently shared transcriptional profiles. It was also observed that the genes of X chromosome had a different range of functions as compared to the genes of autosomes. Take, for example, Nonv phenotypes (nonviable phenotypes that result in lethality or sterility) were much rarer in cases where the dsRNAs were from X chromosomes than from autosomes. The Nonv phenotypes represented, quite frequently, genes that have orthologues in other eukaryotes.

By examining the domain-compositions of genes in the various phenotypic groups the investigators could suggest which domains are shared with a wide range of eukaryotes (in yeasts, plants, metazoa) and which are more limited (only in metazoa), or even restricted to worms only. Those with a wide range were considered ancient domains.

Can information from the tiny *C. elegans* teach us how to reduce human obesity? A team of investigators from various laboratories in Boston, the Washington State University, Pullman, WA and the University of Cambridge, UK (Ashrafl *et al.*, 2003) suggested that this was possible. Ashrafl *et al.* (2003) also performed a genome-wide screening in *C. elegans*. They focused on fat storage and metabolism in this worm. They found a way to tag fats in the worm by feeding them with bacteria that had dyed fats. For that, they grew bacteria on a medium in which the lipids were stained with the vital dye Nile Red. This feeding caused the staining of the lipid droplets in the worms intestine, the principal site of fat-storage in the worm. The investigators then fed the worms with the transgenic bacterial clones that were derived in their previous study (Kamath *et al.*, 2003). These clones had dsRNAs that represented 86 per cent of the ORFs in the *C. elegans* genome. Feeding on such individual bacterial clones is expected to silence the respective genes. In practices, the worms that fed on the various bacterial clones were evaluated for changes in fat deposition. This was a formidable task but it resulted in identifying 305 genes that had a positive role in fat deposition. When these genes were silenced (by RNAi) there were disorders in fat metabolism or in fat deposition; while the silencing of other 112 genes caused more fat deposition or enlarged fat droplet size. By reference to existing data on *C. elegans* genes and genes in mammals with homologous sequences the authors could learn about fat or lipid metabolism in both *C. elegans* and mammals. Take, for example, reduced amounts of stored fat resulted from RNAi of genes encoding worm homologs of enzymatic components of the membrane lipid biosynthetic machinery such as choline/ethanolamine phosphotransferase and CDP-alcohol phosphatidyltransferase. Reduced or disorganized fat deposition patterns were also caused by RNAi of several known components of sterol metabolism, and by RNAi of *C. elegans* genes whose mammalian homologs are involved in gastrointestinal digestion and uptake of food. Notably, increased fat was caused by RNAi of a worm homolog of the mammalian hepatocyte nuclear factor 4-α (HNF4-α). Mutations in the gene encoding this factor are associated with the onset of diabetes of the young. Several similar correlations were revealed and in total over 50 per cent of the *C. elegans* fat-regulatory genes that were identified by Ashrafl *et al.* have mammalian homologs that have not been previously implicated in regulating fat storage. The authors suggested that possibly, homologs of these newly identified *C. elegans* fat-regulatory genes also controlled mammalian body weight.

It appears that wisdom that was derived from the tiny worm may be useful in negating human obesity. In the Proverbs of King Solomon, he suggested deriving wisdom from another small animal: "Go to the ant, thou sluggard, consider her ways and be wise...." (Hebrew Bible, Proverbs, Chap. 6). Only from ants we should learn to store food and to avoid famine in subsequent lean seasons; while from *C. elegans* we should learn how to avoid excessive fat storage in our bodies.

In spite of the success in using RNAi for functional genomics (by reverse genetics) in *C. elegans*, it seemed that RNAi does not inhibit all the genes of this worm. Partial resistance to RNAi was recorded in its nervous system; the neurons in the head region were almost fully resistant to RNAi (see Simmer *et al.*, 2002, for references). A multinational team consisting of investigators of the Hubrecht Laboratory in Utrecht, the Netherlands, from several laboratories in the USA and from Cambridge, UK, found a way to circumvent this resistance. Simmer *et al.* (2002) found that a loss of function of a putative RNA directed RNA polymerase (RdRP) of *C. elegans* that is encoded by the gene *rrf-3* resulted in a substantial enhancement of sensitivity in several of the worms tissues, especially in the nervous system. Since this worm is full of nerve cells, the finding that such cells are rendered RNAi-sensitive in an *rrf-3* mutant background was rather helpful for RNAi Silencing. This finding was soon utilized by a group of investigators from several laboratories in Sussex and in Hampshire, UK. Keating *et al.* (2003) performed a whole-genome analysis of 60 G protein-coupled receptors in *C. elegans* by gene knockout with RNAi and used the RNAi hypersensitive *rrf-3* strain. Such an analysis was of considerable importance because G-protein-coupled receptors (GPCRs) are the largest family of genes in animal genomes and represent more than 2 per cent of the genes in *C. elegans* and in man. The evolutionary conserved seven-transmembrane encoded-proteins of these genes transduce a diverse range of signals. The authors screened the capability of all GPCRs-predicted sequences to bind to small-molecule neurotransmitters or neuropeptides by using RNAi and quantitative behavioral assays. The RNAi silencing of genes was performed, as in studies mentioned previously, by feeding the worms with transgenic bacteria that expressed specific dsRNA. Several genes were detected and were likely to be involved in reproduction while other genes were likely to have a role in locomotion. The findings were consistent with the known action of neuropeptides on the body-wall muscle and the reproductive tract of *C. elegans*.

The use of reverse genetics (functional genomics) by the aid of RNAi in *C. elegans* and the application of this approach in other eukaryotes was reviewed by Robert Barstead of Oklahoma City, OK (Barstead, 2001).

Another large-scale screening of mutants concerns the mobility of a transposon in *C. elegans*. Plasterk and associates (Vastenhouw *et al.*, 2003) of the Hubrecht Laboratory in Utrecht used their long acquaintance with transposons and in collaboration with investigators from the Tennis Court Road (Cambridge University, UK) augmented the collection of *C. elegans* genes involved in silencing. The notion that the natural function of RNAi is to protect the genome from the transposition of transposons was suggested by the Utrecht investigators already in the past (Plasterk and Ketting, 2000). These investigators also noticed that both RNAi and co-suppression are implicated in transposon silencing and share parts of the same machinery (Ketting and Plasterk, 2000). The Tc1 transposon has 32 copies in the genome of *C. elegans*. "Luckily" for this worm, these transposons are jumping in the somatic cells but not in the germline. We could infer that if they would jump around in the germline they would mutate this worm to extinction. Vastenhouw *et al.* (2003) used a procedure to isolate genes involved in RNAi by following the jumping of Tc1 away form an insertion site. They thus used a worm line in which the *unc-22* gene was inactivated by the insertion of a Tc1. Such worms do not move. But when the Tc-1 moves away from *unc-22*, the progeny with the restored *unc-22* will move. The investigators then used an "RNAi feeding library" and screened 14 387 predicted *C. elegans* genes sequences. Mutations in 27 of these genes caused the progeny to regain mobility. This was an indication that the Tc1 jumped away from *unc-22*, meaning that silencing of Tc1 was impaired. Some of the isolated genes affect both transposon silencing and the RNAi (e.g. *mut-16*). Can RNAi inactivate genes be involved in RNAi? In practice this can happen (see: Dudley *et al.*, 2002 for references and discussion). Under which conditions this can happen is not clear yet. Transforming worms that express the *gfp* transgene with a chimera gene that contains a truncated *gfp* gene provides the possibility to investigate cosuppression. Such a chimera transgene will cosuppress the endogenous *gfp* so that the worm will not fluorescent. But if the cosuppression is silenced the worms should regain the ability to synthesize GFP and show fluorescence. The mutants isolated by Vastenhouw *et al.* (2003) were analyzed for their capability to affect this cosuppression. Four of the newly isolated genes were essential for cosuppression. When these were knocked out, there was no cosuppression: the worms regained GFP production in the germline cells. The authors estimated that there are more than 27 genes involved in transposon silencing, cosuppression and RNAi. Many such genes may have escaped from

the screening because they are vital for other function and null mutations in such genes may lead to lethality. In a way this research of
the Plasterk team is an extension of a much earlier study of Plasterk
and associates (Ketting *et al.*, 1999) in which the *mut-7* gene was isolated from *C. elegans*. The *mut-7* encoded a protein that is similar to an
RNase. When this gene was mutated the Tc1 transposon started to be
mobile in the germline cells. This finding led the authors, already in
1999 to suggest that silencing of transposons and RNAi were related
mechanisms.

Development of *C. elegans* Larvae, Heterochronic Genes and the Emergence of MicroRNA

To understand the involvement of small dsRNA and especially
miRNA in the development of *C. elegans* one must become acquainted
with the biology and genetics of this worm. A detailed treatment of
these subjects is beyond the frame of this book. The reviews of Slack
and Ruvkun (1997) and Ambros (2000) as well as an earlier publication of Sulston *et al.* (1983) provide details and literature on the embryonic cell lineage and development of *C. elegans* as well as on mutants
affecting this development. Investigators analyzed the development
of *C. elegans* in great detail. Starting with the fertilized egg, each
cell division was followed and the fate of each of the two daughter-
cells resulting from each division was traced. Thus, for example, it
is known that 17 cycles of cell divisions are required till the mature
gonad cells are established. The number of cell divisions required to
reach the fully differentiated stage of each tissue and organ is known.
The temporal and spatial program is very exact and does not change
among individual worms, unless the worms are exposed to specific
environmental conditions (as feeding conditions) or are being affected
by certain mutations. From the hatching phase to the mature worm
there are four molting stages, L1, L2, L3 and L4. After the last molting the worm differentiates into a sexually mature animal. At each
molting stage specific cells are triggered to divide in the endoder-
mal, the mesodermal and the ectodermal cell lineages. Thus, all the
cell divisions that lead to specific fates are programmed in a precise sequence during the molting stages. Any diversion from this

program can be recorded because the worms are almost transparent and fit nicely in the view-field of a medium-power microscope (the length of the mature worm is about 1 mm). It takes 3.5 days from hatching to maturation. As noted above, the number of cell cycles required to reach the maturation of specific tissues is very different. For the external cuticle of the worm, only eight rounds of cell division are required. For the worms feeding organs nine to 11 rounds are required, while 17 cell divisions are required to attain gonad maturity. After each cell division there are two possibilities — both daughter cells divide further or one of the daughter cells will not divide further but starts to differentiate. The various organ-differentiation patterns differ with respect to these possibilities. In the differentiation of the cuticle it is frequent that only one of the daughter cells will divide further. In the gonad-differentiation it is more frequent that both cells divide further until differentiation takes place eventually. The latter possibility is compatible with the observation that only *two* "stem-cells" for gonads in the hatched worm contribute to the many gonad cells of the mature animal. There are many "stem-cells" in the hatched worm for the making of the cuticle.

Heterochrony means any change in the timing of a developmental event relative to another developmental event, that is, if normally gene A is expressed before gene B, in heterochrony this order is not kept. If gene A and gene B cause different morphogenetic events, the regular sequence of events is not kept in heterochrony. There are numerous examples of heterochrony occurring in nature, such as juvenile features retained in mature animals. Heterochronic genes (as A and B mentioned above) control the relative timing of developmental events. The molecular-genetics of heterochromatic genes was investigated many years ago (e.g. Chalfie *et al.*, 1981) but recent investigations implicated such genes with RNA silencing. I shall take the study of Grishok *et al.* (2001) as an example.

The latter study is a collaborative effort of 10 investigators from six research institutes (including A. Fire, G. Ruvkun and C. C. Mello as participants). These investigators analyzed two heterochronic genes of *C. elegans*: *lin-4* and *let-7*. The products of *lin-4* and *let-7* are dsRNAs of 22 nt and they were termed *small temporal RNAs* or stRNAs. The *temporal* designation is related to the time-limit of the expression of these genes. But the name was temporal by itself — they are now included in the *miRNAs*. *lin-4* is essential for the transition of larval stage one (L1) to larval stage two (L2) whereas *let-7* controls the transition of a later larval stage to adult stage. *lin-4* suppresses the level of the protein LIN-14 in the L1 stage and *let-7* suppresses the level of protein LIN-41 at a later larval stage. After Grishok *et al.* (2001) isolated 14 cDNAs of the *rde-1* genes (out of 23 such homologs in C. *elegans*), they then produced dsRNAs from two of these homologs and termed them *alg-1* and *alg-2* (for argonaute like genes). Following the injection of these two dsRNAs, the progeny had morphogenetic defects as bursting at the vulva and a lack of the adult-specific alae (longitudal stripes that run the length of the cuticle on both sides of the worm) as well as additional abnormalities. As for *alg-1*, even the dsRNA representing a part of the gene sequence caused the vulval bursting. It was also found that *alg-2* is synergistic to *alg-1* but has a weak effect by itself. But neither *alg-1* nor *alg-2* are essential for RNAi. An additional gene of C. *elegans*, *dcr-1*, is predicted to encode a protein related to the *Dicer* of *Drosophila* and the "Carpel Factory" of *Arabidopsis*. As such it was implicated in RNAi (PTGS) and the regulation of development. The RNAi of *dcr-1* was used to assess its role in the developmental control and RNA interference in C. *elegans*. Indeed, *dcr-1* (RNAi) induced developmental abnormalities during larval growth that were similar to those induced by *alg-1/alg-2* (RNAi). In summary inactivation of genes related to the RNAi pathway genes, the *Drosophila Dicer* homolog *dcr-1* and two homologs of *rde-1* (*alg-1* and *alg-2*) caused heterochronic phenotypes similar to *lin-4* and *let-7* mutations. All these three genes (*rde-1*, *alg-1* and *alg-2*) are necessary for the maturation and activity of the *lin-4* and *let-7* stRNAs. This study therefore indicated that there is a common processing machinery that generates guiding-RNA for both RNAi and endogenous gene regulation — at least in C. *elegans*.

The involvement of *lin-4 and let-7* in the RNAi mechanism, as emerged in the example of Grishok *et al.* (2001) was a relatively recent awareness. But these two genes were investigated as regulators of C. *elegans* larval development since many years ago (e.g. Lee *et al.*, 1993; Ha *et al.*, 1996; Feinbaum and Ambros, 1999; Slack *et al.*, 2000; Pasquinelli *et al.*, 2000). Such studies took place in the respective laboratories of Victor Ambros at Harvard University and Dartmouth

College, NH and of Gary Ruvkun at the Harvard Medical School. While the term "small temporal RNAs" (stRNAs) referred to their temporal expression during limited developmental stages, the term *miRNA* was then applied to the small RNA species that were derived from these stRNAs.

A Burst of MicroRNA Findings in *C. elegans*

Three articles on miRNA were published in *Science* (Vol. 294) in the October 26 issue. Two of these three publications (Lau *et al.*, 2001 and Lee and Ambros, 2001) were on *C. elegans* while Lagos-Quintana *et al.* (2001) of the Tuschl laboratory reported on miRNA in *Drosophila*, their favorite experimental animal. We shall deal, in some detail, with the RNA silencing in *Drosophila* in Chap. 5. Many miRNAs were revealed by Lagos-Quintana *et al.* (2001). They were given *miR* numbers (e.g. miR-1 to miR33); these miRNAs had similarity to *lin-4* and *lin-7* of *C. elegans* and their predicted precursors were all stem-loop-shaped.

The "fishing" success of the Bartel Laboratory was even greater than that of Tuschl and associates. Lau *et al.* (2001) revealed 55 previously unknown miRNAs in *C. elegans*. For their "fishing" the Bartel Laboratory focused on the population of small dsRNAs that was a *Dicer* product. They then looked at those dsRNAs that had the appropriate features: a length of about 22 nt, a 5'-terminal monophosphate and a 3'-terminal hydroxyl group. The dsRNAs that were caught in this "net" were sequenced. A considerable number of dsRNAs (330) that were sequenced had matching sequences in the *C. elegans* genome. An additional population (182 clones) had sequences that are homologous to sequences in the genome of *E. coli*. This was expected because the worms were fed with these bacteria. There were 55 dsRNAs that fitted the criteria of miRNAs (i.e. derived from a transcript that will fold into a hairpin RNA). Most of these dsRNAs had a length of 21–24 nt. The authors listed 54 miRNAs and recorded the sequences of all but one of these miRNAs. Like miRNAs of *Drosophila* mentioned above, the predicted precursor RNAs could be folded to imperfect dsRNA. The coding sequences for these predicted precursors were commonly outside of protein-encoding genes. Some were coded by intron sequences. By retrospect it was clear why the miRNAs were not easily fished by computer search of the *C. elegans* genome. The sequences of the predicted precursors are not strictly homologous to any known gene, not even to genes that are silenced by the respective miRNAs. There are commonly one or more base-pair differences between a miRNA and its target sequence. In a total length of ca 21 nt such differences are sufficient to render computer search ineffective. On the other hand, once isolated the miRNAs could be identified by northern blot-hybridizations. Many of the discovered miRNAs had developmental-expression patterns, as was previously revealed for *lin-4* and *let-7*. Moreover, most of the miRNAs of *C. elegans* had orthologs in another nematode species (*C. briggsae*). The *let-7* as well as seven of the novel miRNAs are conserved even in *Drosophila*; most of the latter are even conserved in humans. Not all the novel miRNAs are expressed only during certain periods of the larval development. The miR-1 is expressed throughout the *C. elegans* development and this miRNA is also the most conserved among the miRNAs found in *C. elegans*. In some cases, the sequences for different miRNAs are clustered in the chromosomes. They can thus be expressed together and subsequently diced into different miRNAs. The authors estimated that their yield of miRNAs is only a fraction of the population of miRNAs in *C. elegans* and suggested that there are more than 100 miRNA genes in *C. elegans*.

Lee and Ambros (2001) had the same rationale to look for additional miRNAs (with similarity to the then known *lin-4* and *let-7*). But they used a different search strategy. They applied a computer program to search for the sequence of the *C. elegans* genome and also used cDNA cloning. In their effort to identify additional miRNAs they looked for sequences that exhibited four characteristics: (1) expression of a mature RNA of about 22 nt; (2) location in intergenic (non-protein coding) sequences; (3) high DNA-sequence similarity between orthologs in *C. elegans* and *C. briggsae*; and (4) processing of a ca 22 nt mature RNA from a stem-loop precursor transcript of about 65 nt. By using a computer program they identified three new miRNAs. They then produced a cDNA library of 1.6×10^6 independent *lambda* clones from a size-selected (ca 22 nt) fraction of total *C. elegans* RNA. The resulting clones were screened on the bases of several criteria. This procedure yielded 13 new miRNAs. One of the latter (mir-60) was also identified by the computer search. Some but not all the new miRNAs vary in abundance during larval development of *C. elegans*. Hence, these miRNAs may be involved in the regulation of larvae development. Like Lau *et al.* (2001), Lee and Ambros (2001) also predicted that there could be more than 100 different miRNAs in *C. elegans*.

More recently, Eric C. Lai (2003) of the University of California in Berkeley reviewed the studies on miRNAs in *C. elegans* and other organisms. He emphasized the fact that in the past only coding sequences were taken into account for "making of an organism". Only recently, it was revealed that a "constellation" of non-protein coding RNA genes was revealed as having a major role in this "making". The latter were thus unknown in the past and the investigators did not know about these unknown genes. This led Lai to quote Donald H. Rumsfeld, the U.S. Secretary of Defense who said, **"But there are also unknown unknowns. The ones we don't know, we don't know."**

For this citation Rumsfeld gained the journalist's prize for ambiguity. The truth is Rumsfeld did not deserve the prize. Although what he said "sounds queer", it is clear! Let me divert to the campus of Eric Lai which is in Berkeley, the town named after George Berkeley (1685–1753), the philosopher who claimed that what we sense with our senses are not realities but merely the consequences of our perception.

Lai (2003) reminded us that miRNA emerged from genetic studies, primarily in *C. elegans* where *lin-4* and *let-7* were discovered sequentially. These genes did not encode proteins but were rather transcribed during certain stages of the larval development. The transcripts inhibited the translation of specific protein-encoding genes that are essential for the worm's differentiation. Both these transcripts resulted in short dsRNAs that were first termed small temporal RNAs (stRNAs) and later miRNAs. Both *lin-4* and *let-7* have sequence similarity but not identity, with sequences in their respective target genes (*lin-14* and *lin-44*). This causes *lin-4/lin-14* and *let-7/lin-44* binding and consequently the inhibition of the translation of the protein-coding genes. Lai also indicated that the turning point in miRNA awareness resulted from the studies in several laboratories that provided evidence for the linkage between RNAi and miRNA. It was found that both mechanisms (causing co-suppression and regulation of differentiation) require the *Dicer*. In miRNA *dicer* is required to process *lin-4* and *let-7* RNAs from their respective stem-loop precursors. Moreover, the RNA-induced silencing complex (RISC) was found to be required for RNAi and miRNA systems. More on miRNAs of mammals and of plants will be discussed because since 2001, miRNAs were revealed in other organisms where even more research was devoted to this small RNA rather than in nematodes.

The term miRNA does not define a function but the process by which these small RNAs were derived, all from a DNA sequence in the genome that frequently resides in an intron. The transcript of this sequence has a length of about 70–100 nt in animals and is longer in plants. The sequence of the transcript is such that it folds on itself, forming a hairpin or stem-loop structure in which there is complementation that causes the formation of a double strand. This dsRNA may be with bumps in places where there is no complementation between the two strains. But there are also miRNAs with full complementation of the base-pairs. Those with full complementation may have also full, rather than near-perfect complementation with their target, mRNA. We shall see that those miRNAs that have a near-identical sequence to their target may commonly cause the cutting of the target rather than inhibiting its translation into protein. The siRNAs and the mature miRNAs share the same length of about 21 nt. They also share the requirement for an Argonaute protein motif for their production. But then siRNA are involved only in cutting their mRNA target while miRNA are involved in either inhibition of translation from their target mRNA, or in cutting this mRNA.

The same laboratories of Bartel and collaborators and of Ambros and collaborators who wrote about miRNA worked hard for about 2 years and came up with significant additional information on miRNA in *C. elegans*. The Bartel Laboratory (Lim *et al.*, 2003) used a similar approach as they used previously. This included a computational procedure and comparisons between genomes of *C. elegans* and *C. briggsae* to identify miRNA in these worms. They combined the MiRscan program with molecular identification and validation. By these procedures the number of validated miRNAs in *C. elegans* was increased to 88 and the authors estimated that no more than 35 additional miRNAs remain to be revealed in this worm. The 88 miRNAs represented 48 gene families, almost half of these (meaning 22 families) are conserved in humans. Some of the miRNAs are very abundant, between 1000 molecules per cell, reaching for some miRNAs, over 50 000 molecules per cell.

The Ambros Laboratory (Ambros *et al.*, 2003) used the extraction of total RNAs from worms, the construction of cDNA libraries and northern blot analyses, as they did in their previous study (Lee and Ambros, 2001). By cDNA sequencing and comparative genomics, Ambros and associates also attempted to identify additional miRNAs in *C. elegans*. Their yield was 21 "new" miRNA genes. They also revealed other "new" small RNA species. Among the latter there were 33 members of a class of tiny noncoding-RNA genes that were of the same size as miRNA (20–21 nt) but this new class of RNAs was apparently not processed from hairpin precursors, thus not authentic miRNA. In addition, about 700 distinct small antisense RNAs were found with a length of about 20 nt. These were coded from more than 500 *C. elegans* genes. The investigators assumed that these were endogenous siRNAs. From their findings, the investigators concluded that in *C. elegans* there are various ongoing gene silencing mechanisms, by RNAi, miRNA and possibly by another group of small RNAs. Just like Lim *et al.* (2003b), Ambros *et al.* (2003) also estimated that there are 100–120 miRNA genes in *C. elegans*. About 80 per cent of these genes are conserved in a related nematode, *C. briggsae*, and about 30 per cent have apparent homologs in insects and/or vertebrates. The addition of the "new" miRNA genes of Ambros *et al.* (2003) and of Lim *et al.* (2003b) to the "old" miRNA genes summed up to 96 genes according to Ambros *et al.* (2003); not many more genes are expected. Interestingly, some miRNAs can be grouped in classes of similar or even very similar sequences. The authors suggested that serveral similar miRNAs could target the same mRNA and reduce its translation. This could allow redundant control of target-gene expression by multiple miRNAs. In summary the study of Ambros *et al.* (2003) indicated that a vast complexity of various gene-silencing (either "direct" or via the chromatin region in which the genes reside) is going on in *C. elegans*. If this is so, one should assume that similar manifold silencing is going on in other animals too and possibly in all eukaryotes. Such manifold silencing may be essential for the normal differentiation and physiology of eukaryotic organisms.

A third laboratory, that of Gary Ruvkun of the Harvard Medical School (Grad *et al.* 2003), also looked for additional miRNAs in nematodes. For that they developed a special informatic method to predict miRNAs in these animals. They took into account miRNAs in other animals (*Drosophila* and man) and verified their computer-findings by northern blot hybridization and a sensitive PCR approach. The hairpin structure of four "new" microRNAs is shown in Fig. 7. They further increased the number of assumed miRNA genes and suggested that the genome of *C. elegans* may encode 140–300 miRNAs and potentially many more. Are there really more? This is a yet unknown information and we are led back to the citation of Donald Rumsfeld about the **unknown unknowns**.

While several laboratories were engaged in a wide range search of miRNAs in *C. elegans* as shown above, the laboratory of Ann Rougvie, at the University of Minnesota, MN, investigated in detail a heterochronic gene pathway (Abrahante *et al.*, 2003). The tissue that this laboratory focused on is the stem cell-like hypodermal cells called *seam cells*. The correct maturation-time of these seam cells is regulated by a heterochronic gene pathway. During the L3 larval stage the miRNA *let-7* is transcriptionally activated and it then down-regulates its target genes (e.g. *lin-41*). This allows the seam cell to progress to their terminally differentiated adult-state cells. This terminal differentiation requires a zinc-finger transcription factor, LIN-29, in the hypodermis of worms at the L4 larval stage. If LIN-29 accumulates earlier than normal there is a precocious maturation of the seam cells while if this accumulation is retarded, this maturation does not take place at the correct time, causing additional molting cycles (beyond L4). Abrahante *et al.* (2003) looked for mutations that affect the differentiation of seam cells during the L3 to L4 larval stages. They found a heterochronic mutant and first termed it *lin-57* but then the name was changed to that of a previously revealed mutant *hbl-1*. This name was given because of its sequence similarity to the *hunchback* gene of *Drosophila* (*hbl-1* is an acronym for *hunchback*-like number 1). To identify quickly mutants that affect maturation of seam cells, the authors

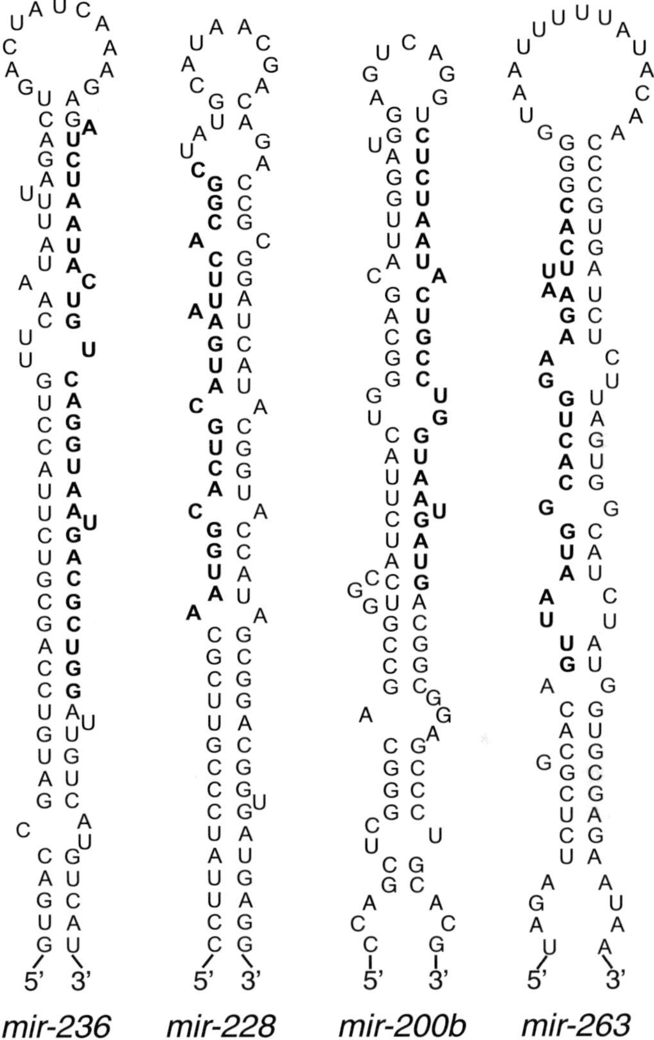

Fig. 7. Identification of microRNA precursors in *C. elegans*. The stem regions to which the antisense probes were designed are identified by bold letters. (From Grad *et al.*, 2003.)

looked at the locomotion of worms. Moreover, rather than using wild-type worms, worms with the *rol-1* mutation were used. The latter have a very different mode of locomotion before the maturation of the stem cells (when they move in a sinusoidal pattern) and after maturation of these cells (then they roll.) As for the *lin-57* mutation, the loss of function caused precocious hypodermal maturation (during the L3 to L4 molt). This early maturation is not completely normal and early mature cells have a greater than the normal number (16) of nuclei per cell. The *hbl-1* mutants also have defects in their vulva and thus an egg-laying defect.

The mutant revealed by Abrahante *et al.* (2003) seemed to be a defective version of the HBL-1 zinc-finger protein. The investigators found that *hbl-1* acts on *lin-29* and it requires *lin-29* in order to exert its function. HBL-1 regulates the timing of seam cells differentiation by preventing LIN-29 accumulation in the hypodermal tissue before the L4 stage. Furthermore, *hbl-1* is also acting to maintain a high level of cell division in seam cells during earlier stages because disruption of *hbl-1* by RNAi reduces the number of these cells at L2. The 3' UTR of *hbl-1* is involved in down-regulation by posttranscriptional regulation. In addition, neuronal cells are also affected. And also with these cells the 3' UTR of *hbl-1* has a temporal regulatory role. But what binds this 3' UTR of *hbl-1*? When this 3' UTR was analyzed it was found to contain several conserved sequences that can form heteroduplexes with *let-7* (a miRNA). This investigation therefore provided strong evidence for the mRNA, encoded by *hbl-1*, to be a target of the miRNA *let-7*. But there are also binding sites for *lin-4* (another miRNA) in the 3' UTR of *hbl-1*. Hence, it is plausible that *lin-4* also participates in the down-regulation of *hbl-1*. The situation is more complicated than summarized above because the various larval cell lines are not equally affected by *hbl-1*. The authors suggested a scheme for the heterochronic gene-regulations that emerged from their study (Fig. 8). Names of all these genes may confuse the novice in this field. Let us therefore remember that LIN-14 and HBL-1 are transcription factor proteins; LIN-28 and LIN-41 are proteins with hallmarks of translational regulators, while *lin-4* and *let-7* are miRNAs.

On the same date (October 29, 2002) Abrahante *et al.* (2003) submitted their article, a team of Frank Slack and associates (Lin *et al.*, 2003) of Yale University, Harvard Medical School and the University of Texas in Austin, submitted their article to the same journal (*Developmental Cell*). The two articles were published sequentially in the same issue and have almost identical titles. The

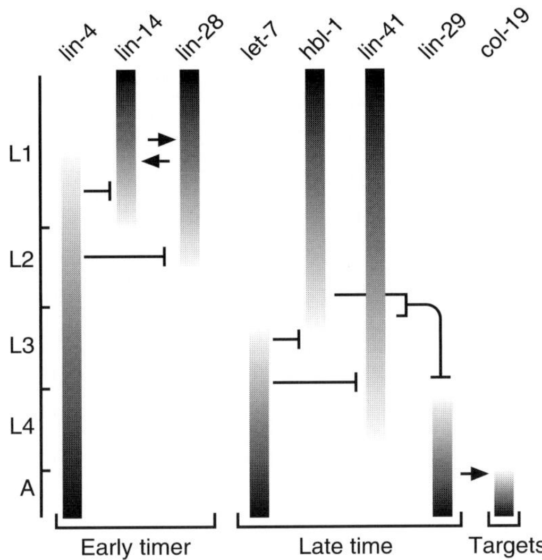

Fig. 8. Heterochronic gene regulatory cascade. Stages of post-embryonic development are indicated on the Y axis. Temporal gene-expression patterns are indicated by filled boxes. Regulatory interactions are indicated by connecting bars (negative interactions) or arrows (positive interactions). (From Abrahante *et al.*, 2003.)

results of the two investigations are also very similar. The difference (though not great) lies in the discussion of the results. The team of Lin *et al.* (2003) stressed in their discussion, the evolutional conservation of the hunchback-gene during temporal patterning. They also claimed that spatial and temporal patterning might share certain features. Functions of key genes and their mechanisms of regulated expression were very probably conserved among rather different animals. They therefore assume that in the future specific miRNAs would be identified in *Drosophila* that bind to and regulate translation of *hb* just as such miRNAs were found that regulate the translation of *hbl-1* in *C. elegans*.

Transcriptional Silencing in *C. elegans*

General background for polycomb and trithorax complexes

We have dealt with two kinds of gene silencing: silencing of the mRNA by its cleavage and inhibition of the translation of mRNAs into proteins. There is also gene silencing by the prevention (or inhibition) of the transcription of RNA from its coding gene-sequence. In this kind of gene-silencing, small RNAs probably participate as guides to silence regions of the DNA in the genome, but the RNA itself is not silenced. Therefore, it will not be considered as RNA silencing.

Activation of genes by their transcription into RNA in eukaryotes depends on a complicated interaction between the relevant DNA sequence and a plethora of chromatin-remodeling factors, histone-modifying enzymes and general transcription factors. We shall see that with some imagination we can look at the process of blocking transcription as in an American football match. First, one factor (protein) gets hold of a DNA site (a football player who holds the ball) but before the first factor increases its grip, an additional factor (or several factors) joins in to ensure a stronger grip on the site. This may go on until a large inhibiting complex of factors is built at and around the silenced DNA site and the transcription from this site is prevented. In many cases, this silencing is carried on through cell (and nuclear) divisions. Nuclear DNA replication requires access of the DNA polymerase complex to dsDNA. How is this done at sites that are complexed with the factors (proteins) that are executing the silencing? There is no answer to this question yet, but obviously the silencing can be maintained through mitosis and even meiosis, causing an epigenetic effect. The programming of "on" and "off" states of chromatin was studied in depth in *Drosophila*. The subject was reviewed in Simon and Tamkun (2002) and this review is recommended for those interested in a broad background. Figure 9 summarizes a model for the binding of regulatory complexes with chromatin. Briefly, there are two main groups of chromatin regulatory proteins that were revealed in *Drosophila*: the polycomb group (PcG) and the trithorax group (trxG). Both PcG and trxG proteins have roles in the maintenance of chromatin states (off or on) by maintaining spatial patterns of homeobox (Hox) genes expression. The Hox genes regulate insect pattern since the very early embryo-stage. Generally, PcG proteins are *repressors* maintaining the "off" state of chromatin while trxG proteins are *activators* of this state and maintain transcription. But there are exceptions. Some PcG proteins behaved (in certain genetic tests) as activators suggesting that the functional division into PcG suppressors and trxG activators is an

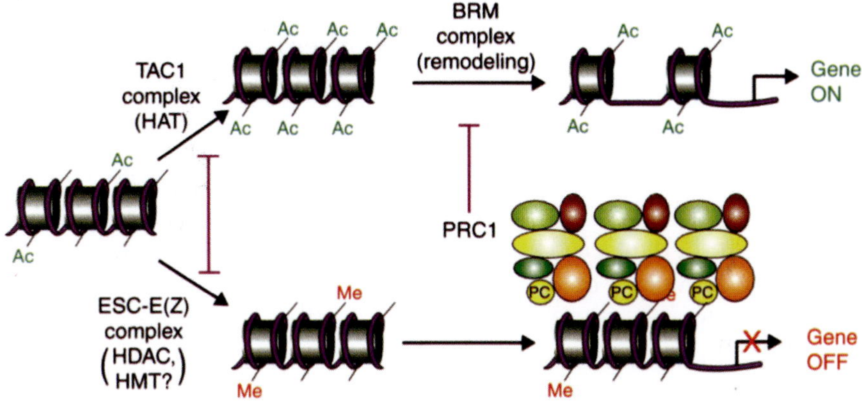

Fig. 9. Model for multistep mechanisms of the trxG and PcG chromatin complexes. The illustration depicts a nucleosome array at a target genes under trxG/PcG control. The trxG activating pathway is shown at the top and the alternative PcG-repression pathway is shown below. "Ac" is acetylation of histone tails and "Me" is methylation. In the model TAC1 acetylates nucleosomes and ESC-E(Z) deacetylates. Distinct histone codes created in the first step then help attract either BRM complex (trxG) or PRC1 (PcG pathway). Vertical bars indicate opposing effects. (From Simon and Tamkun, 2003.)

oversimplification. Both PcG and trxG are built of complexes, for example, the PcG has the PRC1 (polycomb repressive complex 1) and the ESC-E(Z) complex (for histone deacetylation). The trxG has the TAC1 (trithorax acetyltransferase complex 1) and the BRM complex. The situation is more complex because there are additional complexes in PcG and in trxG and some of the complexes are built of several subunits. There are also functional aspects that presently seem perplexing. Take, for example, in order to prevent transcription the PcG does not have to "sit" on the respective promoters. Cases were revealed in which the PcG exerted its suppressive activity by binding to the chromatin several kilobases away from the promoter. There are indications that PcG and trxG can interact, but details of such interactions are not known yet. It is also not known if bindings of PcG and trxG to the chromatin requires a prior chromatin remodeling, such as changes in the tails of the histones that are associated with nucleosomes (e.g. methylation and/or acetylation of certain amino acids).

Homeobox genes in nematodes

Hox genes were found in genomes of animals. They encode home-odomain transcription factors and thus have a role in patterning during development, especially the pattern of the anterior-posterior axis following the establishment of the zygote in the fertilized egg. The organizations of Hox genes in the genomes of nematodes, insects (*Drosophila* and *Anopheles*) and man have similarities that suggest that the Hox genes have a common, very ancient ancestor. Obviously, this organization has diverted considerably so that nematodes have their unique Hox-gene clusters. The Hox-gene clusters in the human chromosome 7 are composed of four clusters that are very close to one another. In *C. elegans* the Hox genes on chromsome II are arranged in four gene clusters in one region, then there is a gap of about 4 Mbp and after the gap, there is another pair of Hox genes. Since the nematodes comprise a very large and diverse phylum, there are nematodes that differ in their Hox-gene organization from *C. elegans.*

Hox genes have roles in specifying cell fates along the body axis in *C. elegans.* Take, for example, among the Hox genes of this worm, the anterior gene *ceh-13* and the posterior gene *php-1* are required in patterning. The elimination of *php-1* causes some defects and even posterior-to-anterior cell-fate transformation. Hox genes are also regulated by the PcG and trxG groups proteins. The trxG of *C. elegans* is LIN-59 which regulates the Hox genes positively. The PcG gene-homologs of *C. elegans* are *mes-2* and *mes-6* (maternal-effect sterility). They function in silencing the X-chromosome in germ-line cells of this worm. But recent studies suggest that they also have a role in regulating Hox genes in *C. elegans.*

Polycomb-like proteins regulate patterning in **C. elegans**

In the past, no role was identified in *C. elegans* for PcG homologs in the somatic Hox-gene regulation. Then, again, two reports were submitted to *Developmental Cell* (on January 23, 2003 and February 11, 2003, respectively). The respective articles were printed in the same issue, one after the other (Ross and Zarkover, 2003, and Zhang *et al.*, 2003). The investigators of the two teams looked at the same organs that are normally located at the caudal end of *C. elegans.* This "tail" contains, in male worms, very specific structures that are required for sensing and mating with the hermaphrodite worms. There are nine pairs of bilateral sensory *rays*, each composed of two neurons and a structural cell (these are the V-rays). More posterior to the V rays are three pairs of T rays. The posterior rays were amply studied in *C. elegans* and provided model systems for the identification of genes that govern developmental patterning, cell-fate specialization and male-specific neural development. Just as the fruit fly has head bristles for the benefit of *Drosophila* geneticists, the worms have caudal rays for the benefit of *C. elegans* geneticists. Both Ross and Zarkower (2003) and Zhang *et al.* (2003) used the caudal rays to answer the question of whether the Polycomb-like group regulates Hox-gene expression in *C. elegans*. Of two protein complexes in the PcG in *Drosophila*: the PRC1 and the ESC-E (Z) complex (for histone deacetylation), the former (PRC1) has no analogs in *C. elegans*. But there are two protein homologs in *C. elegans* for the ESC-E(Z) couples. There are the MES-2 (homolog to E(Z)) and MES-6 (homolog to ESC) in these worms. In addition, there are two other proteins MES-3 and MES-4 in *C. elegans*. Genetically, MES proteins have a repressive function that manifests itself when the worms have the mutant genes, *mes-2*, *mes-3*, *mes-4* and *mes-6*. In such genotypes there are derepressions such as the dereppression of silenced X chromosome. Actually, the absence of MES protein leads to germ-line degeneration and maternal-effect sterility. Till the study of the two teams no role for MES proteins in somatic patterning and regulation of Hox gene expression was found in *C. elegans*. The specific question of Ross and Zarkover (2003) was whether the MES proteins in *C. elegans* will affect the Hox genes *mab-5* and *egl-5* and by that change the differentiation of the caudal V rays. The situation is somewhat complicated but it was of advantage to the investigators. The *mab-5* Hox-gene has an activator: *pal-1*. When the *pal-1* is mutated the male worm loses its mating ability. Males lacking the *mes* genes activity display anterior expansion of tail rays. More specifically worms with a null mutation in *mab-3* have a loss of V rays. The authors looked at 2000 worms and found 10 mutants that restored the V rays. From the analysis of these alleles, the authors concluded that indeed *mes-2*, *mes-3* and *mes-6* are normally active in the male soma during V ray development. Genetic analysis also showed that *mes* genes act upstream of the Hox genes (as the PcG group in *Drosophila*). The *mes* mutations also affected *pal-1* by suppressing it, causing in some cases restoration of mating capability. It was additionally found that the appearance of ectopic rays or deformed rays are all consistent with Hox misexpression in the V ray lineage of *mes* mutant males. It was assumed by the investigators that normally *mes* proteins negatively regulate Hox genes thus preventing wrong patterning of V rays.

The *mab-5* Hox gene also regulates other *C. elegans* developmental processes such as the correct migration of certain neuroblasts during larval development. The authors found that the *mes* genes are essential for the correct negative regulation of *mab-5* in the process of these neuroblast movement. Additional experiments revealed that generally *egl-5* and *mab-5* are regulated by *mes* genes. Using specific anti *mes-2* and anti *mes-6* antibodies these proteins were found in both the germ line cells and soma cells during L2 and L3. Because *mes-2*, *mes-3* and *mes-6* are homologs of the PcG group it appears that basically the PcG/Hox-gene interactions found in flies and other organisms also operate in *C. elegans*.

The Zhang *et al.* (2003) publication focused on another protein, SOP-2, that regulates the activity of Hox-genes in *C. elegans*. To clarify the background we should remember that Hox genes encode conserved sequences of homeobox-containing transcription factors. Thus, they can initiate the transcription of specific genes in cells where the Hox genes are active. Another level of regulation is by the PcG group. In *Drosophila* the latter will repress the Hox genes where the transcription of the target gene should be silenced. Looked at it from the other side: when Hox genes activate transcription without control, homeotic appearance of differentiation is expected. The PcG are causing the repression of chromatin so that the DNA in the affected region will not be transcribed. In our case, the DNA sequences are those coding for the Hox proteins. Thus, if the chromatin in the Hox gene region is *not* complexed with PcG, the transcription of target genes will proceed to where it should not be active. Is this sequence of controlling mechanism going on also in *C. elegans* where previously no real PcG homologs were found to affect Hox gene in the worms soma? Zhang *et al.* (2003) provided evidence that the same principle is also true in *C. elegans* but the role of PcG in other animals was taken over by another protein, SOP-2, that is *related* but not orthologous with any PcG protein.

To characterize the *sop-2* gene a genetic approach was taken. The investigators looked for *sop-2* mutants that will have a homeotic effect. They found a mutation, *sop-2* (bx 91). This was a very helpful mutation because it was temperature dependent. At 15° the worms developed normally, as wild-type worms. At 25° they were scrawny and were arrested at an early larval stage. At 20° the worms with this mutation were generally long, uncoordinated, male abnormal and showed homeotic transformation. The investigators looked especially at the homeotic transformation. In wild-type (male) worms there are cell division patterns and differentiation within two bilateral rows of epidermal stem cells, known as seam cells that were mentioned above. These are regulated by the Hox genes *mob-5* and *egl-5*. The three most posterior seam cells, V5, V6 and T produce nine pairs of sensory rays. The *mob-5* is expressed in the V5 and V6 cell lineages and is required for the differentiation of rays, rather than the longitudinal cuticular ridges (alae). The *egl-5* Hox gene is expressed in the V6 lineage and is essential specifically for rays number 2 to 6. The homeotic change in males caused by *sop-2* (bx91) at 20° was that the seam cells V1 to V4 produced, in the mutant, rays and fan-like cuticular structures in the anterior body region. Can we say that the mutant male worm acquired a moustache? Probably, this will be too anthropomorphic, but clearly male rays appear where they should not exist. It clearly indicated that in *sop-2* (bx91) at 20°, the Hox genes *mob-5* and *egl-5* promoted transcription of ray-forming genes in cells where these Hox genes are normally silenced by the PcG type proteins. This means that the protein encoded by *sop-2* took the role of the PcG proteins. In the absence of *sop-2* proteins (or when they are at a level below the critical one), the Hox genes are expressed at times and locations where they should be silenced in order to lead to normal differentiation. The expression of the Hox genes was followed by tagging the expression of the Hox gene with the GFP. Indeed, the results were as expected: there was Hox-GFP expression in cells where there is normally no such expression. Additional tests showed that ectopic expression is not limited to the differentiation of rays — other differentiation systems such as body size, sex determination and vulva development became ectopic by the *sop-2* (bx91) mutation too. The latter types of differentiation are not known to be regulated by Hox genes. Actually, in the latter mutant ectopic expression of Hox genes starts during embryonic development (when the embryo has reached a certain stage and the main cell-lineages are established). In addition to controlling Hox genes the *sop-2* encoded protein seemed to control the expression of other genes. The *sop-2* gene was isolated and the DNA-sequence was found to differ between *sop-2* and *sop-2* (bx91), by a single base-pair difference that leads to the conversion of proline to serine. A dsRNA with homology to the essential coding region of *sop-2* caused the appearance of phenotypes that were similar to those of *sop-2* (bx91) mutants, but the former were not temperature sensitive. Sequencing also revealed that the protein encoded by *sop-2* is not a member of the PcG group but it has a domain of another group of proteins — the SAM domain containing proteins.

While the Zhang *et al.* (2003) team added very important information on the control of differentiation in *C. elegans* some major questions are still open for further investigation. One unsolved problem is how exactly the suppression of Hox genes is exercised by *sop-2*. Is the SOP-2 protein interacting with other proteins to silence the expression of the gene? Is chromatin remodeling playing a role in this silencing?

Silencing of the Transposable Element Tc1 in the Germ Line of *C. elegans*

As already mentioned above, TEs comprise a major component of many organisms. Organisms in which TEs reach about half or more of the genomes sequence include maize and man (see: Galun, 2003). Most are not full-size and intact TEs but rather "skeletons" that lost their mobility capability. The most abundant TEs in *C. elegans* are the Tc1 elements that belong to Class II TEs which have terminal inverted repeats (TIRs) and code for a transposase that recognizes the TIRs and is active in the transposition. There are 31 Tc1 elements in the *C. elegans* genome. Active transposition (by a cut-and-paste mechanism) takes place in somatic cells but it is suppressed normally in the germ line. If Tc1 is jumping in the germ line we could expect chaos because whenever a Tc1 is inserted in an essential gene this would lead to lethality, or at least misfunction in the progeny. Whatever it is, we can safely assume that the silencing of Tc1 in the germ line had a selective advantage in the evolution of *C. elegans*. Nevertheless, there are "mutator" lines of *C. elegans*. In these worms Tc1 is transposing in the germ line too. The "mutator" mutants were found to be deficient in RNAi. Consequently, Ronald H.A. Plasterk of the Hubrecht Laboratory in Utrecht (Plasterk, 2002) suggested that the silencing of Tc1 in the *C. elegans* germ line is maintained by RNAi. More recently, this laboratory provided evidence for this suggestion (Sijen and Plasterk, 2003). Each of the TIRs of Tc1 is 54 nt long and there are in total 1611 nt in this TE. The investigators looked for transposon-derived dsRNA in wild-type *C. elegans*. Such dsRNA were picked up mainly with probes representing TIRs. This already indicated that the transposon dsRNA has its origin mainly in the TIR sequences. The authors suggested that this resulted from a mechanism in which the Tc1 transcript folds back so that the two TIR transcripts approach each other and form a dsRNA. The finding that the transcription of Tc1 is done by a read-through process supports the suggestion of a fold-back and a dsRNA formation by the TIRs. As for Tc1, TIR derived dsRNA was also found for other TEs — Tc3 and Tc5 — that have a similar general structure as Tc1 but with longer TIRs and there are much fewer Tc3 and Tc5 than Tc1 elements.

The dsRNA in *C. elegans* is cleaved by an RNase IIII-like enzyme (DCR-1) as indicated above. The cleaved fragments then range from 20–24 nt and are termed siRNAs. The authors actually detected Tc1-specific siRNA in wild-type worms although with rather low abundance. Tc1-specific siRNAs were cloned and one clone corresponded to the TIR. It was further found that in all kinds of mutants that are defective in Tc1 silencing these Tc1 siRNAs were not as abundant as the wild types. While in mutants defective only in RNAi the Tc1 siRNA were readily detected.

Finally, Sijen and Plasterk (2003) found no evidence that the silencing of Tc1 was induced by the inhibition of transcription, meaning mediated by chromatin changes. By default, this silencing is thus processed only or primarily by the posttranslational mechanism, meaning that the Tc1 transcript is cleaved.

For summary and overview the reader can refer to the chapter by Ketting *et al.* (2003) of the Plasterk Laboratory, in the book edited by Hannon (2003). This is a relatively recent description of RNAi in *C. elegans.* More specifically, it lists the analyzed genes that are involved in RNAi of *C. elegans* (Table 1).

The phenomenon of RNAi in *C. elegans* is gradually becoming transparent. Transparency is a term that fits this nematode nicely because one of the virtues of *C. elegans* as a model for genetic studies is its transparency; changes in its internal structure can swiftly be observed in whole animals. However, several aspects of RNAi in this organism should be explored further. Among these aspects are the spatial and temporal movements in the RNAi process. By spatial movement I mean the systemic transmission: movement from cell to cell and from one tissue to another, of the RNAi effect. Only recently was a protein (SID) revealed that was implicated with the transport of

Table 1.   Genes involved in RNAi in *C. elegans.*

Locus	RNAi function soma	germ line	Protein domains	References
rde-1	√	√	PAZ; PIWI	Tabara *et al.* (1999)
rde-3	√	√	not cloned	Tabara *et al.* (1999)
rde-4	√	√	dsRNA binding	Tabara *et al.* (2002 and unpubl.)
mut-7	−	√	RNase D	Ketting *et al.* (1999)
mut-8/rde-2	−	√	none detected	B. Tops *et al.* (unpubl.)
mut-14	−	√	DEAD-box RNA helicase	Tijsterman *et al.* (2002)
mut-15	√	√	none detected	R.F. Ketting *et al.* (unpubl.)
ego-1	−	√	RdRP	Smardon *et al.* (2000)
rrf-1	√	−	RdRP	Sijen *et al.* (2001)
rrf-3	√[a]	√[a]	RdRP	Sijen *et al.* (2001)
dcr-1	√	√	dsRNA binding; helicase; nuclease; PAZ	Grishok *et al.* (2001) Knight and Bass (2001) Ketting *et al.* (2001)
sid-1	√	√	transmembrane	Winston *et al.* (2002)

The locations where RNAi defects are observed in mutant strains are indicated for each gene. The one exception in the list is *rrf-2*. This protein is one of the four RdRP family members in *C. elegans,* but no role in RNAi has been observed yet.

[a]loss of *rrf-3* leads to hypersensitivity to RNAi. (Ketting *et al.* 2003.)

dsRNA into cells (Winston *et al.*, 2002; Feinberg and Hunter, 2003). The temporal aspect means the maintenance of the RNAi effect throughout the life cycle of the worm and its transmission to the next generation. There is evidence that RdRP is involved in maintenance and transmission but the details are not fully understood yet.

RNA Silencing in *Drosophila* and Mosquitoes

Since the early years of the 20th century, when Thomas Hunt Morgan shifted his attention to the fruit fly, *Drosophila melanogaster*, and established the "fly room" at the Columbia University, this fly became a main actor in genetic investigations. After about 90 years of serving as a model animal for genetic and developmental research, *D. melanogaster* continues to contribute vastly to our scientific knowledge. An enormous number of genes was characterized and located on the fly's genome. The nucleotide-sequences of all the fly's chromosomes are known and its development from the egg, through the embryo, the larva, the pupa, up to the mature fly, has also been clarified in great detail. *D. melanogaster* has a short life cycle and a great number of stocks can be cultured in a small space. Moreover, cytogenetic methods as *in situ* hybridization of the polytene chromosomes and molecular-genetic techniques, such as genetic-tranformation as well as cell culture protocols, are all well established in this animal.

We shall see below that only a few months after RNAi was discovered in *C. elegans* the first report on RNAi in *D. melanogaster* was submitted for publication (Kennerdell and Carthew, 1998). Once established, the RNAi of this fly yielded a shower of significant results. In this chapter I shall first provide a background for the development of *D. melanogaster* and thereafter review the early history of RNAi in this model animal. Then a selected number of RNAi studies in *D. melanogaster* will be described.

The Main Features of *Drosophila* Development

The following description of *Drosophila* development is highly abbreviated. The description is intended to serve as a basis for understanding experimental procedures that were followed during investigations concerning RNAi. I recommend that readers who are not familiar with the development of insects, consult relevant texts or at least general-biology textbooks such as Raven and Johnson (1996) that include useful illustrations. Rudel and Sommer (2003) reviewed the developmental mechanisms of diverse animals, among them insects and more specifically, *Drosophila*. This review is recommended to those who wish to acquaint themselves with comparative animal development.

There are several kinds of "bodies" during the development of this fly. After the fertilized egg is laid a larva emerges. This is a tubular body consisting actually of an "eating machine". The larva is converted into a pupa and from the pupa emerges the fly body. This can be considered a "flying and reproduction machine". The passage from one kind of body into the other is termed *metamorphosis* (the book of Franz Kafka, 1937, *The Metamorphosis*, is recommended but it will not contribute to the knowledge of insect development).

We shall start with the egg, even before fertilization. While still in the body of the mother, the egg is fed with special *nurse-cells* that help the egg to grow. These nurse-cells move the maternal mRNA into the egg-cell (the oocyte). Because the nurse cells are located at one end of the egg, the flow of mRNA into the oocyte establishes a gradient. It is more concentrated at one end of the egg-cell than at the end that is far away from the nurse-cells. This gradient is the blueprint for the future division into regions of the early embryo. After fertilization the mitoses of the zygotic nucleus proceed quickly, taking about 8 minutes between each nuclear division. There are 12 rounds of nuclear divisions without parallel cell-division. Thus, a mass of about 4000 nuclei, all of them residing in the center of the egg, is quickly formed. These nuclei form a syncytial blastoderm. Then they migrate and space themselves along the cortex of the blastoderm. Subsequently, cell walls grow between the nuclei. Thus, micromeres form a single layer of cells that encompasses a central cavity. Because of the graded distribution of maternal (nurse cells) material, the nuclei at the different locations in the embryo experience different levels of this maternal product. At this phase, as well as in subsequent phases of the embryo, the original orientation of the egg, before fertilization, is not "forgotten" and the blastoderm nuclei/cells experience different levels of the maternal product. Part of the maternal mRNA (e.g. from a gene termed *bicoid*) remains at one end and thus marks what will become the embryo's anterior end; the polarity of the embryo is thus established. The future head and thorax will develop where this mRNA was most concentrated. Then, $2\frac{1}{2}$ hours after fertilization, the *gap* genes come into the play and cause the division of the developing embryo into large blocks. The *gap* genes are activated by the maternal *bicoid* protein and the former are already embryonically active genes. About 30 minutes later, another embryo-gene is activated (*Hairy*) and this gene causes the subdivision of the blocks into seven fundamental embryonal regions. There are six *gap* genes and they were given fancy names such as *Kröppel* and *hunchback*. While the *hunchback* mRNA is transcribed throughout the development of the embryo its translation is controlled by the protein encoded in *manos.* The manos protein can bind to the mRNA of hunchback and thus reduces the translation. Due to this inhibition of translation another gradient is formed with much more *hunchback* protein at the anterior end. The next stage of segmentation is activated by a group of about 16 *segment polarity* genes. Among these is the gene *engrailed* which causes the division of each of the seven regions into halves, resulting in 14 narrow compartments. This is the final subdivision and is maintained throughout the fly's development. There are three head segments, three thoracic segments and eight abdominal segments.

Thus, within 3 hours after fertilization a quick and highly orchestered series of events that include the activation of embryo genes and segmentation, the basic body plan of the embryo is established. During the next (about 20) hours organ formation takes place. Here, other genes such as *tinman* are operative. These genes are already active 5 hours after fertilization. The larva of the fly was called an "eating machine", thus a gut, a heart and a mouth are the essential organs. This "machine" is ready to function at about 1 day after fertilization. After hatching the larva does not need to look far for its feeding. In nature the mother fly places the egg at the source of food (the fruit) and starts to feed right away. In the laboratory the investigator puts the eggs into milk bottles (or other appropriate containers) filled with feed. As the larva feeds it grows quickly but because it has a chitinous exoskeleton that has a limited stretching ability the growth is limited. Hence, after about 1 day, the exoskeleton is shed. Before the hardening of the new exoskeleton that replaces the previous one, the larva expands and continues to feed. Then the exoskeleton is shed again and this is repeated two more times. Thus, three stages, termed *instars*, occur within a period of about 4 days.

While the task of the larva is to store energy for future development, there is another process going on in the larval body: a dozen groups of cells called *imaginal disks* are set aside and divide actively without apparent differentiation. These are "buds" which will come to light only in the next phase of the fly's development.

At the beginning of this next phase a hard shell forms and the larva is transformed into the *pupa*. From the outside the pupa looks like a resting body but very active changes take place inside it. There the larval cells break down and furnish the nutrients for the formation of the organs of the final body of the fly. Then the cells of the imaginal disks assemble and differentiate into the organs of the adult fly. The "flying-machine" will then emerge.

The Homeotic Genes of *Drosophila*

We have discussed the homeotic genes while dealing with RNA-silencing in nematodes. Historically, these genes were first revealed in flies. As these genes encode specific transcription factors they trigger the transcription of other specific genes. When the homeodomain genes operate orderly the correct organ will be developed in the appropriate site. But when the homeodomain gene is mutated strange things will happen. Take, for example, one such mutation, *bithorax*, will result in the formation of an extra pair of wings because it has an alteration in thoracic segments. The result of another mutation, *Antennapedia*, will cause legs to grow out of the head, in place of antennae. It was found that several homeotic genes are mapped together on the third chromosome of *Drosophila*, in a cluster called *bithorax complex*. These genes actually affect the body parts that comprise the rear half of the thorax and the whole abdomen. Conspicuously, the *order* of the genes, in the cluster on chromosome three, is the same as that of the body parts controlled by them. Thus, the genes at one end of the cluster control the segments of the thorax; further genes control organs in the anterior part of the abdomen and genes at the other end of the cluster control the organs at the posterior part of the abdomen. The *antennapedia* complex of homeotic genes control organ formation at the anterior part of the fly.

The homeotic genes were actually discovered in *Drosophila* since the middle of the 20th century. The basic role of the homeotic genes in *Drosophila* is very similar to that of *C. elegans*. In the fruit fly the general body pattern is set very early in the development of the embryo, leading to the establishment of 14 segments. By that the overall pattern is established (see: Coen, 1999). This pattern provides the spatial clues where the various homeotic genes should be set into action. As the homeotic genes code for transcription factors, each of these genes will produce a protein that will bind to a promoter of a gene (or genes) that is more directly involved in the construction of specific organs and cause the respective transcription. Here, again is a warning. We may say that the protein encoded by a specific homeotic gene activates the transcription of a *leg-forming-gene*. But this is a vast simplification. We do not know at all how a leg (or a wing etc.) is structured. It is obviously a rather complicated process of cell division, cell expansion and changes in the characteristics of cells, in space and time. Surely many genes participate in this process of structuring a leg (or a wing, etc.). For the geneticist a *leg-forming-gene* stands for a gene that when it is mutated or missing (null mutation) no leg is formed. The geneticist is aware of the complication and is using the term *leg-forming-gene* only as a convenient short term. The warning is thus directed to the innocent reader who should keep in mind that a *single leg-forming-gene* does not exist.

The homeotic genes have a common general structure and code for homeodomain proteins with the same basic components. At the amino-end of these proteins there is a variable region that determines the specific activity of the protein. Then, towards the carboxyl end of this protein there are 60 amino acids that constitute the *homeodomain*. This domain has four α-helices. The coding region of this homeodomain is termed *homeobox*. One of the helices of the homodomain has the capability to bind specifically to the DNA of a target gene that is activated (by promoting its transcription) by the respective homeotic gene. The homeotic genes of *Drosophila* were termed HOM genes too. There are parallel homeotic genes in vertebrates, and in mice they were termed HOX genes. Due to similarities in nucleotide sequences the HOM genes were instrumental to reveal homeotic genes in other animals. Thus, the HOX genes in mice and man were identified. They also have the same general arrangement in these mammals as in the fruit fly and their relative location on the genome parallels the head-to-tail roles in the developing animal. Though two main clusters are found in flies (*Antennapedia* and *bithorax*) there are four such clusters in mammals. The similarity in nucleotide sequences and the order in which they are lined up in the genome that parallels the order of body components in which they are active, is maintained in very different animal phyla. This clearly indicates that homeobox genes appeared very early in animal evolution. Moreover, similar genes were revealed in flowering plants, thus we may place the evolution of archaic homeobox genes very early in the evolution of eukaryotes.

What was learned from developmental studies in *D. melanogaster* and other model animals can be summarized as follows. Soon after the egg is laid, the maternally pre-formed mRNAs that were deposited into the egg cell during orgenesis from one (anterior) end are being translated into proteins. Because of the flow coming from one end, a *gradient* of pattern forming proteins is established. Thus, a kind of "topographical map" is established. This lays the foundation for a division of the early embryo into segments. The homeotic genes that are residing in all the cells of the embryo start to be expressed only according to the topographical map in which the cells are located. Once being expressed and their respective proteins are translated the homeoboxes of these proteins "look" for the DNA binding sites in their target genes. They then initiate the transcription of the latter genes which are causing the differentiation of specific organs. Thus, there are three "layers" of positive regulators. We saw in nematodes and shall also see that there are also essential negative controlling entities as the miRNAs that can exert their impact by cleaving specific mRNAs or suppressing the translation from specific mRNAs. There is also negative control via the chromatin such as the effect of Polycomb proteins. These negative controls assure that genes that are *not* required, at a certain location and at a certain phase of development, will *not* be expressed.

The Early History of RNAi in *D. Melanogaster*

Wnt proteins comprise a large family of compounds of major importance in animal development. The *wnt* genes were discovered in divers animals as nematodes (*C. elegans*), mice and the fruit fly (*D. melanogaster*) and probably exist in all metazoa. The genes encode a secreted family glycoproteins, usually 350–400 amino acids in length. These have at least 18 per cent sequence-identity and a conserved pattern of 23–24 cysteine residues. The functions of Wnt proteins and their signaling mechanisms were reviewed in detail by Cadigan and Nusse (1997). This signaling is involved in several embryogenesis phenomena as embryonic induction, generation of cell polarity and specification of cell-fate. For its signaling the Wnt protein requires a cell-surface receptor and a mechanism of relaying the signal to the cell nucleus, In *Drosophila*, the Wg (wingless) is the best known Wnt family member. *Drosophila* homozygous for the wg (*wnt1*) mutations have segment polarity defects. The intercellular signaling that involves Wg in *Drosophila* is rather complicated. In order to understand the first RNAi report in this fly we shall only point to some of the interactions. Let us look at two neighboring cells that belong to two adjacent parasegments of the larval epidermis. One expresses the Wg protein and the other, that is posterior to it, produces the En (Engrailed) protein. Wg signals to maintain En expression; the En expressing-cell activates Wg by secreting the Hh protein. The Wg protein is secreted from this cell with the assistance of the Porc protein (an ER transmembrane protein). The secreted Wg probably acts through the Dfrizzeled-2 (Dfz2) receptor of the En cell. A cascade of events takes place in the En expressing (target) cell. The Wg, through some relays, can inhibit some target genes among them the *en* gene. From the En cell a complex signal then goes out to regulate-back the expression of the Wg protein. To further complicate the issue the Wg protein consists of different functional domains, each with a different function in the patterning of the embryonic epidermis that will form the cuticle of the fly's larva. Thus, Wg is also required for the patterning of the adult eyes, legs and wings. The laboratory of Richard W. Carthew of the University of Pittsburgh, PA, was engaged in the development of eyes (Zheng *et al.*, 1995) and wings (Zhang and Carthew, 1998) of *Drosophila*, more specifically, in the interaction of the *frizzled* encoded cell-surface transmembrane-protein (Fz) and the interaction of the *Dfz2* encoded receptor-protein (mentioned above), with the Wg secreted signaling-protein. Eight months after the publication of the first report on RNAi in *C. elegans*, the Carthew laboratory was ready with its report on RNAi in *Drosophila* (Kennerdell and Carthew, 1998, submitted it on October 20, 1998). The latter investigators asked whether Fz and Dfz2 transduce the Wg signal in the embryonic epidermis. They used dsRNA that corresponded to the *Fz* and to the *Dfz2* gene. These dsRNAs were synthesized *in vitro* by annealing the ssRNAs (sense and antisense) and were injected into the syncytial blastoderm embryos of the fly. But first, to test the system, two other genes were chosen for RNAi application: *ftz* and *eve*. These genes are required for the segmentation of the embryo. Their transcription occurs $1\frac{1}{2}$–2 hours after egg laying, meaning 10–60 minutes after the injection of the dsRNAs. The injected RNA was predominantly double stranded but also contained some ssRNA. The injected annealed RNA of either *Ftz* or *eve* into wild-type embryos clearly interferred with cuticle patterns similar to the respective mutants *Ftz* and *eve*. The antisense ssRNAs and the sense ssRNAs of these genes had a much weaker interference activity (by one order of magnitude). Even a level of about 30 molecules of dsRNA could cause a detectable interference. The dose of injection was reflected in the severity of the interference and the high doses caused interference as the respective null mutations of these genes. The effect of dsRNA was very specific; no gross malformation other than those caused by the *frz* and *eve* mutations were observed. The injection of dsRNA of the Ftz caused the respective reduction or elimination of the Ftz protein in the embryos. The investigators injected dsRNA corresponding to other genes that are expressed in the embryos and obtained similar results, indicating that injecting dsRNA into embryos is an efficient way to silence the respective genes. Thus, the dsRNA of *wg* was injected into the fly's embryos. This caused a defective cuticle pattern as that of *wg* mutants. But the effect was localized: only the larval regions that corresponded to the

 RNA Silencing

region where the injection was affected. The rest of the larvae stayed normal.

The localized effect of dsRNA, representing the *wg* gene, was also observed when other dsRNAs, representing additional genes, were injected into the fly's embryos. The investigators then focused on the *fz* and *Dfz2* genes. They prepared dsRNA representing the 5′ UTR of these genes and mixed the two dsRNAs. The injected mixture caused changes in the cuticle pattern. Again, the changes were restricted to the site of injection into the larvae. Furthermore, if only either the dsRNA representing *fz* or the dsRNA representing *Dfz2* was injected, there was no change in cuticle pattern. This suggested that these two genes function redundantly — either of them is sufficient for normal cuticle patterning. This possibility was examined further. In an additional experiment the investigators used a *fz* mutant and injected the ds *Dfz2* RNA into it. In this case the cuticle pattern was affected in a similar manner to the effect of injecting a mixture of ds *fz* RNA and ds *Dfz2* RNA into wild-type embryos.

As indicated above, the DFz2 protein acts as a receptor for the Wg signal; thus, the former protein should act downstream of the latter (Wg) protein. This was verified by injection experiments. Similarly, ZW3 kinase was previously suggested to act downstream of Fz and DFz2. This sequence of activities was verified by dsRNA injection too.

In summary this pioneer study by Kennerdell and Carthew (1998) furnished the following information:

- Like in *C. elegans*, dsRNA representing a given gene in *Drosophila* can silence this gene.
- As in *C. elegans*, dsRNA of *Drosophila* is by far more efficient than ssRNA in silencing.
- The RNAi approach is instrumental in analyzing gene function and gene interaction in *Drosophila*.
- The RNAi mechanism is rather efficient, as a small number of dsRNA molecules (e.g. 30 molecules per cell) will cause a silencing effect.

There was also a significant difference between *C. elegans* and *Drosophila*. In the latter animal the effect of dsRNA was not mobile: only the site of injection was affected. In addition, while in *C. elegans* the effect of RNAi was stable and could even be transferred to the next sexual generation, no such temporal extension of RNAi effect was revealed by Kennerdell and Carthew (1998). These authors thus extended their study to achieve a stable impact by RNAi. Kennerdell and Carthew (2000) introduced, by genetic transformation, a sequence of DNA into the fly's genome. This DNA was constructed in a way that after transcription it will form a hairpin by complementation between the upstream and the downstream regions of the transcript with the loop of a few nt in between. Such a hairpin-loop will have an extended dsRNA sequence. If this dsRNA is complementary to a given region of a gene the expected result would be silencing of the respective mRNA. To first test their idea the authors used embryos that carried a fused *engrailed-lacZ* transgene in their genome. Such embryos can be stained by X-Gal and the bands of the stained embryos are clearly visible. When these embryos were injected with the dsRNA of 745 nt that corresponded to a coding region in the LacZ sequence, the bands were not visible. Injection of the antisense ssRNA did not abolish the bands. The injection of a hairpin-loop of the LacZ sequence did abolish the bands almost completely. This indicated that the effect of the hairpin-loop was similar to the effect of the dsRNA. RT-PCR tests of the abundance of mRNA for *LacZ* confirmed that the mRNA was similarly affected: it was strongly reduced by the hairpin-loop construct.

The investigators then performed their "real" experiment. They used the same flies with the *engrailed-lacZ* transgene and transformed the latter flies with a construct that would, when activated, transcribe an RNA that will fold into a hairpin-loop, representing a fraction of the *lacZ* gene. The transcription of the hairpin-forming RNA was initiated in the F_1 generation when the transformed flies were mated with flies that had the appropriate activator. The results were that this hairpin-loop transcript caused, in most F_1 embryos, the same inhibition of pattern as direct injection of dsRNA. The authors were encouraged by a report of Martinek and Young (2000) who obtained similar results with another gene *period* that could be silenced by an inverted-repeat RNA.

An early study on gene silencing that is probably related to RNAi in *Drosophila* was done by Jensen *et al.*, (1999a, 1999b) at the CNRS in Villejuif, France. These investigators asked whether an active TE in *Drosophila* can be silenced by a transcript that has homology to the TE transcript. They were dealing with the *I* TE. This is a non-LTR retroelement (it does not contain long terminal repeats) with similarity to the

LINEs of mammals (see: Galun, 2003, for details). When *I* invades *virgin* genomes of *Drosophila* (the term was coined by Jensen *et al.*, 1999a and 1999b) it first has a very high frequency of transpositions, causing a high mutation rate, chromosomal non-disjunction and female sterility. But the high rate of transposition is transient. The number of *I* elements per genome reaches a final level, then after about 10 generations, transposition ceases. The transposition of *I* is not by cut-and-paste but rather by a replicative process, hence transposition also increases the number of *I* elements in the fly's genome. The investigators were aware of previous studies on co-suppression in plants and animals and therefore had the idea that if a part of the *I* element is to be transcribed in a fly, this partial transcript may cause co-suppression of the transcription of *I* and thus silence this element. They thus made a construct that contained a sequence of 969 nt from the 5′ end of ORF2 of the element. They also made a similar construct but with stop codons in order to abolish full translation. The constructs were introduced into flies and the transgenic flies were propagated for several generations. Then they were crossed with flies that contained an active (full length) *I* element. When control flies (without the transgene or with a transgene that cannot be transcribed) were crossed with an *I* containing fly there was a close to 100 per cent dead embryos probably due to massive transposition. But the transgene drastically reduced embryo mortality. The level of reduction was variable but strongly correlated with the number of transgene inserts in the flies. Four such inserts gave much higher protection (silencing of the *I* introduced by mating) than a single insert. Also, a number of sexual propagation cycles were required before the trasngene was capable of silencing the incoming *I*. Clearly, the co-suppression of *I* was not dependent on a translation from the transgene. This approach was first followed with a sense transgene (Jensen *et al.*, 1999a) but thereafter also with the antisense transgene (Jensen *et al.*, 1999b). When the authors submitted their second report they were already aware of the study by Kennerdell and Carthew (1998) that showed RNAi in *Drosophila* by a specific dsRNA. The possibility that the introduction of the transgene will trigger a dsRNA that will repress the transcript of the *I* element was therefore discussed but without a definite conclusion.

One day before the publication of the article by Kennerdell and Carthew (1998), namely on December 24, 1998, an article by Misquitta and Paterson of the National Cancer Institute, Bethesda, MD, was accepted for publication in PNAS (Misquitta and Paterson, 1999). These authors also intended to silence a specific gene from *C. elegans* by employing the RNAi system known at that time. They were engaged for the differentiation of embryonic muscles in *Drosophila*. They had previous information that particular sets of muscle precursor cells will establish a muscle pre-pattern in each hemi segment and then recruit fusion-competent mesodermal cells to build the characteristic muscle groupings. This is termed muscle "founded" differentiation. In *Drosophila* it was assumed that *nautilus*-expressing cells are the "founder-cells". Such a possible mode of differentiation could be analyzed if there would be a way to eliminate the *nautilus* expression. But no *nautilus* mutant was available. The authors therefore chose the means to prevent the *nautilus* expression. In one way they aimed an ablating agent to such cells. They targeted the toxin ricin to those cells by using the *nautilus* promoter. Indeed, the *nautilus* expressing cells were killed. This caused a severe disruption of muscle formation. The other way was to inject dsRNA into the embryos. The injection was performed to the posterior end of the eggs, slightly off their center. They made dsRNA from the mRNA of *nautilus* and found an effect that was similar to that of using ricin ablation of *nautilus* expressing cells. The efficiency of the RNAi approach was also demonstrated by preventing the expression of *lacZ* in transgenic flies that normally expressed this marker gene. In many additional RNAi studies the authors also found that dsRNA corresponding to parts of the *nautilus* transcript caused the same inhibition of embryonic muscle differentiation. Furthermore, the dsRNA approach was successfully used with additional genes as *twist* and *engrailed* and the respective distorted phenotypes were obtained. When the dsRNA of *white* (eyes) was injected into the embryos the adult eyes were affected but only in less than 3 per cent of the injected animals. [A note was added by B.M. Paterson to the page proofs of the PNAS publication: "While my paper was in press, similar results were reported by Kennerdell and Carthew".] Whatever it is, the two publications by Kennerdell and Carthew (1998) and of Misquitta and Paterson (1999), represented independently performed studies and opened the way for the utilization of the RNAi approach in *Drosophila*.

In the fall of 1999, the stage was ready to consider PTGS and RNAi comprehensively. This was done by Phillip A. Sharp of MIT in Cambridge, MA, in a *Perspective* article (Sharp, 1999). On the basis of studies on PTGS in plants, a study in *Trypanosoma brucei* (Ngo *et al.*, 1998) and the publication of Kennerdell and Carthew (1998) on RNAi in *Drosophila*, Sharp summarized the available information on the suppression of gene expression by specific homologous dsRNAs. It is probable that while phrasing his *Perspective* article he already possessed the results of the RNAi study of his own laboratory that was published a few months later (Tuschl *et al.*, 1999). The Sharp article not only presented a good summary of the then available information but also made a rather accurate prediction of how PTGS, RNAi and dsRNA are involved in the specific silencing of genes. He suggested the following diagram:

dsRNA containing exons of gene x

↓

RNAi or PTGS

↓

agent causes gene-specific degradation of mRNA from
gene x in both the nucleus and cytoplasm; gene-specific
agent or inducing agent can be duplicated by a normal cell
and transmitted to other cells

We shall see below, how accurate the diagram of Sharp really was. What was performed with dsRNA in plants, nematodes, *Trypononsoma* and *Drosophila* during 1998 and the early months of 1999, as summarized by Sharp (1999), probably triggered a special effort by Sharp and associates (Tuschl *et al.*, 1999). These investigators intended to obtain information on the molecular mechanisms by which specific mRNAs are degraded in the presence of dsRNAs that are homologous to the respective mRNAs. Sharp and associates assumed that a cell-free system should be useful to obtain such information.

The source of the cell-free system was a lysate prepared from 2 hour-old *Drosophila* embryos. These were collected, dechorionated and washed. Then the material was dried and ground in an appropriate buffer. The lysate was then centrifuged (14 500 g) and the supernatent was frozen in liquid nitrogen in aliquots and stored at −80°C until use. This lysate, derived from the syncytial blastoderm of fly embryos, can translate transcripts into functional proteins. Moreover, the level of synthesized protein is correlated, within a certain range, to the level of available mRNA.

The authors asked whether the embryo lysate would be useful to show the specific degradation of a given mRNA by a dsRNA that was homologous to a part of this mRNA. To visualize the effect of degradation of the mRNA they used mRNAs of two different luciferases: the *Renilla reniformis* (sea pansy) luciferase (Rr-Luc) and the *Photinus pyralis* (firefly) luciferase (P_p-Luc). These two luciferases are encoded in different DNA sequences and the luminescence of each requires a different substrate. The dsRNAs were about 500 nt long. They were transcribed from PCR products of the respective genes (Rr-Luc and P_p-Luc). In both cases the transcribed region started 100 nt downstream of the start of translation. The investigators first produced the respective ssRNA and the single-stranded antisense RNA (asRNA). These were annealed to produce the dsRNAs. Because lysates may differ considerably in their translation capability, internal controls were always included. Thus, when they tested the Rr-Luc, the investigators also added the P_p-Luc mRNA as a control, and *vice-versa*. The tests consisted of first adding the dsRNA and after an incubation period, the mRNAs for the luciferases was added. A typical experiment is presented in Fig. 10. After a further incubation the level of luciferase activity was recorded. The first incubation (e.g. 10 minutes) was required and the investigators found that a minute amount of dsRNA can drastically reduce (to about 30 per cent of the normal level) the translation of the respective luciferase. The level of mRNA was also lowered. The interference was specific, meaning that the dsRNA from the P_p-Luc gene reduced only P_p-Luc production and not the production of Rr-Luc. Moreover, dsRNA of another gene (*nanos*) did not affect the interference. As in *in vivo* experiments performed earlier, the length of dsRNA was important. Thus, when the length was only 49 nt the dsRNA did not reduce the level of luciferase. The level was slightly reduced by dsRNA of 149 nt and a 505 nt already resulted in maximal reduction of the luciferase. The usefulness of the embryo lysate of *Drosophila* was clearly indicated by the experiments of Tuschl *et al.* (1999). The specifity of dsRNA interference did not exist in other

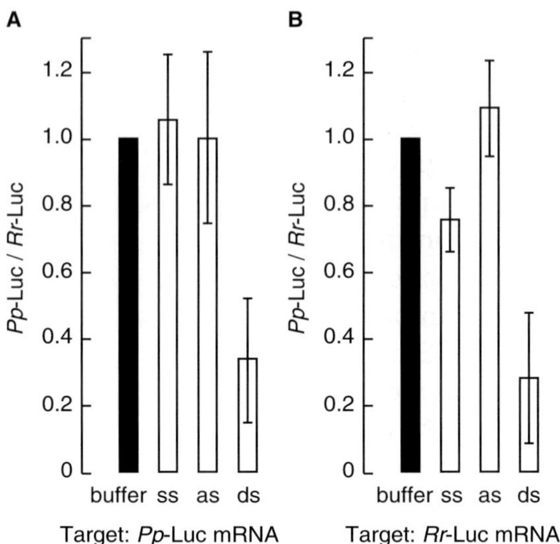

Fig. 10.   Gene-specific interference by dsRNA *in vitro*. (A) Ratio of luciferase activity after targeting 50 pm Pp-Luc mRNA with 10 nM ssRNA, asRNA or dsRNA from the 505 bp segment of the Pp-Luc gene. (B) Ratio of luciferase activity after targeting 50 pM *Rr*-Luc mRNA with 10 nm ssRNA, asRNA or dsRNA. An *Rr*-Luc to *Pp*-Luc ratio of one indicates no gene-specific interference. (From Tuschl *et al.*, 1999.)

in vitro systems that were analyzed by these authors: wheat germ extracts and rabbit reticulocyte extracts, but see Tang *et al.* (2003) in Chap. 12.

The effective collaboration of Sharp and associates and the efficient *in vitro* system that they developed, yielded one additional contribution to unravel the RNAi in *Drosophila* (Zamore *et al.*, 2000), but then the members of the group went on to establish their own respective teams.

While Zamore, Tuschl, Sharp and Bartel were engaged in utilizing their *in vitro* system to elucidate the molecular mechanism of RNAi in *Drosophila*, another team in Cold Spring Harbor, NY, had the same aim in mind but developed for this purpose a cell-culture system (Hammond *et al.*, 2000). It was clear that there was a difference in the efficiency of the journals in which the results were published. The team from Cold Spring Harbor submitted their manuscript to *Nature* on November 26th, 1999; it was accepted on January 26th, 2000 and it appeared in print on March 16th, 2000 (a total of almost 4 months from submission to publication). The team of Zamore *et al.* (2000) submitted their manuscript on March 2nd and it was accepted (after revision) on March 10th. It appeared in press on March 31st (a total of less than one month). The moral is that if you desire swift publication, approach a friendly neighbor journal rather than an overseas one.

It should be noted that during the few months, between the submission of the article on the establishment of the *in vitro* system in *Drosophila* and the submission of the report on the further utilization of this system (Zamore *et al.*, 2000), there was a great increase in information on PTGS and RNAi especially in plants and nematodes. This information included the identification of genes required for the initiation of dsRNA silencing, its intercellular movement and its transmission to the next sexual generation. One specifically important information was the dsRNA related occurrence in plants of RNA fragments with a rather uniform size of about 25 nt (Hamilton and Baulcombe, 1999). An important finding of Zamore *et al.* (2000) was that ATP is required for the *in vitro* RNAi. This was shown by the same *Rr*-Luc luciferase system described above but under the condition in which ATP was either left in its regular level or when ATP was depleted before the test of dsRNA interference. Furthermore, by the use of inhibitors of protein synthesis such cycloheximide and puromycin, the translation of mRNA by the *in vitro* system was reduced about 2000-fold while the RNAi proceeded at normal efficiency. On the other hand, chloramphenicol, an inhibitor of mitochrondrial protein synthesis did not affect the translation in the embryo lysate (and also had no effect on RNAi). Also, the 7-methylguanosine cap was not affecting the RNAi process. As was found in plants (by Hamilton and Baulcombe, 1999) and in *Drosophila*, fragments of 21–23 nt of dsRNA were found. Zamore *et al.* found that these short RNA species represented about 15 per cent of the input dsRNA. To justify the length of these short RNAs they noted that 22 nt corresponds to two turns of an A-form RNA-RNA helix. This ~22 nt size is a real riddle. Any sequence that is shorter than 19 nt will lack specificity; such a sequence may have several (non-specific) RNA targets in the cell. On the other hand, a sequence of 23–25 nt is already very specific. The chance that a homologue to such a sequence exists in several sites in the genome, but in different genes, is extremely low. More than

25 nt are superfluous for specificity. The experimental system also indicated that the dsRNA is cleaved directly into the short species and there is no single-stranded intermediate. The cleavage of the dsRNA into the 21–23 nt species did not require ATP. Finally, the results led Zamore *et al.* (2000) to the suggestion that the 21–23 nt fragments that are derived from the dsRNA cleavage are guiding the subsequent cleavage of the mRNA.

The teams of Hammond (Cold Spring Harbor), Bernstein (State University of NY at Stony Brook), Beach (University College, London) and Hannon (Cold Spring Harbor) had similar aims as those teams mentioned above: to study the biochemistry of the RNAi system in *Drosophila*. Hammond *et al.* (2000) used cell-culture as well as cell extracts in their investigation and the *Drosophila* Schneider 2 (S2) cells in their study.

The first step was to use S2 cells that were transiently transfected with a *lacZ* vector and consequently had β-galactosidase activity that could be detected by X-gal. When the cells were co-transfected with a dsRNA that corresponded to the 300 nt of the *lacZ* sequence the expression of *LacZ* was drastically and specifically reduced. Note that contrary to the syncytial cell-free extract no pre-incubation was required for the elimination of β-galactosidase/LacZ activity. Only the dsRNA reduced *LacZ* activity. The ssRNA, either sense or antisense, did not affect *lacZ* activity. Once the investigators were satisfied that the cell-culture system is functional in RNAi they proceeded to test whether RNAi would also be active against endogenous gene expression. For that an endogeneous "reporter" gene is required. The investigators chose *cyclin E*. The activity of the latter gene is essential in *Drosophila* for progression into the S phase of the cell-division cycle. The cultured S2 cells are mostly in the G2/M phase during their log-phase division. But when a dsRNA representing the first 540 nt of *cyclin E* was transfected into these cells, the cells were arrested in the G1 phase. The RNAi effect of the dsRNA was dependent of the length of the dsRNA: 540 and 400 nt were very effective, 200 or 300 nt were less effective and 50 or 100 nt were not effective at all. Parallel to the cell-cycle arrest there was also a reduction of the *cyclin E* mRNA. A modest reduction in *fizzy* mRNA was also recorded when a dsRNA corresponding to this gene was transported into cultured S2 cells. The *fizzy* gene activity is required for the promotion of the anaphase stage. The results clearly showed that RNAi may be a generally applicable method for probing gene function in cultured *Drosophila* cells.

The reduction of an endogenous transcript could result from either effects that happen during or after transcription. To explore these two possibilities, Hammond *et al.* (2000) turned to a cell-free system where the cells were first transfected with the appropriate dsRNA and extracted. Then a corresponding mRNA was added. The degradation of this mRNA was subsequently assayed. For these tests the investigators used dsRNAs and mRNAs of *cyclin E* and *lacZ*. Clearly, the extract degraded only the mRNA of *cyclin E* when dsRNA representing *cyclin E* was transfected into the source-cell. This dsRNA did not affect the mRNA of *lacZ*. The same was recorded when the dsRNA represented *lacZ*; in this case the cell-extract degraded the mRNA of *lacZ* but not of *cyclin E*. The extract thus should have formed a complex that can degrade a specific mRNA. The authors termed this enzymatic complex RISC (**R**NA-**I**nduced **S**ilencing **C**omplex). This (as yet hypothetical) RISC could degrade mRNAs provided they had a minimal length. The authors found that both sense and antisense mRNAs were degraded by the presumptive RISC. Clearly, degradation was only recorded if the transfected dsRNA had homology to the mRNA. The RISC did not degrade dsRNA. Thus, the enzyme that is active in cutting long dsRNA into short dsRNA is probably different from the RNase in RISC. The authors made an additional prediction: the short (21–23 nt) RNAs that result from the long dsRNA are then joining the RISC and guiding it to a homologous site of the mRNA where the cleavage will take place. The RISC should thus be a ribonucleoprotein particle. Indeed, further purifications and analyses indicated that the enzymatic complex contained short RNA species but it was not clear whether these species were ssRNAs or dsRNAs.

The Bernstein *et al.* (2001) team that consisted of almost the same team as Hammond *et al.* (2000), continued the study of the biochemistry of RNAi in *Drosophila*. The former assumed that indeed the degradation of specific mRNAs in this fly as well as in plants (PTGS) is accomplished by a multicomponent nuclease that is guided to the site of cleavage. The ~22 nt dsRNA "guide" fragments were detected previously in plants as well as in *Drosophila* systems that consisted of either a cell-free extract from embryos or S2 cells. The authors then asked whether there is a nucleoprotein complex that can cut the long dsRNA long into ~22 nt dsRNA fragments. Simple fractionation already indicated that there were two different nucleases in the extracts. Several enzyme candidates were analysed with respect to their ability to cut dsRNA into ~22 nt fragments. One of them was a complex of dual RNAse III and an amino-terminal helicase domain (CG4792). This latter complex could carry out the cleavage into the recorded ~22 nt dsRNA species. The authors termed this enzyme *Dicer* (*Der*). As expected, *Dicer* was capable to cut dsRNA but not ssRNA. Also, the *Dicer* required a minimal length of dsRNA. Fragments of 200 nt or less were hardly cut (diced). The authors used anti-*Dicer* antibodies and immuno-precipitated a nuclease activity from *Drosophila* embryo extracts and S2 cell extracts. This *activity* produced the dsRNAs of ~22 nt from the long (about 500 nt) dsRNA. The production of guide RNAs required ATP. The authors followed the RNAi approach by applying dsRNA that is homologous to the code for *Dicer* enzymes to S2 cells. The lysate of such cells had a 6- to 7-fold reduction in *Dicer* activity. This and other tests indicated that *Dicer* is active *in vivo*.

Another team of investigators from the cradle of *Drosophila* genetics, the Columbia University, NY, asked whether results obtained by previous *in vitro* studies can be repeated *in vivo*. Yang *et al.* (2000) used a similar approach as Tuschl *et al.* (1999), but without homogenization. The former investigators monitored the degradation of a mRNA in individual fly embryos. This could be performed by injecting the mRNAs of two luciferases, Rr-Luc and P_p-Luc, into the individual embryos. The two luciferase reporter genes have unrelated nt sequences. This permitted the use of one of the respective mRNAs as a control while the other was used to monitor the degradation by its cognate dsRNA. The mRNA was obtained either by the injection of a DNA plasmid or as final mRNA (capped and polyadenylated). The investigators commonly used 0–1 h embryos. The dsRNA (0.1 nl) was injected first. Then after 20 minutes, the mRNA was injected. The embryos were then assayed with a dual-luciferase system after a 3-hour incubation. Thus, the information could be obtained from many individual embryos. It sounds straightforward

but requires a highly-trained person and rather precise and sophisticated instruments. Basically, the investigations by Yang *et al.* (2000)
provided similar results as those by Tuschl *et al.* (1999), Zamore *et al.*
(2000) and Hammond *et al.* (2000) although there are some differences.
In the individual embryo system the length of the dsRNA that is effective in RNAi can be shorter — 80 nt is enough. Yang *et al.* (2000) found
that the RNAi induced gene silencing can be saturated by a rather low
level of dsRNA but it is also inhibited by excess unrelated dsRNA.
Also, the study of Yang *et al.* (2000) clearly suggested that the antisense
strand of the dsRNA determined the target specificity. Excess complementary sense or asRNA competed with the RNAi reaction. The
conclusion that the antisense strands of dsRNA fragments are determining mRNA target-specificity was an obviously important step for
understanding the biochemistry of the RNAi in *Drosophila*.

The RNAi System is Substantiated in *Drosophila*

By the end of 2000, the RNAi mechanism in *Drosopila* was amply investigated by *in vivo* and *in vitro* methodologies. A phase of research then
started, in which this mechanism was studied in greater detail and
suggestions based on the studies from 1998 to 2000.

An example of investigations that substantiated previous results and
added important information is the report by Elbashir *et al.* (2001a) that
was submitted by Thomas Tuschl after he moved to the MPI for Bio-
physical Chemistry in Göttingen, Germany. Several aspects of RNAi
in *Drosophila* were investigated by the Tuschl team, mostly based on
the *in vitro* system that was developed in Sharp's laboratory at the
MIT. First, the minimal length of the dsRNA that is required for dicing the dsRNA into 21–22 nt fragments was more precisely identified. Whenever the length of the dsRNA was less than 39 nt it lost
its capability to cause RNAi (the assay was based on degradation of
the mRNA of luciferases). By employing 5'-capped single stranded
(mRNA) targets and different dsRNAs, the investigators found that
the cleavage of the target is predominantly at a specific site. It is 10 nt
downstream from the 5' end covered by the dsRNA. If the dsRNA is
long enough (e.g. 52 nt) there may be an additional (weaker) cleavage 23–24 nt downstream of the former cleavage site. Furthermore,

the investigators found that the choice between cleavage of sense and antisense target RNA is not arbitrary. It is determined by the direction of the dsRNA that provides the "guide" for this cleavage.

Detailed analysis of the cleavage products of dsRNAs clearly indicated that they resulted from a reaction that is characterized by an RNase III cleavage. Chemically synthesized 21 nt and 22 nt dsRNAs were also capable of causing the cleavage of target (ss) RNAs. When these short dsRNAs were applied at a 10 nM concentration there was only a low level of cleavage of the target mRNA but at 100 nM the cleavage was "readily detectable". Further, a 10-fold increase in concentration (i.e. 1000 nM) did not increase the cleavage further. These authors termed the 21 nt and 22 nt dsRNAs with the overhanging 3′ ends: short interfering RNAs or *siRNA* and the complex of protein and RNA, a small interfering ribonucleoprotein particle of siRNA (rather than RISC).

The model for *dsRNA-directed mRNA cleavage* that was proposed by Elbashir *et al.* (2001a, 2001b, 2001c) is as follows:

- An RNase III-like enzyme cuts dsRNA into fragments that are predominantly 21–22 nt long and have short overhanging 3′ nucleotides.
- The dsRNA processing proteins that bind to the dsRNA in either the 3′ to 5′ or the 5′ to 3′ direction, or a subset of them, remain bound to the fragments of the dsRNA. The dsRNA is separated into two ssRNAs, one of which is guiding the cleavage of the mRNA.
- It is possible that the cleavage of the dsRNA and the cleavage of the mRNA are performed by the same endonuclease.

We shall see that the third point of the model was not supported by later studies. Quite a different aspect of RNA silencing in *Drosophila* was studied by Vladimir Gvozdev and associates at the Moscow State University in Russia (Aravin *et al.*, 2001). These investigators studied two groups of genes on the X and Y chromosomes of the fruit fly. These are the *Stellate (Ste)* repeats on chromosome X and the *Suppressor of stellate, Su(Ste)* repeats on chromosome Y. The interaction between the repeated genes is involved in the maintenance of male fertility of *Drosophila*. The *Su(Ste)* has similarity to *Ste* but its ORF is heavily damaged, although it is 90 per cent homologous to the ORF of *Ste*. The *Su(Ste)* has additional differences, relative to *Ste* (e.g. an inserted TE). When the X-linked *Ste* is overexpressed, as in the case of an absence of Y-linked *Su(Ste)*, the fly will be male sterile, meaning that the *Su(Ste)* repeats suppress the overexpression of *Ste*. The authors presented evidence in favor of a homology-dependent RNAi-related mechanism of *Stellate* silencing that was caused by the *Su(Ste)* repeats. It seemed that both strands of the *Su(Ste)* repeats are transcribed and a dsRNA that is involved in the silencing of the mRNA of *Stellate* is formed. Mutation-relieving of *Su(Ste)*-dependent silencing of *Stellate* has another effect. It leads to a derepression of LTR and non-LTR retrotransposons that reside in the fly's genome. Actually, dsRNA fragments of 25–27 nt that are homologous to sequences of *Ste* and *Su(Ste)* were detected in the testes of normal flies. This study required a considerable effort: each analyzed sample required hand-dissecting and isolation of 100 testes. The authors conclude their discussion by stating: *"The future studies of RNAi-related phenomenon may result in unexpected findings …."* Expectation of the unexpected is a very reasonable expectation, but this statement is reminiscent of the "unknown unknowns" of US Secretary of Defense Donald H. Rumsfeld that was mentioned in Chap. 4.

In the chapter on RNAi in nematodes, I mentioned the pioneer study by Grishok *et al.* (2001) who revealed the small temporal RNAs (*stRNAs*) was highlighted. These stRNAs were subsequently termed *miRNAs* and because of their important role for normal development in plants and animals, more details will be covered in Chaps. 10 and 12. Hutvagner *et al.* (2001) including Thomas Tuschl who moved to Göttingen and Phillip Zamore who moved to Worcester, MA focused their attention on the expression of *let-7* of *Drosophila*. The RNA entity of *let-7* is a 21 nt dsRNA that we came across while dealing with *C. elegans*. It was then termed small temporal RNA. The question was whether *let-7* has a precursor and if it has one, when, during the development of *Drosophila*, the precurser (*let-7L*) and the mature stRNA, will appear. RNAs were thus extracted from 11 stages of the fly's development (e.g. early-mid-late embryos; 1st, 2nd and 3rd larval instars; pupal stages and adult) and the levels of *let-7* and of a 72 nt precursor (*let-7L*) of *let-7* were analysed. Both *let-7L* and *let-7* were first detected at the late 3rd instar stage.

The *let-7L* was also found in the early pupae but then disappeared
and only the mature *let-7* was retained. The *let-7L* is a (72 nt) ssRNA
that is capable of folding into a hairpin with a loop of 5 nt. The two
strands that form the stem of the hairpin are not 100 per cent complementary; several pairs of nucleotides do not match. The synthetic
let-7L was digested to the mature *let-7* by embryo lysates. The *let-7* that
was produced *in vitro* had the same 5′ ends as the natural *let-7* RNA.
The authors suggested that the same *Dicer* that is required to convert
a dsRNA into a siRNA also cleaved the hairpin-loop precursor (*let-7L*)
into the mature *let-7*.

The RNAi system in *Drosophila* was further clarified by an additional study by the Hannon and Hammond team in Cold Spring Harbor.
Hammond *et al.* (2001a, 2001b) used their cultured fly cell system to analyse the RISC complex of *Drosophila*. This complex, as indicated above,
was known to be a multicomponent nuclease complex that includes a
ssRNA guide (of ~21 nt) that leads the RISC to the mRNA that is the
target for degradation. It is also known that about a dozen genes affect
the dsRNA response in various organisms. Among these was the family
of Argonaute genes (e.g. *AGO1* found in plants). The investigators thus
wished to link the biochemical information to the genetic information.
In other words, they intended to identify, by their *in vitro* system, the
proteins that are components of RISC in *Drosophila*. They thus purified
the RISC complex. In cell-free extracts the RISC is bound to ribosomes.
Thus, the RISC was first isolated together with the ribosomal fraction.
It was then separated from the ribosomes (by extraction with a high
salt concentration). Further purification identified a complex of about
500 kD. A protein that belongs to the Argonaute family was identified
in this complex. The gene for this protein was termed *AGO2* of which it
contained the PAZ and the PIWI domains. Further identification of the
AGO in the RISC complex was performed by AGO2-specific antibodies.
Furthermore, to test whether AGO2 is essential for RNAi in *Drosophila* S2
cells, the investigators used the RNAi system to silence the production
of endogeneous AGO2. The assay was based on the ability to silence the
Renilla luciferase mRNA. Indeed, when the AGO2 was silenced the ability to silence the mRNA for the luciferase was also abolished. It should
be noted that *Dicers* also contain a PAZ domain. The RISC and the *Dicer*

thus seemed to share a component but are most probably two different entities (see the *Epilogue* for more details).

The Tuschl team used the *Drosophila* embryo-lysate to obtain more details on the requirements from the siRNA to efficiently trigger the RNAi effect (Elbashir *et al.* 2001c). Their assays were based on the degradation of the mRNA of firefly luciferase (P_p-Luc). The latter was evaluated by the levels of luminescence. The mRNA of another luciferase (Rr-Luc) was used as an internal control, as indicated above (Tuschl *et al.*, 1999). A low ratio of P_p-Luc/Rr-Luc was an indication of high degradation of the mRNA of P_p-Luc. The interference was caused by different dsRNAs. The latter were constructed from sense and antisense RNA sequences that were derived from the coding sequence of the P_p-Luc gene. These sequences were of different lengths and different overhangs, in the annealed dsRNA, at the 5′ and the 3′ ends. Figure 11 provides an example of the experiments that were aimed to determine the optimal total length and the optimal length of overhangs of dsRNA for the RNAi effect. From the tests presented in Fig. 11 and from additional tests, the authors concluded that 21 or 22 nt is the optimal length of siRNA and that a 3′ overhang of 2 nt is more effective than a 1 nt overhang. The effectivity was strongly reduced when there was a 3 nt overhang. It was also found that the degradation of the target mRNA is highly sequence-specific but not all the positions of the nucleotides in the siRNA are equally important: a mismatch between siRNA and target sequences is the most detrimental in the center of siRNA. It was also found that the position of the cleavage site in the target (mRNA) is defined by the 5′ end of the guiding siRNA rather than by its 3′ end. The detailed information that was obtained by Elbashir *et al.* (2001a, 2001b, 2001c) required probably a lot of painstaking work but it had an important purpose. This information was very useful for further studies in which synthetic siRNA would be used to target (and silence) specific genes.

The results of Elbashir *et al.* (2001a, 2001b, 2001c) were soon supported by a research conducted in Heraklion, on the Greek Island of Crete. Boutla *et al.* (2001) also tested the RNAi efficiency of short dsRNAs. These synthetic siRNAs were 22 nt long and most of them consisted of two 5′ phosphorylated RNA strands but hydroxylated-end

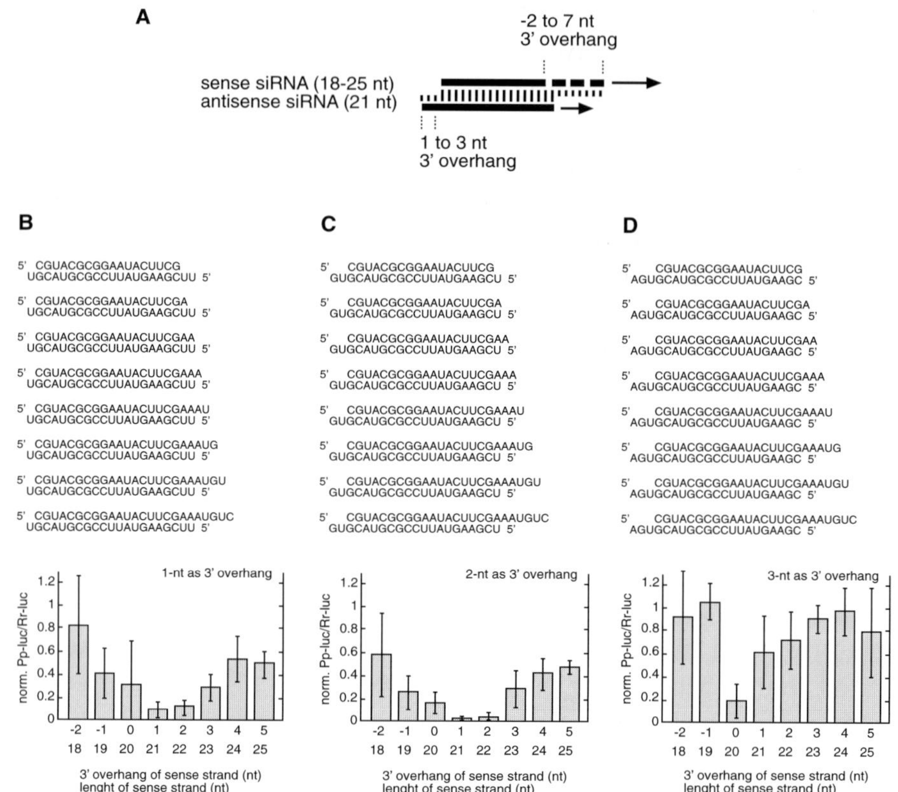

Fig. 11. Variation of the length of the sense strand of siRNA duplexes. (A) Representation of the experiment: three 21 nt antisense strands were paired with eight sense siRNAs; the siRNAs were changed in length at their 3′ ends; the 3′ overhang of the antisense siRNA was: (B) 1 nt, (C) 2 nt or (D) 3 nt, while the sense siRNA overhang was varied for each series. The sequences of the siRNA duplexes and the corresponding interference ratios are indicated. (From Elbashir *et al.*, 2001.)

forms were also tested. There was a difference in methodology. The Greek investigators used an *in vivo* system. They injected the synthetic siRNAs into embryos of the fruit fly. The siRNAs had homology to either of two genes *Notch* or *Hedgehog*; silencing of either of these genes caused a clearly identified phenotype. Grossly, the results of Boutla *et al.* (2001) were similar to those of Elbashir *et al.* The main conclusion of the former was that short dsRNAs (siRNAs) of 22 nt in length are also effective silencers of genes that have homology to the sequences in the respective synthetic siRNAs.

Further confirmations and extensions of the findings of Elbashir *et al.* were reported in two consecutive *Cell* publications: Lipardi *et al.* (2001) and Nykänen *et al.* (2001). The former publication of the Bruce Peterson team at NIH in Bethesda, MD, also employed embryo extracts of *Drosophila* and found that there could be a "chain reaction", meaning that the mRNA that was the target of the initial cleavage by siRNA yielded dsRNA that could serve as secondary siRNA. This mRNA-dependent siRNA formation was apparently dependent on RNA-dependent RNA polymerase (RdRP). Lipardi *et al.* (2001) suggested a model for RNAi in *Drosophila* (Fig. 12). Note that the model of these investigators contains *Dicer*-cleavage stages but no RISC. Moreover, in their model siRNA primers are not inducing directly the

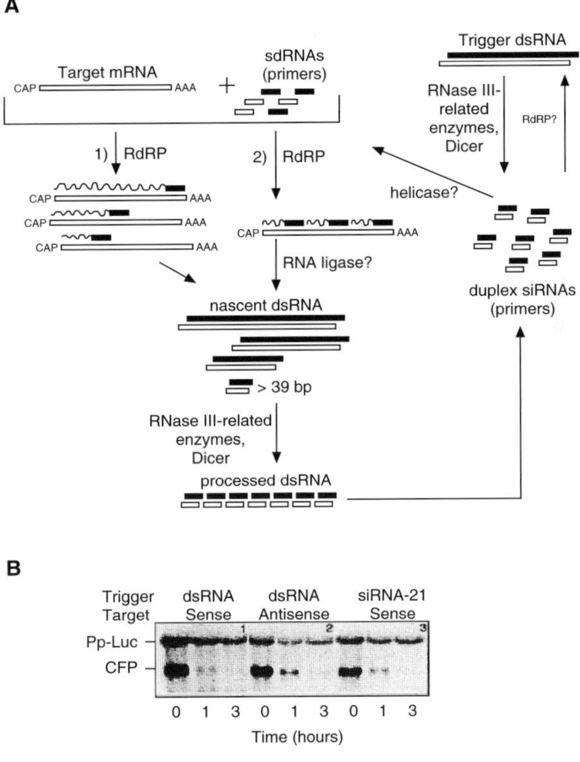

Fig. 12.   (A) A model for RNAi. (B) Observation on the predications stated in the model: changes in mRNA levels imposed by different triggers. (From Lipardi *et al.*, 2001.)

degradation of mRNA but are first serving to produce longer dsRNAs that are then processed into shorter dsRNAs. The team of Phillip Zamore in Worcester (Nykänen *et al.*, 2001) focused on the biochemistry of RNAi. They detected four steps in the RNAi process. The first step is the cleavage of long dsRNA into siRNA. This step requires ATP. The next step consists of incorporating siRNA into an inactive protein/RNA complex. The third step again requires ATP for unwinding the siRNA duplex and conversion of the inactive complex into an active one. The last step is again ATP-independent and it consists of recognition of the target sequence and cleavage of this target. The model of Nykänen is illustrated in Fig. 13. Note that the latter model does include the RISC complex and its ATP-dependent activation.

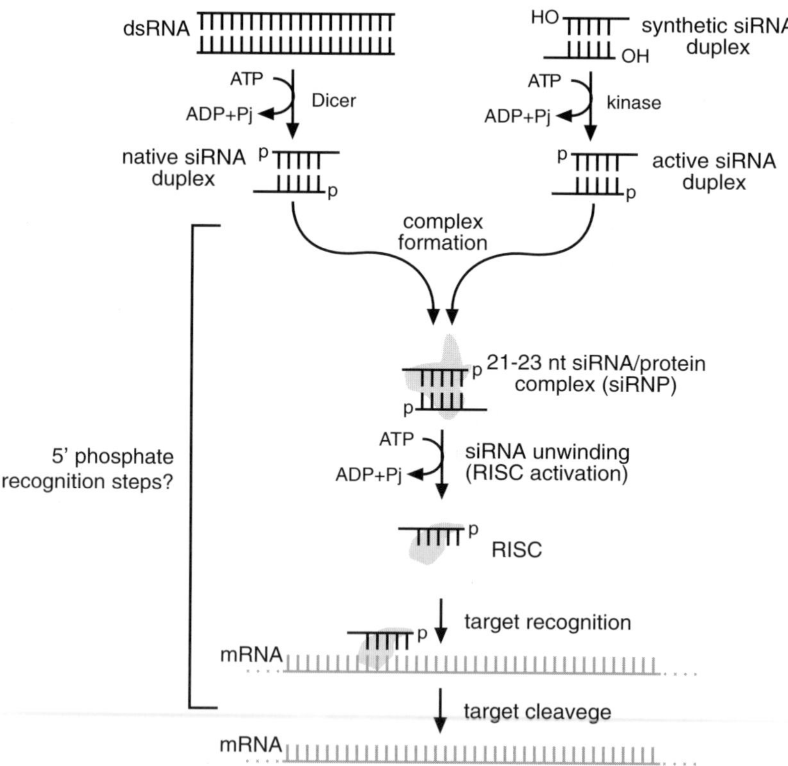

Fig. 13. A model for the RNAi pathway. The authors did not know if only one or both siRNA strands are present in the same RISC complex yet. (From Nykänen *et al.*, 2001.)

A further clarification of the RNAi mechanisms in *Drosophila* and its relation to the silencing of transcription emerged from the study of Pal-Bhadra *et al.* (2002) of the laboratory of J.A. Birchler, the University of Missouri at Columbia. These investigators found that a mutation in *piwi* that encodes a component of *Dicer* blocks the RNAi (PTGS). It is also causing transcriptional gene silencing (TGS). This indicated that the two types of silencing are connected, but the exact biochemistry of this connection stayed enigmatic.

Williams and Rubin (2002) of the University of California in Berkeley focused their attention on the *AGO1* gene. This gene (as well as another member of the *Argonaute* family, *AGO2*) is expressed throughout embryonic development in *Drosophila*. The authors confirmed that AGO1 is not required for the conversion of long dsRNA into the ~21–22 nt siRNA (processed by *Dicer*) but AGO1 expression is essential for a later stage of the RNAi process, probably just following the pairing of the "guide" (a single-stranded sequence of the siRNA) with the target (mRNA). Following this process the mRNA could be either cleaved by the RISC complex or the guide RNA may serve as primer for a RdRP to produce more dsRNAs that is then again substrate for the dicing into siRNAs.

Soon after the publication of Williams and Rubin (2002) the Zamore laboratory (Schwarz *et al.*, 2002) provided information that indicated that siRNA does not serve as a primer for RdRP in the RNAi of *Drosophila*. Moreover, the latter investigators found that in *Drosophila* as well as in man the RdRP is not required at all for the RNAi process. In this respect *Drosophila* and probably all mammals differ from plants, hyphal fungi and nematodes in which PTGS, quelling and RNAi, respectively, do involve a stage of RdRP. The role of siRNA in flies and mammals seems to be restricted as guides for the RISC complex to bind at the correct sequence of a mRNA. How does the RISC surveys all the mRNAs to bind only to the correct mRNA and only to the correct site of its sequence is still not known.

How much can we learn about human behavior from the behavior of insects? King Solomon, as I have already mentioned, indeed told us that we can derive wisdom from these animals. Akira Ishizuka, Mikiko Siomi and Haruhiko Siomi (2002) of the Tokoshima University in Japan also told us that wisdom on RNA silencing can be derived from insects; more precisely from *Drosophila* flies that have a neurological defect caused by a mutation in the *dFMR1* gene. The Fragile X syndrome (FXS) of humans affects approximately 1:4000 males and 1:8000 females worldwide, making it the most commonly known monogenic cause of mental retardation. The syndrome is characterized by general mental retardation with notable deficits in language and executive function. In most cases the syndrome is correlated with a considerable increase in the number of repeats of a trinucleotide (CGG) at the 5′ untranslated region of the gene *FMR1*. These repeats lead to hypermethylation of chromatin and consequently to transcriptional arrest. The mice *FMR1* gene is very similar to the human gene. When *FMR1* is knocked out in mice the result is abnormal dendritic spines in the brain, reminiscent of a maturation delay. The *FMR1* protein is an RNA-binding protein and is associated with polyribosomes and specifically it has affinity to the two ribosomal proteins L5 and L11. *FMR1* is thought to be related to posttranscriptional regulation of gene expression in a manner critical to the correct development of neurons. It is assumed that the *FMR1* protein acts as a negative regulator of translation but how this happens *in vivo* is enigmatic. In *Drosophila* there is a single gene *dFMR1* that is homologous to *FMR1*. The fly gene and the mammalian gene share a number of topographical landmarks, including two RNA-binding motifs. The *Drosophila* gene also has an RGG box (i.e. a motif known to bind specific DNA sequences) and the dFMR1 protein is ribosome-associated. Moreover, genetic studies showed that the *dFMR1* (in *Drosophila*) has a role in the regulation of synapse growth and function, and this protein is probably acting in this fly as a translational suppressor. All the above information on FMR1 and *dFMRI* prompted Ishizuka *et al.* (2002) to examine the role of the *FMR1* protein in the fruit fly as a model in order to obtain significant insights into the function of this protein in mammals.

Because the RNAi system in *Drosophila* was already well studied and it appeared that key components of this system are shared by *Drosophila* and mammals, Ishizuka *et al.* (2002) focused on the RNA silencing system. By several isolation and identification methodologies, these investigators found that the dFMR1 protein is present in a complex that is isolated from the S2 cells of the fruit fly. This complex also contains the AGO2 protein and has RNase capability. The investigators provided arguments in favor of the possibility that FMR1 joins the pre-formed RISC complexes (before the specific ssRNA guide is added). It was also found that dFMR1 interacts with a *Drosophila* homologue of the p68 RNA-helicase. This helicase is active in the unwinding of the siRNA that preludes the final "maturation" of the active RISC particle. In short the study of Ishizuka *et al.* (2002) provided convincing data that show the involvement of *dFMR1* in complexes and components that are active in the RNAi of *Drosophila*, especially with the miRNA induced inhibition of translation. On the other hand, it is yet a long way till the specific impact of mutation in *FMR1* on human neural disorders is clarified.

Only 10 days after the Ishizuka *et al.*'s (2002) manuscript was submitted to *Genes and Development*, another report on the Fragile X-related protein in *Drosophila* was also submitted, to the same journal by the laboratory of Hannon and Hammond of the Cold Spring Harbor Laboratory (*Genes and Development* is published by the Cold Spring Harbor Laboratory Press). The latter publication (Caudy *et al.*, 2002) preceded directly the publication of Ishizuka *et al.* (2002). The investigators from the Cold Spring Harbor Laboratory (Caudy *et al.*, 2002) did not intend to provide insight into the Fragile X Neural Retardation in humans (as was the aim of Ishizuka *et al.*, 2002). The aim of Caudy *et al.* (2002) was to address the underlaying mechanisms of siRNA-guided degradation of mRNA and they investigated the degree to which the mechanims of siRNA and miRNA overlap. The miRNAs were mentioned in the chapter on RNA intereference in nematodes and will be discussed again in the chapters on RNA silencing in mammals and in plants. The investigators thus intended to determine the composition of RNAi-effector complexes.

One of these complexes is the RISC. The laboratory of Hannon and Hammond analysed the RISC of *Drosophila* in the past (Hammond *et al.*, 2001a, 2001b) as noted above. These investigators found then that the Argonaute-2 (Ago-2) protein is a core component of RISC. Caudy *et al.* (2002) intended to look for additional components of RISC. For that, they performed a large-scale biochemical purification procedure. The rate of RISC activity and the level of the Ago-2 protein were followed during purification. After several purification steps, a number of additional proteins indeed emerged consistently and co-purified with active RISC and Ago-2. Among these "new" proteins were two RNA-binding proteins, VIG and the *Drosophila* homolog of the human Fragile X Mental Retardation protein, *FMRP*. The *Drosophila* homolog was termed *dFXR* in the Caudy *et al.*'s (2002) publication. The VIG protein is encoded from within an intron of another gene, *Vasa*. VIG has one important characteristic; it has an RGG box. Hence, it can probably bind RNA. It is an evolutionary conserved protein with homologs in animals, plants and yeast. The other additional protein found in RISC, the dFXR (dFMR1 of Ishizuka *et al.*, 2002, and note that different authors may use slightly different designations for the same gene) is an ortholog of the human FMRP that was mentioned above. The latter is encoded from a locus on chromosome X that is epigentically silenced in individuals with the Fragile X syndrome. This protein is expressed in neural cells and in cells of several other tissues. The human genome encodes several FMR family members and these contain the RGG box. It was found that when the genes for VIG and dFXR were silenced by the RNAi procedure, the capability to perform the RNA silencing was impaired. This could mean that RISCs devoid of the VIG and dFXR proteins are not functional.

The investigators also looked into the possibility that miRNA is involved in the RISC complex that contains VIG and dFXR. Indeed by immunoprecipitation one known miRNA (miR2b) seemed to be included in the RISC complexes that contained VIG and dFXR. Turning back to the human syndrome and assuming that in humans the wild-type (FMR) protein is a component of this RISC, how does a mutant FMR causes the specific known syndrome? The authors made several suggestions but this subject is open to further studies.

A Hypothetical Deal with Mosquitoes

Caplen *et al.* (2002) of the NIH in Bethesda, MD, asked whether the RNAi mechanism that exists in *Drosophila* is also operational in the mosquito *Aedes albopictus*. Actually, these investigators used mosquito cells that can be transformed by appropriate plasmids. Consequently, viral sequences coded by these plasmids are then expressed in the mosquito cells. Two viruses were investigated: the Semliki Forest virus (SFV) and the serotype 1 dengue virus (DEN1). The SFV is a mosquito-borne pathogen that infects mammals and can cause human encephalitis, a lethal disease. There are other "relatives" of SFV that belong to the *alphavirus* genus. The genome of *alphavirus* is a ssRNA that functions as a mRNA. A polyprotein is produced and then cleaved and leads to further processes that finally results in viral replication. The genome of DEN1 also consists of a ssRNA and its replication follows a somewhat different route.

Caplen *et al.* (2002) first assessed the RNAi capability of *A. albopictus* cells. For that the cells were cotransfected with a plasmid and dsRNA. The plasmid was for the expression of the reporter gene *gfp* that should express the GFP. The dsRNA was either a part of the mRNA for GFP or control dsRNA. The dsRNA with a sequence that had homology to the code for GFP significantly reduced the fluorescence (but did not eliminate it). Also, the level of mRNA for GFP was reduced. The RNAs interference was specific to the kind of dsRNA that was introduced into the mosquito cells. When the investigators turned to the inhibition of viral replication they found that the appropriate dsRNA should be introduced into the cells 18 hours before the plasmid that bears the viral sequence. Adding dsRNA and plasmid simultaneously did not cause a significant RNAi reaction. The 5' of the SFV RNA encodes 4 nonstructural proteins that are essential for the future replication of the virus. The investigators found that not all the four dsRNAs that are homologous to these genes were equally effective in RNAi respectively. In fact, only dsRNA for two out of the four reduced the respective expression (SFV *nsp-2* and SFV *nsp-4*). Similar tests were performed with cells infected with the DEN1 virus. Again, the dsRNAs corresponding to different parts of the viral genome had different impacts on viral replication but some dsRNAs drastically reduced this replication. Albeit, in no case was the DEN1 replication stopped.

In summary there is potential for the use of an RNAi system in mosquitoes that may lead to more effective silencing of pathogenic viral RNA. For practical application of dsRNA to eliminate pathogenic viruses from mosquitoes there are still many hurdles: how to introduce (stably?) the dsRNA, how to apply this to wild mosquito populations, and so on. I suggest the following deal with *Aedes* mosquitoes. We shall supply the mosquitoes with armaments (dsRNAs) to fight their own pathogens; but in turn we shall also add to their genome dsRNA-coding sequences that will abolish the replication of human viruses in these insects. The willingness of the *Aedes* mosquitoes to accept the deal will increase if they are exposed to an environment that is loaded with insect pathogens. There are obviously dangers with such a deal. One danger is that the population of *Aedes* will increase so drastically that their quantity will cause nuisance and they become intolerable. Here we have an advantage. We may turn to additional insect pathogens against which we did not furnish dsRNA protection.

Returning to the studies on mosquitoes, it appears that the RNAi system is also functional in these insects but probably less effectively than in fruit flies. What about other insects? The simple (and correct) answer is that we do not know. We do not even know how many insect species exist. Are there less than 1 million insect species or are there more than 5 million? It may be misleading to reach a general conclusion on RNAi in insects that is based on *Drosophila*. One feature of *Drosophila* has already taught us that the empiricistic approach should be used with caution and only as a guide for further experimentation: the *Drosophila* chromosomal telomers are uniquely composed of TE. Such transposable-element "building stones" were not reported in other insect genera or other metazoa.

The Utilization of RNAi to Analyse the Expression of *Drosophila* Genes

Maurizio Gatti and associates of the "La Sapienza" University in Rome (Somma *et al.*, 2002) used dsRNA-mediated interference to ablate genes that are required for cytokinesis in *Drosophila*. They thus analysed the phenotypes of dividing S2 cells that were treated for 24 or 72 hours with various dsRNAs that were derived from genes having putative involvement with cytokinensis. Twelve different dsRNAs were put to test. These dsRNA represented 12 genes that were implicated in either mitosis or male-meiosis, or in both. The dsRNAs of most of these genes (*anillin, acGAP, pavarotti rho1, pebble, spaghetti squash, synthaxin1A* and *twinstar*) disrupted cytokinesis in the S2 cells. The actual reduction of the encoded proteins could be followed in about half of the respective dsRNAs treatments because antibodies were available for western blot hybridization. Indeed, in five out of six such genes the respective dsRNAs eliminated the encoded protein after 72 hours of dsRNA treatment (in two genes, the elimination was noticeable after 24 hours of treatment). There were several cytological phenotypes that were elicited by specific dsRNAs. These phenotypes suggested interactions between central spindle microtubules, the actin-based contractile ring and the plasma membrane. The investigators thus found evidence for interdependence in the structure of the central spindle and the contractile ring.

Schmid *et al.* (2002), a team from the California Institute of Technology, in Pasadena, CA, were exploring the regulation of axon guidance and synaptogenesis during *Drosophila* development. Some of the genes involved in this process were revealed previously but there was the question of redundancy. Redundancy is problematic in genetic/morphogenetic studies because if two or more genes control the same morphogenetic process, then mutating (or knocking-out) one of them will not cause any phenotypic change. In the case of axon guidance in the fruit fly five genes of the family of neural-receptor-tyrosine-phosphatases (RPTPs) were implicated with this guidance. But single mutations of only three of these (*Dlar, Ptp52F, Ptp69D*) cause lethality. The null mutation of two other genes (*Ptp10D, Ptp99A*) caused no embryonic phenotypes. Moreover, their offsprings were also viable and fertile. Past experiments indicated that combinations of more than one mutation of these five genes resulted in a more complicated picture. Here RNAi techniques could be very useful. If the dsRNA representing two or more of the RPTPs could be applied together, a combinatorial analysis would be possible. The investigators found that indeed it is possible to inject into fly eggs (blastoderm stage) a mixture of several dsRNAs representing several members of a gene family. By appropriate staining of neurons and glia in the developing embryo and confocal microscopy, the impact of specific dsRNAs as well as combination of such dsRNAs could be followed and the effect of gene silencing on the central-nervous-system's axon pathways could be evaluated.

Alternative splicing is an emerging genetic issue. This issue was discussed in my book (Galun, 2003) on TEs because alternative splicing is operating in the transcripts of several TEs. This issue was reviewed

(Graveley, 2001) and its impact on human genetics and pathology is becoming evident. Briefly, alternative splicing means that a given transcript can be spliced in more than one way. This will then result in two or (many) more mRNAs from the same gene. It is now estimated that about *half* of the human genes encode at least two alternative spliced mRNAs. The presently known champion of alternative splicing is the *Drosophila* gene *Dscam*. This gene has the potential to generate 38 016 different proteins via alternative splicing. Geneticists are encountering problems when they intend to determine the function of specific proteins that are encoded by a gene that undergoes alternative splicing. Eliminating all transcripts from a gene by mutations will not pinpoint specific proteins. Here could be another case where RNAi could help. The dsRNA could be derived from specific (postsplicing) exons and the elimination of specific proteins could thus be achieved. But there could be a problem. If there is *transitive* RNAi in *Drosophila* this approach will not be helpful. In the transitive RNAi of nematodes, a *Dicer*-cleaved dsRNA will produce a siRNA. The latter will anneal to comparable location on the mRNA and serve as a primer for RdRP. A dsRNA region will then be established upstream of the annealing location. This dsRNA is subject to secondary *Dicer* cleavage and additional siRNAs will be formed. This could lead to a situation that dsRNA derived from a sequence in exon 2 will cause the cleavage of this exon as well as of the exon that is upstream of it: exon 1. Celotto and Graveley (2002) of the University of Connecticut Health Center, in Farmington, explored the possibility of using RNAi as a tool to dissect the functional relevance of alternative splicing. They used the *Drosophila* gene *Dscam* — the champion of alternative splicing. A scheme of a part of this gene with the axon 4 cluster is shown in Fig. 14. The gene *Dscam* contains 95 alternative exons and thus can

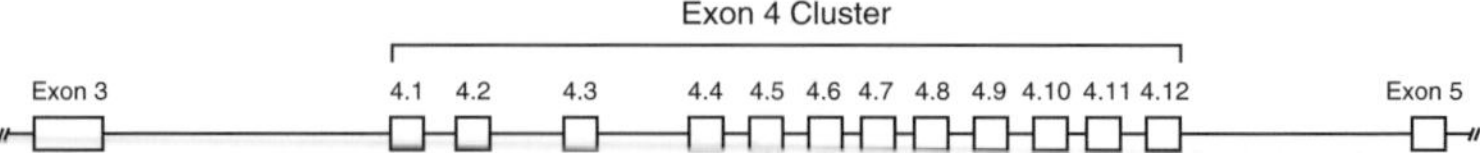

Fig. 14. Organization of the *Drosophila Dscam* exon 4 cluster. Twelve variants of exon 4, called exon 4.1 through exon 4.12, lie between constitutive exons 3 and 5. The exon 4 variants are alternatively spliced in a mutually exclusive manner such that only one exon 4 variant is included in each *Dscam* mRNA. The exon 4 variants range in size from 159–171 nt. (From Celotto and Graveley, 2002.)

potentially produce 38 016 different mRNA isoforms. The investigators focussed on the exon 4 cluster that contains 12 alternative exons: 4.1 to 4.12. Incidentally, the *Dscam* gene is related to axon guidance, handled by Schmid *et al.* (2002) as noted above. The *Drosophila* cell adhesion molecule (*Dscam*) gene actually encodes an axon guidance receptor with similarity to the human *Down*–syndrome cell adhesion molecule. Terms may be confusing because we have the alternative *exon* and coding for *axon* guidance receptors. But we can ignore the neurology of the fruit fly and summarize the results of Celotto and Graveley (2002). These authors indeed found that RNAi can be used in *Drosophila* cells to selectively degrade specific alternative spliced mRNA isoforms. This could be done by treating the cells with dsRNA corresponding to an alternatively spliced exon. It seems that there is no transitive RNAi in *Drosophila*, probably due to the lack of an RdRP that leads to this process in other organisms (e.g. plants, nematodes). This tempted the authors to recommend the RNAi technique for the analysis of alternatively spliced mRNAs in *Drosophila*. Whether it is applicable in other animals such as mammals is still an open question.

Early Reports on MicroRNA in *Drosophila*

More than a year before Xu *et al.* (2003) submitted the first report on the suppressive effect of miRNA on *Drosophila* genes, an earlier report was submitted by two investigators from Dallas, Texas (Kalidas and Smith, 2002). In the earlier publication the term *miRNA* was not used but in effect a kind of miRNA was constructed to be introduced stably into the fly's genome for RNAi-mediated gene silencing. The investigators from Texas elaborated the arguments why previous attempts to introduce effective DNA constructs that will result in specific and drastic RNAi effects in *Drosophila* failed in the past. One of the problems encountered was the difficulties in cloning the appropriate constructs in bacteria: inverted repeat RNAi constructs are difficult or impossible to recover from bacteria. Kalidas and Smith (2002) therefore looked for help from introns and splicing. Their strategy was to separate two inverted repeats corresponding to a segment of the target transcript by an intron. Transgenic flies were to be produced by introducing the respective cDNA with an appropriate insect vector.

After the respective transcript from this transgene is formed in the nucleus it will be spliced by the fly's own cells and the introns will be removed, leaving after splicing, an inverted repeat RNA sequence without any spacer. This sequence will then fold-back to form a hairpin without any loop. Such a hairpin would be cut by the cell's *Dicer*, resulting in siRNA. This principal strategy was applied with three genes: the *lush* gene, the *white* gene and the heterotrimeric G protein gene, *dGqα*. The *lush* gene is a member of the invertebrate odorant-binding family and it is expressed highly in a small number of cells of the antennae. To assure the transcription at the correct site and time, the cDNA encoding the hairpin RNA contained the endogeneous promoter of *lush*. Normally, *lush* is first expressed at the late pupal stage. Flies that were homozygous to the transgene of the engineered cDNA produced very little LUSH protein (about 1–2 per cent of normal).

The *white* gene encodes an ABC transporter that is required to localize pigments in the insect's eye. The null mutants of *white* have unpigmented (white) eyes. The derived *white* construct was able to completely suppress the expression of this gene: the eyes of the transgenic flies were not pigmented, just as those of the null mutation of *white*. The transgenic RNAi to target the *dGqα* is a more complicated issue because the transcript of this gene is naturally spliced into several mRNAs. But in summary, gene silencing was also achieved with this gene.

There were two early publications on miRNA in *Drosophila*: Brennecke *et al.* (2003) from the Laboratory of Stephen Cohen, at the EMBO Laboratory in Heidelberg, Germany, and Xu *et al.* (2003) from the laboratory of Bruce Hay at Caltech, in Pasadena, CA. The team of Cohen and associates (Brennecke *et al.*, 2003) studied the *bantam* gene. As briefly indicated above the development of flies is the result of an elaborated process of gradients affecting the expression of patterning genes and of genes that promote cell division (at specific locations) as well as of genes that arrest cell division and even cause cell death (apoptosis). The investigators studied the *bantam* locus and detected a code for a miRNA. Flies with or without this miRNA code were analyzed. The investigators found that the *bantam* region does not have the capacity to encode a protein. The sequence that encoded the miRNA in *Drosophila* was also found in the genome of the mosquito *Anopheles gambiae*. The respective miRNA sequence was found in normal larvae of flies, starting from third instar. It was absent in *bantam* mutants that had a specific deletion in this locus. Further studies indicated that the *bantam* mature miRNA of 21 nt stimulated at specific times and locations, cell proliferation and prevented apoptosis. The pro-apoptotic gene *hid* was found to be the target of silencing of this miRNA.

The team from Caltech (Xu *et al.*, 2003) focused on another miRNA of *Drosophila: mir-14*. These investigators reported that *mir-14* is a cell-death suppressor. In the absence of *mir-14*, there is an uninhibited expression of the gene *Reaper* that will result in cell death. The *mir-14* was also found to suppress cell-death that is caused by additional genes (*Hid, Grim*). Flies that lack *mir-14* can survive but are defective: they are stress-sensitive and have a short lifespan. But *mir-14* seems to affect another gene-activity. Deletion of *mir-14* resulted in flies with increased levels of triacetylglycerol and diacetylglycerol while the addition of codes for *mir-14* into the fly's genome had the opposite effect. Thus, *mir-14* is probably involved in fat metabolism.

Two additional studies on specific miRNAs of *Drosophila* were published simultaneously: one (Bashirullah *et al.*, 2003) concerns a study on *mir-125* by the laboratories of Gary Ruvkun (of the Mass General Hospital in Boston) and Carl Thummel (of Salt Lake City). The other publication from the teams of Victor Ambros and Edward Berger (Dartmouth, New Hampshire) reported the investigation of the roles of several miRNAs (e.g. *mir-34, mir-100, mir-125 and let-7*).

The study of Bashirullah *et al.* (2003) handled mainly the impact of two miRNAs of *Drosophila: let-7* and *miR-125 (mir-125 was later designated as miR-125)*. The *let-7* is very conserved among animals. It was detected in nematodes but was later found wherever it was searched from. It is expressed at a relatively late state of development, meaning in flies at the last instar or later. The *Drosophila miR-125* has similarity to the nematode *lin-4* which is expressed at an early phase of larval development in nematodes. The *miR-125* was found to be expressed in *Drosophila* during the pupal and adult stages. It was observed that the expression of *let-7* and of *miR-125* in *Drosophila* is correlated with the increase of the level of the steroid hormone ecdysone. The investigation provided evidence that indicated that in spite of this temporal correlation this hormone is not the direct inducer of *let-7* and *miR-125* expression. One phenomenon that supported this notion was that there was an increase in *let-7* and *miR-125* under conditions which prevented the impact of ecdysone. These two miRNAs are encoded by sequences that are rather close to each other on the chromosome; both are transcribed from the same region (of about 0.5 kb). It is reasonable that their transcription is regulated by the same promoter. The investigators suggested that not ecdysone but rather another affector causes the initiation of transcription of both *let-7* and *miR-125*. So what is the direct inducer of *let-7* and *miR-125*? The investigators speculated that it is a different hormone but did not provide its identity.

The team that published the report of Sempere *et al.* (2003) did an extensive survey of 24 miRNAs in *Drosophila*. They followed the change of expression of these miRNAs during the development of the flies (e.g. embryo, L1, L2, L3, pre-pupa, pupa and adult). The presence of these 24 miRNAs was followed by extraction and northern blot hybridization with the appropriate probes. More detailed analyses were then performed with part of these miRNAs: *miR-1, mir-34, mir-100, mir-125* and *let-7*. The miRNAs *mir-100, mir-125* and *let-7* increase at the same stage in the fly's development: during the mid L3 stage. The simultaneous change in level is not unexpected because all these three miRNAs are encoded by the same ~800 nt coding fragment of the fly's genome. Contrary to the conclusion of Bashirullah *et al.* (2003) the claim of Sempere *et al.* (2003) was that the up-regulation of the three miRNAs (*mir-100, mir-125* and *let-7*) as well as the down-regulation of another miRNA (*mir-34*) are affected by the hormone ecdysone. The team of Berger and Ambros and the team of Ruvkun and Thummel will soon hopefully come to an agreement about the identity of the hormone that is the direct regulator of the expression of the three miRNAs.

RNA Silencing in Protozoa

The term Protozoa was given to a group of diverse organisms from very different phylae. Among these are obligatory parasites that are pathogenic to metazoa, including man. Obviously, the RNA silencing was investigated in only a few protozoan genera such as *Plasmodium*, *Toxoplasma*, *Tetrahynema*, *Paramecium*, *Trypanosoma* and *Leishmania*. All these genera are considered phylogenetically "ancient" eukaryotes. Clearly, if we accept the assumption that these genera are "ancient" this does not mean that they did not change during the last one billion years. Clear logic tells us that they did change. To give just one example we shall take the genus *Trypanosoma*. The species of this genus are obligatory parasites and require guts of blood-sucking insects and blood cells of mammals to complete their life cycles. How did trypanosomes manage before mammalian blood cells become available? There must have been predecessors that had a very different lifestyle from the trypanosomes of today. Only a handfull genera out of a great number of protozoa were investigated with respect to RNA silencing. But even within this small number of investigated genera there was a surprise: some protozoa lack RNA silencing. After a few years of study of RNA silencing in plants, fungi, insects and mammals some investigators concluded that RNA silencing is an ancient system that exists in *all* eukaryotes!

David Hume (1711–1776) had already insisted that we cannot derive a general truth with respect to the problem of *induction* from repeated experience. For that he gave the example of the rubber ball. You throw the ball to the floor and it bounces. You repeat this several times and each time the ball bounces. But you cannot be sure that after one million times when you have thrown the ball to the floor it will bounce again one more time. The example of Lord Bertrand Russell (1872–1970) on the problem of *induction* was bolder. He featured a cock which expected his farmer to supply him with the daily feed each morning. The feeding was repeated many times but the cock's expectation was not an assurance for an endless continuation. One morning the farmer came with a knife rather than with feed. In an example closer to our theme Russell wrote about the man who saw only white swans and thus insisted that there are no black swans. The man was wrong; black swans are native in Australia.

Back to the reality of RNA silencing: no RNA silencing was revealed in the silencing in some *Trypanosoma* species and neither in the species of the genus *Leishmania* that were investigated. Consequently, in the primitive protozoan parasite *Giardia lamblia* that infests the human small intestine, we cannot predict whether this flagellated protozoa has or lacks the RNA silencing mechanism. Scientists are allowed to make guesses and even assumptions have to be treated with caution, unless checked. In numerous articles and reviews it was stated that RNAi evolved in eukaryotes as a defence against retrotransposons. African *Trypanosoma* species have transposons and have RNAi. But *T. cruzi* also has retrotransposons while no RNAi was found in this species. How does *T. cruzi* cope with its transposons?

Most of the available information on RNA silencing in protozoa is about *Trypanosoma*. Thus, in this chapter we shall deal mostly with this genus. But there is also some information of RNA silencing (negative and positive) in other protozoan genera. I shall summarize this latter information. A relatively recent review on RNA silencing in protozoa and "other nonclassical model organisms" with references till 2002 was written by Ullu and Tschudi (2003).

Indeed, as noted above the protozoa include very different organisms that share only two characteristics: they are not autotrophic and are unicellular. When dealing with specific genera of protozoa I shall furnish basic biological and molecular characteristics of the genus. The information will be rather brief. Readers who are interested in additional information should consult the respective texts on these organisms.

To obtain a rough picture on the phylogenetic relationships among the genera of protozoa that will be mentioned below (e.g. *Trypanosoma,*

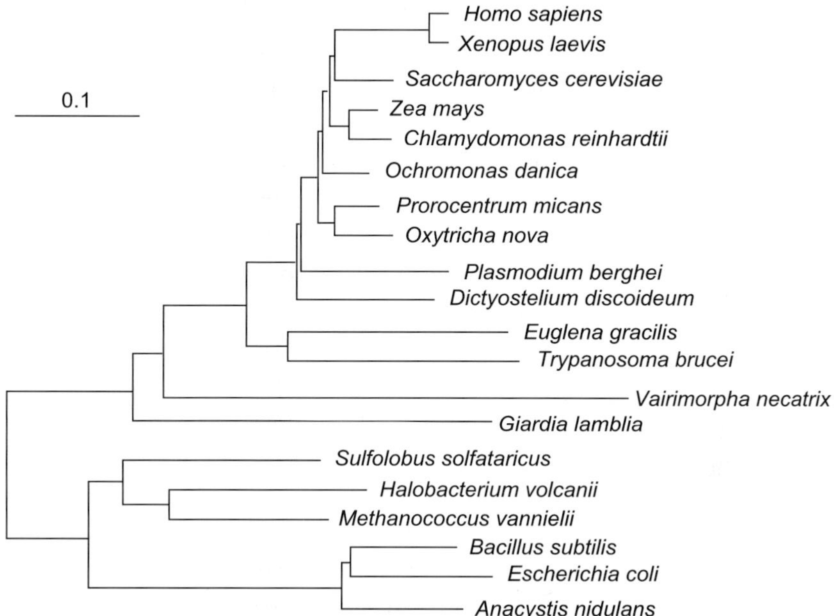

Fig. 15.   A multi-kingdom tree inferred from 16S-like rRNAs. The distance that corresponds to 10 changes per 100 positions is indicated. (From Sogin *et al.*, 1989.)

Plasmodium) as well as between these protozoa and other eukaryotes, one should look at a phylogenetic tree. Such a tree can be based on the nucleotide sequences in 16S-like ribosomal RNAs. A multi-kingdom tree of diverse eukaryotic and prokaryotic organisms was constructed by Sogin *et al.* (1989) as shown in Fig. 15.

RNA Silencing in the Trypanosomatids

The Trypanosomatideae is a family of protozoa that belongs to the order Kinetoplastida. This order is named after their unique organelle, the *kinetoplast*. The latter is a component of the mitochondrion that consists of a large concentrated DNA network in which there are minicircles and macrocircles. The minicircles have elaborated structures combined with guiding-RNAs that are involved in the RNA editing. None of the genera belonging to Trypanosomatidae are free living. Several of the genera as the two that will be discussed (*Trypanosoma* and *Leishmania*) are transmitted exclusively by insects. There are other genera of this family that are associated with plants. There are species of the Trypanosomatida that harbor autotrophic symbionts. Did some of the genera that are now obligatory parasites have ancestors who maintained themselves by such symbionts? The feasibility for such an ancestry is supported by the existence, in some parasitic trypanosomes, remnants of plastids. Albeit, the discussion of this possibility is beyond the scope of this book.

We shall discuss in this chapter the African trypanosomes (*Trypanosoma brucei-ssp.*), the south American *Trypanosoma cruzi* and species of the genus *Leishmania*. Each of these three types of Trypanosomatids apparently initiated its parasitism of mammals independently. They also choose their respective different insect vectors as well as evolve a different survival strategy and obviously a radically different disease pathology in the mammalian hosts. As for the antigen variation manifested by the changing of the surface antigens to evade the immune response of their mammalian hosts, this surely is a relatively late evolution that followed the adoptive immune system in host organisms. The trypanosomatids are thus "old" but by no means stagnant. Clearly, a fight is going on in which the photozoan parasites are using an arsenal that has been accumulating for several hundred million years and humans are trying to fight back using new biological tools that become available in the recent few decades. Can RNA silencing become effective tools in this war? A discussion of the toolkit that evolved by molecular parasitology against trypanosomatids was provided by Beverley (2003).

The African **Trypanosoma** *species*

There are several species of African *Trypanosoma*, among them is *T. brucei* that causes the human sleeping sickness. The vector of *T. brucei* is the Tse-Tse (*Glossina*) fly. About 60 million people in Africa (sub-Sahara) are exposed to this pathogen and about half a million people are actually infected. Chemotherapy is problematic but practical, and no vaccination is yet available. *T. brucei* is also the cause of the animal disease Nagana that is also transferred by the Tse-Tse flies. The Nagana disease is of major economical importance, eradicating cattle in large areas. The *T. brucei* has several nasty subspecies as ssp. *rhodesiensis*, ssp. *gambiense* and *Trypanosoma congolense* of which the latter species also causes the Nagana disease in cattle, inflicting a constant impediment of the livestock industry in central and eastern Africa.

All the above mentioned *Trypanosoma* species as well as the Trypanosomatids in general have the capability to evade the immune response of their mammal hosts due to the antigen variation. *Trypanosoma* species invest a large proportion of their small genome (about six fold larger than the genome of *Escherichia coli*) in genes that encode variants of surface glycoproteins (VSGs). There could be up to 1000 such genes. The trypanosomes have different positions of their flagellum according to the stages in their life cycle: (1) basal flagella — *trypomastigote* (2) median flagella — *epimastigote*, (3) apical flagella — *promastigote* and (4) no flagella — *amastigote*. Also, the mitochondrial genome of trypanosomes is relatively large — about 30 per cent of the total DNA of the cells is mitochondrial DNA.

Briefly, the sucking Tse-Tse flies introduce the parasites into their host and the parasites then multiply in the lymph and blood cells. There, the parasites multiply to enormous quantities (10^9 or more per cell). When a "virgin" fly sucks blood from an infected person (or animal) the parasites enter the insects gut and from there move to the salivary glands. From the salivary system the parasites are transferred to another mammalian host on which the fly is feeding. The same fly may suck bood from different hosts. When this happens different *Trypanosoma brucei* may reach the same fly and matings between the parasites is possible. But such matings are apparently not obligatory for the propagation of the parasite. While in higher eukaryotes the regulation of transcription of specific genes plays a major role in the regulation of gene expression, the Trypanosomatids (as well as some other protozoa) use a different strategy for this regulation. Although they also use RNA polymerase II for the transcription of protein-coding genes, the promoters of individual genes do not play a role in Trypanosomatids. Moreover, the nuclear genome is organized typically in long arrays of ORFs that are transcribed into long polycistronic pre-mRNAs which are then processed by coupled transsplicing and polyadenylation to generate mature, monocistronic mRNAs. Each of the latter bears a common 39 nt mini-exon on its 5′ end. This peculiar transcription does not lead to the regulation of transcription of individual mRNAs. It is thus assumed that gene regulation in these parasites is mainly posttranscriptional. It is unclear what the exact mechanism of posttranscriptional gene-regualtion is but the involvement of RNA silencing was suggested by some investigators. In other words, there is probably a major role of RNA metabolism that takes place posttranscriptionally. These considerations were put forward by the first investigators who started to study the effect of dsRNA on mRNA degradation in *Trypanosoma brucei* (Ngo *et al.*, 1998). The team of Elizabetta Ullu and Christian Tschudi (Yale University, School of Medicine, CT) was studying *T. brucei* for several years and were engaged in the molecular biology of this parasite, handling the peculiar processing of its polycistronic transcript. Therefore, the notion that in trypanosomes the level of mature mRNA that is available for translation into proteins, is primarily the result of degradation rather than the rate of mRNA synthesis, was familiar to these investigators. Hence, soon after the report on the degradation of mRNA in nematodes by dsRNA (Fire *et al.*, 1998), Ngo *et al.* (1998) reported their experimental work with *T. brucei.* This work also clearly indicated that in this trypanosome dsRNA can cause specific degradation of mRNA.

The first clue for RNAi by dsRNA in tryponosomes resulted from the introduction of specific DNA plasmids into the cultured protozoa by electroporation. For the construction of the plasmids the code for the α-tubulin gene was utilized. When the plasmid contained the code in two tail-to-tail orientations the treated trypanosome cells changed drastically. It should be noted that unlike in mammals where long dsRNA may cause unspecific silencing, there is no unspecific silencing in trypanosomes. The altered morphology of these *T. brucei rhodesiense* had multiple nuclei and kinetoplasts and instead the elongated form the cells became almost spherical (FAT). Further transfections with various variants of the plasmid indicated that plasmids that caused the *in vivo* formation of dsRNA representing the α-tubulin mRNA 5′ untranslated region caused this morphological change and the arrest of cytokinesis (but allowed multiplication of nuclei and mitochondria). In further experiments the investigators used a more direct approach. Rather than expecting the formation of dsRNA *in vivo* that should result from the introduction of the respective plasmids, they introduced the dsRNA itself. When dsRNA that represented the 5′ untranslated α-tubulin pre-mRNA was introduced, the same morphological changes and cytokinesis-arrest phenomena were observed. Introducing ssRNAs that represented either of the two double strands did not cause these morphological changes.

The investigators also evaluated the level of α-tubulin mRNA after the dsRNA treatment. Northern blot hybridization indicated that the dsRNA caused a 85 per cent reduction in this mRNA. The effect was transient. The reduction in mRNA was already recorded 1 hour after dsRNA treatment and then stayed low for several hours. But after 20 hours the α-tubulin mRNA returned to normal level. Not only the 5' untranslated region was effective in degrading mRNA of α-tubulin mRNA, other (coding) regions of the α-tubulin transcript were also effective when applied as dsRNA. Furthermore, the investigators provided evidence that dsRNA caused the degradation of the mature α-tubulin mRNA rather than the degradation of the pre-mRNA.

The dsRNA induced mRNA degradation was not confined to the α-tubulin gene. Similar effects were recorded after the introduction of short dsRNA representing the code for the actin gene as well as for the paraflagellar rod, into the trypanosomal cells.

As for dsRNA representing sequences from the α-tubulin transcript, this dsRNA reduced not only the level of the respective mRNA but also about 80 per cent of the α-tubulin synthesis. Like mRNA the reduction of α-tubulin synthesis was transient and returned to normal after about 16 hours. Degradation of mRNA for the paraflagellar rod, on the other hand, was permanent and treated cells did not recover.

Ultrastructural analyses revealed that dsRNA induced defects in the microtubules of the flagellar axoneme and the flagellar attachment zone could cause the inhibition of the cell-cleavage furrow and thus arrest of cytokinesis.

It should be noted that the effects of dsRNA could be lethal to the trypanosomal cells: once they became malformed ("FAT") they never regained their normal reproduction. This was probably true after either means of dsRNA application: the direct application of synthetic dsRNA or the application of a plasmid that will produce the desired dsRNA *in vivo*. On the other hand, the degradation of certain mRNAs was transient by either method of dsRNA treatment.

Can the dsRNA effect be rendered heritable rather than transient? To answer this question the Ullu and Tschudi team recruited Elizabeth Wirtz (of the Rockefeller University, NY). The latter brought her toolkit for introducing a hybrid transgene into *T. brucei*. The Wirtz team previously developed effective plasmids that could integrate transgenes into *T. brucei* and cause the respective expression under inductive conditions (e.g. Wirtz *et al.*, 1999). One such plasmid was the *pLew79*. The latter plasmid vector integrates into the non-transcribed ribosomal DNA (rDNA) spacer due to its flanking sequences and utilizes the tetracycline-regulable procyclic acidic repetitive protein (PARP) promoter to activate the transcription of the transgene. The joint efforts of the Yale and Rockefeller laboratories (Shi *et al.*, 2000) indeed showed that by using the appropriate toolkit and combined ingenuity, stable cell lines that constantly express the dsRNA effect (upon tetracycline induction) could be established. Figure 16 shows the schemes of plasmids that were used by Shi *et al.* (2000). The structure of these plasmids included two opposing transcript regions in a way that *in vivo* these two (sense and antisense) regions will form a stem-loop RNA structure. In practice the investigators engineered into their plasmids the sense and the antisense sequences of the α-tubulin mRNA, with a spacer (converted to loop) between these two sequences (e.g. *pLewFAT* of Fig. 16). The linearized plasmids were

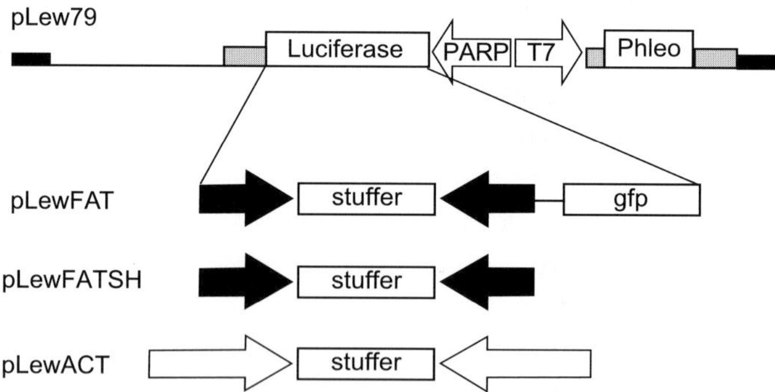

Fig. 16. Structures of the plasmid constructs used for the expression of dsRNA in trypanosomes. The scheme of the basic plasmid vector, *pLew79*, is shown at the top. The solid arrows are α-tubulin mRNA 5′ UTR and the open arrows are actin mRNA sequences. The grey boxes represent sequences containing pre-mRNA processing signals. The drawing is not drawn to scale. (From Shi *et al.*, 2000.)

transfected by electroporation into procyclic trypanosome cells. Since the plasmids contained a selectable gene the transformed cells could be selected on selectable medium (phleomycin) and cloned. Whenever such cells were induced by tetracycline the α-tubulin sequences were transcribed and the morphology of the cells changed to the FAT phenotype. This change was accompanied by a drastic reduction of α-tubulin mRNA (10-fold reduction). The level of applied tetracycline affected the FAT phenotype and the reduction of α-tubulin mRNA.

Similar results were obtained with the mRNA of actin. In the latter case a sequence of 426 nt, immediately downstream of the actin translation-initiation, was used (in sense and antisense) to derive, *in vivo*, the stem-loop, in which the stem will furnish the desired dsRNA. In this case the effect of tetracycline-induced production of the dsRNA had an even more drastic and quicker affect: there was a 10- to 20-fold reduction of actin mRNA at 8 hours after tetracycline application. The surprise was that the elimination of actin did not seem to harm the cultured cells. In addition, these investigators found that the T7 RNA polymerase can also be used to build effective dsRNA producing plasmids.

Phillippe Bastin, Keith Gull and associates of the University of Manchester, UK, were studying the ontogeny of the flagellum of trypanosomes for many years. They thus recruited the newly discovered dsRNA effect on flagellum differentiation (Bastin *et al.*, 2000). They focused on the intraflagellar structure termed *paraflagellar rod* (PFR) which is composed of two major proteins, PFRA and PFRC. The investigators transformed trypanosome cells with various plasmids. One of these was capable of rendering the cells inducible (by tetracycline) to produce a dsRNA from a region of the transcript for PFRA. Such an induction caused the quick elimination of new mRNA for PFRA as well as new PFRA protein. These reductions paralleled with cell paralysis. Using such RNA silencing the investigators could follow the process of flagellum construction, finding that the PFR is constructed by a polar assembly at the distal end of the flagellum.

I mentioned above that Shi *et al.* (2000) used mainly the derivatives of the *pLew79* plasmid for the transformation of *T. brucei* but also turned to the T7 promoter in the construction of plasmid vectors. The latter construction was the main vehicle of transformation of a team from the Johns Hopkins School of Medicine in Baltimore, MD, (Wang *et al.*, 2000b). The main plasmid of the latter investigators was *pZJM* (that is also based on plasmids constructed by the Wirtz Laboratory) and after integration in the trypanosome genome could be induced by tetracycline to cause, *in vivo*, the synthesis of a dsRNA of a defined gene's transcript. Various genes could thus be represented in the *pZJM* plasmid and cause the specific silencing of the appropriate gene whenever the cells were exposed to tetracycline. As the vector used by Shi *et al.* (2000) the *pZJM* vector plasmid, after linearization, was also flanked with sequences of the spacer in the rDNA. These flanks caused the vector to be integrated into this spacer. Such an integration was the norm because in trypanosomes there is a commonly homologous recombination with transgenes. The investigators thus integrated sequences of about 500 nt in opposing directions into their vector plasmids, and the linearized plasmids were introduced into the trypanosome cells. When the cells (after selection on antibiotics to assure that they integrated the vectors) were induced with tetracycline, the respective genes were indeed silenced. The investigators looked at eight target genes such as genes involved in DNA replication in the mitochondrion. Indeed, a specific silencing of the target genes was achieved and the investigators could follow the phenotypic effects of these silencings. The effect was not restricted to genes involved in DNA duplication. When a plasmid that included opposing sequences from the gene for ornithine decarboxylase (ODC) was used the respective dsRNA caused a deficiency in putrecine and thus a reduction in polyamines. Such polyamine-deficient trypanosomes do not grow but persist in an elongated form for up to 8 weeks. When putrecine is added the parasite cells can be rescued. Not only could the sequence from a single gene be integrated into the *pZJM* plasmid, but also the investigators constructed plasmids with two such sequences, each in opposing directions. The respective *pZJM* could then silence (after tetracycline induction) two different genes.

The team of Ullu and Tschudi at the Yale University (Djikeng *et al.*, 2001) proceeded their RNAi studies in *T. brucei* by asking if, in these protozoa, the long dsRNA that are derived *in vivo* from transgenes are also diced to small dsRNAs (siRNAs) as they are in other organisms (plants, nematodes, flies and mammals). For this search the investigators used a strain of trypanosomes with a transgene that, after induction, will produce dsRNA (of several hundred nt) with homology to the actin gene (line ACTI). The cells were induced and the RNAs were extracted and fractionated by centrifugation. Indeed, small RNAs were detected in sedimented fraction (100 000 xg). These RNAs were complexed with high-molecular-weight RNA. The small RNAs were cloned and sequenced. Indeed, 20–30 nt RNAs were detected and by sequencing they were found to have actin mRNA sequences. This strongly suggested that a *Dicer*-like enzymatic complex also exists in this low branch of eukaryotic animals. There was also a bonus-finding. There were short dsRNA in the RNA extract of 24–26 nt that had sequences with homologies to either of two retrotransposons that exist in *T. brucei*: INGI and SLACS. This finding clearly indicates that even in *T. brucei* there is a defense mechanism against the transposition of TEs that is based on dsRNA silencing.

The Ullu and Tschudi team went on to investigate further the RNAi mechanism in *T. brucei* (Djikeng *et al.*, 2003; Shi *et al.*, 2004). They found that 10–20 per cent of the siRNA is homologous to a dsRNA co-sedimented with polyribosomes. The siRNAs seemed to be associated with translating ribosomes. This assumption lead the investigators to assume that siRNA is recognizing its target mRNA while the latter is in the process of translation. In their recent report on RNAi in *T. brucei* the Yale team focused on the Argonaute protein (Shi *et al.*, 2004). Argonaute genes were previously found to be genetically associated with the RNAi of very diverse organisms. In most cases the organisms had several Argonaute genes (e.g. four genes in *Drosophila*, ten genes in *Arabidopsis* and in nematodes there are probably about 20 such genes). Unicellular organisms seem to have only one Argonaute gene (e.g. *Tetrahymena*). Shi *et al.* (2004) found a single gene for Argonaute (AGO1) in the *T. brucei* genome. This genomic sequence encodes two motifs, PAZ and Piwi, that were found in the Argonautes of metazoa. The AGO1 protein was a component of the RNP particle that is associated with polyribosomes. Furthermore, silencing of the AGO1 by the respective dsRNA considerably elevated the transposition of the *T. brucei* TEs. The latter finding indicated that dsRNA mediated RNA silencing is keeping the retrotransposons of this protozoa at bay.

A close relative of *T. brucei* is *T. congolense.* The latter species is a parasite of livestock in eastern and central Africa. As a subject of trypanosome research *T. congolense* has the advantage that its entire life cycle can be reproduced in axenic cultures, *in vitro. T. congolense* served the investigations of a team composed of investigators from Hokkaido, Japan and the University of Iowa (USA) for several years. Clearly, *T. congolense* is of no veterinary relevance in either Hokkaido or Iowa. Nevertheless, effective tools to investigate this parasite and especially its antigenic variation was developed by this international team. Consequently, Noboru Inoue (Hokkaido), John Donelson (Iowa) and associates initiated a study of RNAi in *T. congolense* (Inoue *et al.*, 2002) that had similarity with the studies of Ullu, Tschudi and associates on *T. brucei.* In fact the Hokkaido/Iowa team used similar plasmids and transgenic trypanosomes as those used by the Yale team (also based on plasmids originally developed by Elizabeth Wirtz and associates, the *pLEW* plasmids). The integrated plasmid could thus be induced to transcribe, after tetracycline induction, a transcript that *in vivo* will form specific dsRNA. The *T. brucei* plasmids had to be modified to function properly in *T. congolense.* Inoue *et al.* (2002) also succeeded to cause in *T. congolense* the FAT phenotype by a plasmid that caused transiently the formation (*in vivo*) of a dsRNA with homology to the α-tubulin. One of the abundant proteins in trypanosomes is the ribosomal protein PO (RPPO), one of the proteins of the large subunit of the ribosome. The investigators amplified a region of the RPPO gene (by PCR). This region was then introduced in an appropriate plasmid that was moved into the protozoa. When treated with tetracycline, death started at 48 hours after the protozoa were transfected (by electroporation). The death was not affecting all protozoa and after a few days there was recovery. This phenomenon suggested that RPPO is essential for the functionality of the protozoa.

Additional reports on RNA silencing in *T. brucei* came from Israel. Shulamit Michaeli (Bar Ilan University, Ramat Gan, Israel) is engaged for several years with the study of the molecular biology of trypanosomatids, primarily focusing on transsplicing and on the C/D and H/ACA-like small nucleolar RNAs (snoRNAs). The codes for these snoRNAs are clustered and repeated in the trypanosomes genome. These codes are transcribed (as other genes in these protozoa, see above) as polycistronic transcripts by RNA polymerase II. The transcripts are then processed into mature snoRNAs. Additional information on the transsplicing in Trypanosomatids will be provided in a section below. Some snoRNAs are involved in pre-rRNA cleavage and modifications. Thus, ribose methylation is guided by C/D snoRNAs. Michaeli and associates (Liang *et al.*, 2003a, 2003b) asked whether the snoRNA genes can be silenced in three trypanosomatide species: *Leptosomas collosoma, Leishmania major* and *Trypanosoma brucei*. Indeed, silencings by antisense methods were achieved in *L. collosoma* and *L. major.* I shall not detail these results. The silencing of snoRNAs in *T. brucei* was investigated by the dsRNA procedure. In the latter species Michaeli and associates followed the same basic procedures reported above: using the (tetracycline) inducible synthesis (*in vivo*) of dsRNA from opposing T7 promoters. They utilized the *pZJM* expression vector (mentioned above) that contained coding sequences of various snoRNAs in a way that the respective dsRNAs will be formed in the transformed trypanosomes. Consequently, the snoRNAs were indeed silenced in the transgenic cells that were induced by tetracycline. The silencing reached various levels. The C/D snoRNA was silenced more than the H/ACA RNA. Reduction of the levels of snoRNAs reduced the rate of modifications of rRNA that is guided by these snoRNAs.

A further study of the Michaeli team (Mandelboim *et al.*, 2003) on RNA silencing in *T. brucei* can serve as a reminder to the novice. It deals with the elaborate way this tiny organism deals with the expression of its genes. Obviously, tiny does not mean simple. I have mentioned above the transsplicing of transcripts in trypanosomes. A short description of the present knowledge of transsplicing will be presented in a special section below, illustrated with a scheme. In the transsplicing reaction the spliced leader is donated to pre-mRNA from a small RNA termed SL RNA. For the transsplicing the actions of small ribonucleoproteins (snRNPs) is required. These snRNPs are carrying several small RNAs: U2, U3, U5 and U6. There is also a need for splicing factors: U2AF35, helicases, nuclear RNPs and serine/arginine rich proteins. The exact roles of these factors are not clear yet. The snRNAs that participate in transsplicing have different "cap" structures. Take, for example, the SL RNA has a complex "cap" ("cap 4") which consists of methylation of the sugar groups in the tetranucleotide AACU. The correct modification is essential for transsplicing. The U2 and the U4 possess a trimethylated guanosine (TMG) "cap". The U6 has an inverted "cap" whereas the U5 lacks TMG and has a 5'-phosphate terminus. The SL RNA is transcribed from a distinct promoter by RNA polymerase II. The SL RNA is a component of a particle termed SL RNP. Actually after transcription the SL RNA undergoes modifications. It is noteworthy that SL RNA was found in the nucleus as well as in the cytoplasm. It is probable that essential modifications of the SL RNA take place in the cytoplasm. Then the SL RNA is returned to the nucleus for further modifications. Why this shuffling? Unlike most other animals that elaborated their form vastly during the last billion years, Trypanosomes maintained their overall form but they did elaborate one process considerably. This process is the transsplicing which became so elaborate that talented investigators had to work very hard in order to understand this process. Of the several snRNAs the U6 snRNA has a unique biogenesis. The RNA is transcribed by an RNA polymerase III and acquires a α-monomethyl "cap" structure. This U6 probably does not go out of the nucleus — it stays in it. The modification of U6 (methylation and pseudo-uridylation) takes place in the nucleolus. In addition there are Sm proteins. These bind to the U snRNAs and SL RNA. The structure of the trypanosome Sm proteins differs from homologous Sm proteins of higher eukaryotes (e.g. man).

Michaeli and her associates (Mendelboim *et al.*, 2003) focused their attention on two Sm proteins: SmE and SmD1. They asked what would be the consequences, with respect to transsplicing and splicing, when the levels of these proteins would be specifically reduced by RNA silencing. To obtain answers to this question the investigators used an appropriate strain of procyclic *T. brucei.* The strain has all the required ingredients for the RNA silencing and the selection of transformed cells. Like most eukaryotic cells, *T. brucei* cells like "company": when diluted beyond a certain level they will succumb rather than multiply. Thus, feeder-cells are required to maintain the multiplication of diluted cultures.

The silencing of the genes that encode SmE and SmD1 was achieved by either of two methods. In the first method the gene encoding the SmD1 protein was silenced. For that the investigators constructed plasmids that will transcribe RNA sequences that will fold after integration. For silencing the SmE encoding gene, the method of silencing was done by the T7 opposing promoters. As in previous silencing of trypanosomes the linearized plasmids were integrated into the rDNA spacer and the transgene contained a selectable gene (phleomycin resistance) for cloning of transformed cells. Due to the appropriate construction of the vector plasmid, the transformed cell can grow and multiply happily as long as they are not treated by the inducer, tetracycline. After tetracycline induction there was an arrest of growth and a reduction of mRNAs for the SmE and SmD1 genes. Notably the silencing of SmD1 was more prolonged than the silencing of SmE. The latter silencing was lost after 10 days. Possibly silencing by the stem-loop producing plasmid is more stable than silencing by the T7-opposing promoter system. Silencing of specific Sm genes caused the reduction of U1, U2, U4 and U5 while the U6 and the spliced leader-associated RNA (SLA1) remained unchanged. On the other hand, the level of SL RNA was dramatically elevated. But this elevated SL RNA was not the normal species: it lacked the typical modification of natural SL RNA. It also retained a polypyrimidine tail. This SL RNA was retained in the cytoplasm. This indicated to the investigators that the SL RNA that was accumulated in the cytoplasm represent an intermediate in the normal processing of the SL RNA before its final formulation that happens after re-entry into the nucleus. The results also showed clearly that Sm proteins bind U1, U2, U4 and U5 but not U6 and SLA1. In addition the results indicated that unlike the tails of the "Three Blind Mice", tails of SL RNA are cut twice. Once the long tail is cut in the cytoplasm and 5 nt of it is retained. Then, upon re-entering the nucleus the rest of 3' tail is cut off. In summary the investigators suggest the following scenario for the biogenesis of the "mature" SL RNA. The methylation on "cap 4" nt takes place in the cytoplasm, only after Sm assembly. "Cap" modifications of the SL RNA may take place both in the cytoplasm and the nucleus while the ψ modification takes place in the nucleus early in the biogenesis. The different "cap" modifications may assure accurate transport in and out of the nucleus. The SL RNA biogenesis seems to be related to snRNP biogenesis. Future studies will hopefully further clarify the process of transsplicing in trypanosome. Such a clarification should lead to therapeutic interventions to reduce the pathogenicity of these deadly parasites.

Some remarks on transsplicing

Transsplicing exists in several investigated protozoa and is not confined to these organisms; it was also revealed in some metazoa. In this process alternative exons may be spliced in the same pre-mRNA, resulting in different mature mRNAs. The following remarks are based on the review of Shulamit Michaeli and associates (Liang *et al.*, 2003b) where details and many references are presented. Splicing of pre-mRNA to render it to mature mRNA is the norm of eukaryotic genes (although there are typical eukaryotic genes in which there is no splicing of the transcript). This is a rather elaborated process in which the 3' end of one exon is cleaved; the intron between this exon and the next downstream exon is removed and the 3' and of the upstream exon is then connected to the 5' of the next downstream exon. Many protein encoding genes are involved in this elaborated process, some of which have the role to identify the border between exons and introns. Wrong identifications will lead to non-functional spliced products. On the other hand, there is also *alternative splicing*. In the latter case the same pre-mRNA can be spliced in different ways. If the pre-mRNA is from a gene that encodes proteins, then the same gene can lead to different mature mRNAs and consequently to different proteins (see: Galun, 2003, Chap. 6).

It is common in prokaryotes that several enzymes are coded by a polycistronic region of a bacterial genome. Frequently, these enzymes are of the same metabolic pathway. Hence, the promoter that activates the transcription of this polytcistronic transcript regulates the levels of several mRNAs of this pathway but not so in most eukaryotes (e.g. metazoa and plants). In these latter organisms it is usual for each protein to have its "own" separate transcript and a "private" promoter on this transcript. There are obvious exceptions: some viruses (e.g. hepatitis A virus) have a long transcript that is translated into a polyprotein which is then cleaved into individual proteins. Members of this polyprotein may actually serve as proteases that are active in the cleavage of the polyproteins into individual proteins.

As already noted above in transsplicing we witness a different process. Here I shall describe the transsplicing of trypanosomes although the same basic process was also found in other organisms (e.g. nematodes, euglenoids, trematodes and even chordates). In trypanosomes the entire chromsome may be transcribed as a polycistronic RNA. While RNA polymerase II is active in this transcription no promoter for this polymerase was identified in the polycistronic RNAs of trypanosomes yet. A general scheme for transsplicing in trypanosomes is provided in Fig. 17A and a proposal for the

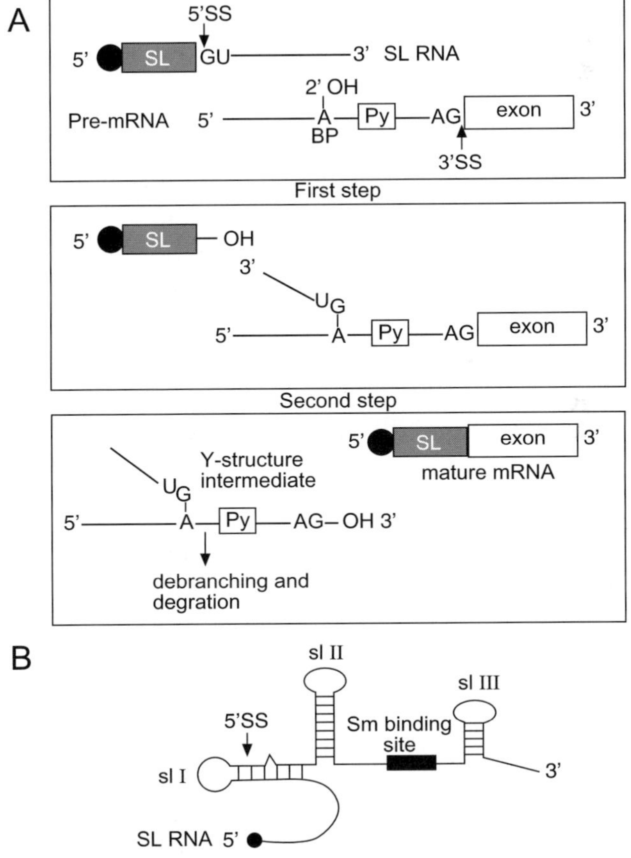

Fig. 17.   Scheme of the mechanism of *trans* splicing. The 5′ splice site GU on SL RNA and the 3′ splice site AG on the pre-mRNA are indicated. (A) The mechanism and (B) The result. BP is branch point; Py is poly-pyrimidine tract. (From Liang *et al.*, 2003.)

involvement of components in this transsplicing is shown in Fig. 17B. An important component of the transsplicing is the spliced leader RNA (SL RNA) as shown in Fig. 17A. It has three conserved stem-loop secondary structures (slI, slII & slIII) as well as a unique "cap" at its 5′ end. Between slII and slIII there is an Sm binding site. In SL RNA we meet small sl structures but we should recall that stem-loops, small and large, are characteristic of a variety of RNA species as ribosomal RNAs and tRNAs. They are probably the way of RNAs to attain defined secondary structures.

There are several species of snRNAs that are involved in splicing. Of these the U1 of trypanosomes is involved in "regular" splicing and U2, U4, U5, U6 and SLA1 are involved in transsplicing. These have homologs in metazoa where they differ to some extent from trypanosome snRNAs. Thus, the U2 of trypanosomes is shorter than the metazoan U2 and also does not contain some motifs (GUAGUA) that exist in the U2 of yeast and mammals. The U5 of trypanosomes also differs from that found in eukaryotes especially in its 5′ end. The SLA1 and its function were revealed only recently and it seems to have a role in guiding modifications in other RNA species (e.g. pseudouridylation on SL RNA).

Base-pairing was detected between the U5 and SL RNA and there is genetic evidence that indicates that such base-pairing is important for correct transsplicing. A similar base-pairing exists between U6 and SL RNA.

The SL RNA is transcribed by RNA polymerase II by an external promoter. It is transcribed with a poly (T) of various lengths. The "cap 4" is added during transcription and it probably takes place in the nucleus. As noted above, the SL RNA undergoes essential modifications in the cytoplasm. The SL RNA has a binding site for Sm protein. The binding of this protein may have a role in further SL RNA modifications. Then the fully modified SL RNA is ready to re-enter the nucleus. Interestingly, the SL RNA in amastigote *Leishmania donovani* has a long poly(A) tail. It sounds peculiar: tail-less (amastigote) cells do have a long SL RNA tail each!

The Sm proteins for which there are binding sites in snRNAs differ in trypanosomes from the Sm proteins of other eukaryotes. Antibodies for the latter Sm do not recognize Sm of trypanosomes.

The polypyrimidine tract (Py in Fig. 17) and the AG dinucleotide, downstream of it are essential signals in transsplicing of trypanosomes. How exactly trypanosomes recognize the branch-point (commonly an adenosine, up-stream of the Py) is not clear. It is also not entirely clear yet how the poly(A) tail is added to the final mRNAs.

In summary there is a set of snRNAs of trypanosomes that are crucial for transsplicing. The U1 is probably not one of these but the latter may serve the "regular" splicing in trypanosomes. The SL RNA interacts with U5 and probably U6 too. The SLA1 interacts with the exon sequence and guides the pseudouridylation of SL RNA. Many additional aspects of transplicing of trypanosomes await verification and/or clarification. Once an *in vitro* system for transsplicing of trypanosomes becomes available, more information on the details of this interesting mechanism is envisioned. Clearly, a thorough understanding of this mechanism in trypanosome that does not exist in the hosts of these parasites can lead to specific and efficient therapeutic intervention.

Apparent lack of RNA silencing in **Trypanosoma cruzi** *and in* **Leishmania**

The protozoa *Trypanosoma cruzi* is the causative agent of the Chagas disease that is a severe human health problem in Latin America. It has a lifecycle that is composed of three stages. In one stage epimastigote cells multiply extra-cellularily in the mid-gut of the bug vector. From the mid-gut these protozoa migrate to the hind-gut where they differentiate into non-dividing trypomastigotes which are infective and pass through mucous membranes and skin cuts of a mammalian host (e.g. man). There, they enter certain cell types and change to amastigotes (i.e. devoid of flagellum) which multiply intracellularily. Then the amastigotes are transformed to trypomastigotes and enter the circulatory system when their host cells lyze. They can then invade new host cells or enter the blood feed of the reduviid bug and the cycle starts again.

John Donelson of Iowa, whose collaboration with Japanese investigators was mentioned above, studied RNA silencing in *T. cruzi*. He collaborated with Brazilian investigators (DaRocha *et al.*, 2004) and used several techniques that were functional in RNAi of *T. brucei* and intended to silence several genes. No RNA silencing could be established. In no case was the respective mRNA reduced. The investigators assumed that some components of the RNAi are missing in *T. cruzi*.

The protozoa *Leishmania donovani* is the pathogen causing one of the several Leishmaniasis diseases. The scientific name of this parasite is composed of two people who described it about 100 years ago: Leishman and Donovan. There are a number of *Leishmania* species that cause Leishmaniasis and this is not a single disease. Clinically there are three major types of leishmaniasis: cutaneus leishmaniasis, mucocutaneous leishmaniasis and visceral leishmaniasis. The third of these is life-threatening. Leishmaniasis is spread worldwide. About 350 million people are at risk and about 12 million people are actually infected. HIV patients seem to have an elevated risk for this disease. The insect vectors of *Leishmania* protozoa are sand flies (of various species). A detailed review on Leishmaniasis was provided by Herwaldt (1999). Due to its medical importance *L. donovani* was a subject of many studies that intended to investigate the genetics and molecular biology of this parasite. After the success of RNA silencing in *T. brucei* Robinson and Beverley (2003) of the Washington University Medical School in St. Louis, MO, used similar methodologies to search for RNA silencing in *Leishmania*. They developed satisfactory procedures for transient and stable DNA transfection. They then applied plasmids with similar constructions to those effective in *T. brucei* to *L. major* and *L. donovani* but achieved no RNA silencing. Also introducing a dsRNA of 24 nt that represented the α-tubulin gene was not effective in the reduction of α-tubulin. The investigators concluded that their data suggested that typical RNAi strategies may not work effectively in *Leishmania* and raised the possibility that *Leishmania* is naturally deficient for RNAi activity.

RNA Silencing in *Toxoplasma gondii*

Toxoplasma is a successful obligatory intracellular parasitic protozoa of the family Apicomplexa in that it infects warm-blooded vertebrates worldwide very commonly but usually does not kill its host. This is also true in man, provided that the host is immunocompetent. In immunodefficient people such as AIDS patients and people treated by immunodepressors (as in chemotherapy and after transplantation) the disease caused by *T. gondii* infestation can be deadly. These patients may suffer from fatal pneumonia and taxoplasmic encephalitis. AIDS patients should therefore detach themselves from the company of cats that frequently host *T. gondii* and where the sexual stage of this parasite takes place.

A summary of the lifecycle of *T. gondii* is shown schematically in Fig. 18. More information on the genetics and biochemistry of *T. gondii* development was provided by Boothroyd *et al.* (1997). Briefly, there are two cycles in *T. gondii*. The asexual cycle takes place in many warm-blooded animals (e.g. cats, dogs, man). It involves two stages of the parasite: rapidly dividing tachyzoite and more slowly dividing (and encysted) bradyzoite. When there is a pressure by the animal's immune system the balance is in the direction of slow division and even encystment. The transmission of *Toxoplasma* during the asexual cycle is commonly done by feeding on undercooked meat. The sexual cycle occurs (only?) in the gut epithelium of cats. These gametes are undergoing fertilization and oocysts are produced.

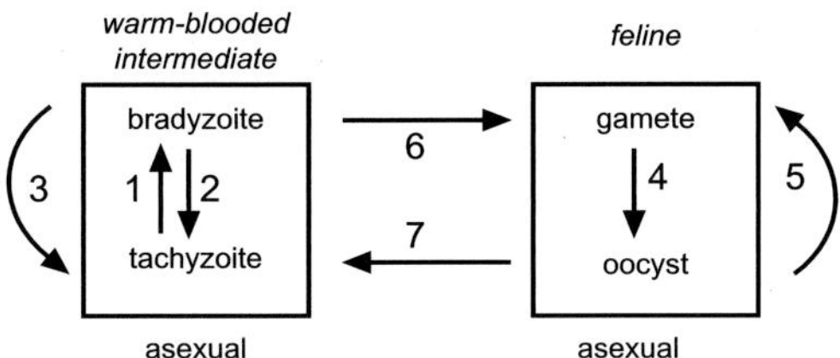

Fig. 18.   The life cycle of *T. gonadii*. The asexual cycle can occur in a large number of warm-blooded animals and is shown on the left. It involves an equilibrium between the rapidly dividing tachyzoite and the more slowly dividing bradyzoite (arrow 1 and 2). The balance in this equilibrium is believed to be determined by environmental factors, especially stress provided by the host's immune response. Encystment results from a strong immune response whereas reactivation occurs when the pressure is relaxed. Transmission through the asexual cycle is made by ingestion of uncooked meat and other foods, containing infectious encysted bradyzoites. The sexual cycle is shown on the right and involves schizogony gametogenesis and fertilization in the gut epithelium of felines. Cat-to-cat transmission through the sexual cycle is by ingestion of oocysts in faecal contamination. Cross-over between the two cycles is represented by arrows 6 and 7. (From Boothroyd *et al.*, 1997.)

The latter are extrucal with the faeces. Research of *Toxoplasma* is facilitated because this parasite can be cultured on mammalian-cell monolayers.

A team at the University of Windsor, Canada (Al-Anouti *et al.*, 2003) investigated the possibility to suppress specifically the expression of a *T. gondii* gene by homologous dsRNA. The investigators focused on the gene that encodes uracil phosphoribosyltransferase (UPRT). There is a way to kill parasites that have an active UPRT. In the presence of this UPRT the addition of 5-fluoro-2'-deoxyuridine (FDUR) will finally be converted in the parasite to 5-fluoro-deoxyuridine monophosphate. The latter is lethal to the parasite because it inhibits the synthesis of the vital thymidine monophosphate. Hence, parasites in which UPRT is silenced will survive the addition of FDUR in the culture medium. The investigators followed two strategies. In one strategy the dsRNA that had a sequence that was homologous to the UPRT gene was produced by an engineered palsmid (similar to the plasmids mentioned above in dealing with *Trypanosoma brucei*). The plasmid was introduced into the parasites, leading to stably transformed *T. gondii*. The result was the down-regulation of the expression of UPRT and capability to grow in the presence of FUDR. The down-regulation of the UPRT could also be achieved by the introduction of *in-vitro* synthesized dsRNA with homology to the UPRT gene into the parasites. It was shown that such dsRNA specifically degraded mRNA of the UPRT gene, resulting in lowering the UPRT activity in the parasites.

RNA Silencing in *Plasmodium*

The protozoa *Plasmodium* (especially *P. falciparum* and *P. vivax*) are the major agent causing malaria. This disease is a major cause of human mortality worldwide. Although the numbers provided by the World Health Organization (WHO) should be accepted with caution, the numbers that it publicized are really alarming: about 40 per cent of the world population is at risk of malaria; there are about 400 million infected people and malaria causes 2–5 million deaths annually. In areas of high endemicity the prime victims are children below the age of five and pregnant women.

The malaria disease is initiated when sporozoites are introduced into humans when female *Anopheles* mosquito are sucking blood from the humans. The sporozoites "donated" by the mosquito migrate to the liver where they invade hepatocytes within 30 to 60 minutes of inoculation. There, the parasite develops into exo-erythocytic forms during a period of 5–10 days (depending on the species of *Plasmodium*). As many as 30 000 merozoites can be produced. During this liver stage there is no clinical disease. When the liver cell finally raptures the merozoites enter the circulation. They invade erythrocytes where they undergo asexual amplification, the infected erythrocytes rapture and the plasmodia invade new erythrocytes. This erythocytic stage is responsible for the disease symptoms during which many erythrocytes are destroyed causing fever and anemia. The infected of erythrocytes have a "side-effect"; they may stick to the side of blood vessels, causing occlusion of these vessels. Then the parasites may develop into sexual forms: male and female gametocytes. When a "new" female mosquito sucks blood from an infected host (e.g. a human), the gamete-containing blood cells enter the mosquito and mate in the mosquito midgut. The resulting sporozoites migrate to the salivary gland and upon the next blood-meal the parasite is introduced into the next host (victim). Interestingly, the *Plasmodium* cells have two extra-nuclear DNAs. One is circular and resembles mitochondrial DNA while the other has similarity to plastid DNA. Could the latter be a remnant of very ancient ancestors of this obligatory parasite when no warm-blooded hosts were yet available, and these ancestors were fed by the sun energy rather than human (or other animal) blood?

Already in 1880 the malaria (*Plasmodium*) parasites were found in the blood of human victims of malaria and by 1900 it became clear that the disease is transmitted by mosquitoes. Only about 50 years later was the liver stage of the parasites lifecycle discovered. More details on the lifecycle and especially on the immune defence of *Plasmodium* are provided by Doolan and Hoffman (1997). A recent review on the genomics and molecular biology of *Plasmodium* was provided by Aravind *et al.* (2003). It may be noted that various mosquito genera specialize in the transmission of *Plasmodium* to certain hosts. Thus, *Anopheles* mosquitoes transmit the parasite to humans, monkeys and rodents; *Culex* and *Aedes* mosquitoes predominate in the natural transmission to birds.

Two research teams pioneered in the study of RNA silencing by dsRNA in *Plasmodium falciparum*. McRobert and McConkey brought the parasite to the University of Leeds, England, to perform their experiments on *in vitro* grown parasites that grow in blood cells where they could be synchronized. They asked if a dsRNA with a sequence homologous to a specific *P. falciparum* gene would silence this gene. They chose the gene dihydroorotate dehydrogenase (DHODH) which is an essential enzyme in pyrimidine synthesis that is required in *P. faliparum*, since in malaria parasites *de novo* synthesis of pyrimidines is required. The investigators synthesized dsRNA *in vitro* and introduced dsRNA by electroporation. The DHODH dsRNA promptly inhibited the growth of the parasites (about 24 hours after electroporation). The dsRNA of a similar enzyme from trypanosomes served as a control and did not inhibit the growth of *P. falciparum*. The mRNA for the DHODH was also reduced (to about 50 per cent) by the application of the respective DHODH dsRNA. Another gene, chorismate synthase (CS), was also silenced by the same approach.

Following McRobert and McConkey (2002) a research on RNA silencing of *Plasmodium* took place in a malaria-area: New Delhi, India. The Indian investiagors (Malhotra *et al.*, 2002) focussed on two *falcipain* genes of *P. falciparum* that encode cystein proteases. There are several proteases in *P. falciparum* that are required for the digestion of the haemoglobin of the host. But the specific role of the individual proteases is not known. The investigators therefore prepared dsRNA respresenting two of these proteases: *falcipain-1* and *falcipain-2*. The investigators thus introduced either of the two dsRNA into the parasite or introduced both dsRNAs. There was a notable effect of these dsRNAs. Each of them affected the morphology and reduced the metabolism of haemoglobin in the parasite. When the two dsRNAs were introduced simultaneously, the effect was even greater. Also, the levels of mRNA for these genes were reduced. It was also revealed that the application of dsRNA caused the formation of siRNA (of about 25 nt) that had homologous sequences to the respective *falcipain* genes.

RNA Silencing in *Amoeba*

The amoebas belong to the phylum Rhizopoda. They are widespread in very different habitats such as saltwater, freshwater and soils. Some are parasites of animal. Among the latter is *Entamoeba histolytica*. The reproduction of amoebas is by fission: direct division into two cells of about equal volume. These protozoa have no cell-wall, nor flagella or any form of sexual reproduction. Their mitosis is similar to more advanced eukaryotes. The species *E. histolytica* is a mammalian parasite that causes amoebic dysentery. This parasitism is only pathogenic with respect to "lodging", the amoebas feed on bacteria but the amoebas release toxins that are harmful to the "host". The amoebas can form cysts that are resistant to digestion by their host. Mitotic divisions can take place within the cysts; when there are four or more amoebas within a cyst it may break and release amoebas within the digestive tracts of the animal host (e.g. man). The cysts are dispersed in faeces and may re-enter another host orally, either directly or by carriers (e.g. insects) or contaminated food. In developed countries about 5 per cent of the population may be hosts to *E. hystolytica* while in developing tropical areas, up to 50 per cent of the population may be infected with this parasite.

David Mirelman of the Weizmann Institute of Science, Rehovot, Israel (i.e. my neighbor and former Dean) and associates are veterans in the study of infectivity and molecular biology of *Entamoeba histolytica*. In a recent study they focused on the amoebapores (AP) that are important virulence factors in this amoeba. The AP is a small protein of 77 amino acids. There are three isoforms of this protein. In the recent study (Bracha *et al.*, 2003) the investigators intended to look for the specific role of AP–A. The present view is that following lectin-mediated recognition and intimate adherence between the trophozoite (the active form of the amoeba) and its target cell, the AP molecules are inserted into the membranes of the latter without depending on the interaction with a specific membrane receptor. Antibodies against AP are therefore unable to inhibit its toxicity. The study intended to change the level of AP–A (the most abundant AP) by the insertion into the amoeba of plasmids that harbor sequences of the *ap-a* gene. Note that they did not construct plasmids in which the gene sequences were put head-to-head so that the transcript will form a stem-loop that would be cut, *in vivo*, into dsRNA. The approach of Mirelman and associates was similar to that of Don Grierson (mentioned in the Introduction) who introduced a transgene that included a part of the PG gene (that encodes the enzyme polygalacturonase) into tomato plants in order to reduce the level of the pectin-degrading activity in the tomato fruits (and consequently extend the shelf-life of the fruits). By constructing the appropriate plasmids that also contained a selectable gene for resistance to G418, the investigators could indeed reduce the levels of mRNA for AP–A as well as the level of the protein. Increasing the levels of G418 caused more plasmids per amoeba and thus a stronger reduction. The effect depended on the regions of the *ap-a* gene that the investigators used in the construction of the plasmid. The sequences from the 5′ regulatory region of the gene were effective in reducing the expression of *ap-a*. Actually, effective sequences and sufficiently high doses of G418 (causing a high number of plasmids per cell) completely eliminated the mRNA of *ap-a* and the AP–A protein. Nuclear run-on analyses and other analytical procedures indicated that reduction of mRNA resulted from reduced transcription rather than from degradation of mRNA that was transcribed. Clearly, the mechanism of silencing was of the TGS type. The insertion of the plasmid with the above mentioned transgene did not cause the formation of siRNAs (with sizes of 22–25 nt and homologues to the *ap-a* gene and/or its flanking sequences).

Will the "regular" PTGS also work with *Entamoeba histolytica*, and will dsRNAs with homologies to specific genes cause posttranscriptional silencing in this or other amoebas? We have to wait for answers.

RNA Silencing in Ciliated Protozoa: *Paramecium* and *Tetrahymena*

The ciliates branched off from other eukaryotes about 1 billion years ago, hence before the branching-off of fungi. They comprise a very diverse group of protozoa. The genetic distances between genera of ciliates can be enormous; two genera may be more distant from one to another than maize from rat! Since RNA silencing was only investigated in two genera that are considered among the phylogenetically advanced ciliates, we shall deal only with these genera: *Paramecium* and *Tetrahymena*. These two genera belong to the order

Hymenostomatida. These two ciliates are rather elaborated animals with several "organs" (like the cilia, "mouth" and excretion device) but are all confined in one cell. Several of the species of *Paramecium* and *Tetrahymena* can be cultured easily in the laboratory (although there are also parasitic *Tetrahymena* species) and hence were favorable subjects of investigations. We shall see that due to unique molecular-genetic features (e.g. in the macronucleus of *Tetrahymena*), these ciliates were studied intensively (see: Prescott 1994, for review and literature).

Paramecium

This genus has usually one micronucleus and one macronucleus. The former contains the whole genome of this protozoa and serves as a kind of germ-line. The latter is diploid with five pairs of chromosomes. The macronucleus serves the "daily needs"; it does not contain all the genes but rather only genes required for the vegetative growth and metabolism. Its DNA is not organized into regular chromosomes. *Paramecium* has a *gullet* (a kind of mouth) that is lined with cilia, into which the food is introduced and passed to foodvacuoles. The food is digested in the vacuoles. After the food is digested it is passed to a "wastebasket", the *cytoproct*, from where it is extruded from the cell. There are anterior and posterior contractible vacuolar systems that can change the specific weight of the ciliate and drive it up or down in the water.

The cells of *Paramecium* usually divide asexually by transverse fission. This reproduction can go on for many generations but then a sexual reproduction is required. Cells from two different mating types approach each other for this conjugation. In preparation for conjugation the micronucleus undergoes meiosis; (usually) four haploid micronuclei are then produced. Each partner sends one micronucleus to its mate and also receives a micronucleus from its partner. In each of the conjugated cells the resident micronucleus and the received micronucleus fuse. After the fusion the diploid a micronucleus divides again. One of the resulting micronuclei stays as a micronucleus while another is converted into an "Anlage", in preparation of its development into the "mature" macronucleus. The "old" macronucleus then degenerates. The mitosis of the micronucleus differs from that of higher eukaryotes: the mitotic stages of the chromosomes take place in an enveloped nucleus (as in some fungal organisms); there is no breakdown of the nuclear envelope. In the macronucleus there are no visible chromosomes during mitosis. There is an amitosis; the DNA is packed in long bundles of chromatin strands. Moreover, the DNA may not be divided equally in the two macronuclei after division. Although the micronucleus does not seem to be transcribed during the vegetative growth its presence is required. When the micronucleus is removed the *Paramecium* may gradually degenerate. Both micronuclei and macronuclei have nucleosomes. Only the macronucleus has nucleoli. The total amount of DNA per macronucleus is several hundred-fold greater than the amount in the micronucleus. It seems that those gene sequences that are retained in the macronucleus are replicated to many copies.

The DNA fragments in the macronucleus have their "own" telomeres with repeated GT nucleotides that differ to some extent at the 5' end from the telomeres of the 3' end. The 5' leaders and three trailers in ciliates genes have uncommon boxes. These were studied more in *Tetrahymena* than in *Paramecium*. There are introns in some of the *Paramecium* genes but the introns are fewer and shorter than in higher eukaryotes.

The coding in ciliates differs from that in other eukaryotes. In *Paramecium* and *Tetrahymena* TAA and TAG are coding for glutamine rather than serving as stop codons. Only TGA serves as a stop codon in these two genera. It is noteworthy that in another ciliate genus, *Euplotes*, TAA and TAG are stop codons while TGA codes for cysteine. The preference of usage of triple nucleotides as codons for amino acids is rather different from that known in other organisms.

In ciliates there is a unique process of elimination of "internally eliminated sequences" (IES) from the genome in the micronucleus to the sequence in the macronucleus. Thus, in a process that has functional homology to the splicing of pre-mRNA, certain IES are cutout and the remaining DNA sequences are spliced. Additional peculiarities in ciliates that involve excision of IES, excision of transposon-like sequences, gene scrambling, etc. were reviewed by Prescott (1994).

The first report on homology-dependent gene-silencing in *Paramecium* came from a team of investigators in Gif-sur-Yvette, France (Ruiz *et al.*, 1998a) who were studying the regulated exocytosis of trichocytes, secretory granules thought to be involved in defense against predators of *Paramecium*. The trichocyst crystalline core and a number of cytoskeletal arrays are built from a set of closely related polypeptides. This team showed previously that the trichocyst matrix proteins (TMP) and the polypeptides that form the innermost cytoskeletal network of the *Paramecium* cortex are encoded by mulligenic families. The genes are apparently co-expressed and the various polypeptides are then co-assembled.

The investigators introduced into the macronucleus, by microinjection, coding sequences of various genes. The DNA was introduced without the cis-flanking sequences. Commonly, such an introduction will cause in the macronucleus, modifications such as the addition of "telomeres", and cause the formation of alien pseudochromosomes that replicate in parallel with the replication of the DNA of the macronucleus. When coding sequences for TMP were injected all the respective genes were silenced. The same was found for a single-copy gene *ND7* (required for exocytotic membrane fusion and trichocyt release). The reduction of the respective protein was evaluated by western blot hybridization and the reduction in mRNA was detected by northern blot hybridization. The microinjections also caused the formation of phenotypes that were expected to result from these gene silencings. This study was performed *before* the publication of Fire *et al.* (1998). The experimental procedure did not cause the production of dsRNA in the macronucleus and was thus similar to the procedure employed by Bracha *et al.* (2003) in gene silencing in *Entamoeba histolytica*. Ruiz *et al.* (1998a) assumed that their gene silencing is related to PTGS but they did not provide evidence for mRNA degradation in their experimental system. Indeed, such evidence was furnished for gene silencing in *Paranecium* by Galvani and Sperling (2001) of the same Center in Gif-sur-Yvette as Ruiz *et al.* (1998a). In the later study (Galvani and Sperling, 2001) several plasmid constructs were engineered, in which sequences from two genes were included: the *T4a* gene, a member of the TMP multigene family encoding secretory proteins and the *ND7* that is a single copy gene required for exocytotic membrane fusion. When silencing took place it had a clear phenotypic consequence, leading to exocytosis-deficiency. Several sequences of the gene were effective in silencing and all were effective but not the 3′ non-coding region of the gene. The constructs apparently caused, *in vivo*, the formation of dsRNAs with homology to part of the gene sequences. The silencing was indeed of the PTGS type, as was revealed by the lowering of mRNA that was traced by nucleus run-on assays. Interestingly, a transgene that contained the 3′ non-coding sequence of a gene inhibited the silencing. Could it be that such a sequence reduced the *in vivo* formed dsRNA? In a continuation of the above mentioned study Galvani and Sperling (2002) added two more *Paramecium* genes (*NSF*, encoding a general membrane factor, and *ICL1a*, encoding centrin, a component of the infraciliary lattice) to the list of genes that can be silenced in this ciliate protozoa. Only the method of introducing dsRNA into the *Paramecium* cells was changed. Bacteria with plasmids that would produce these dsRNAs, *in vivo*, were fed to the protozoa, rather than injecting the plasmids into the macronuclei. This rendered the experimental procedure much simpler. Probably, Galvani and Sperling (2002) followed the studies on RNAi in nematodes. They consequently asked whether feeding dsRNA-producing bacteria that caused RNAi in nematodes will have a similar effect in *Paramecium*. They received an affirmative answer.

Tetrahynema

This genus of ciliates is a relative of the genus *Paramecium*. Like *Paramecium* species *Tetrahymena* species also live in various habitats as fresh water and salty water and some are even parasites of certain metazoa. Most earlier studies were conducted with the species *Tetrahymena pyriformis* (pear-shaped Tetrahymena). These early studies concerned the morphology, mating and especially metabolism. The name *Tetrahymena* is relatively recent (given in 1961) while this genus was one of the very first microorganisms to be put under the microscope (by Antony van Leeuwenhock, in 1676; possibly with lenses supplied by his neighbor, the philosopher Baruch Spinoza). As *Paramecium*, *Tetrahymena* also has rows of cilia (sometimes termed kinetics) that cover the surface and run about parallel to the long axis of the cell. There is also an oral apparatus at the anterior end and an anal pore (cytoproct or cytopyge) on the posterior ventral side. By electron microscopy it was revealed that the somatic cortex of *Tetrahymena* is a rather elaborate structure in which the cilia are anchored. Similar to *Paramecium, Tetrahymena* also has micronuclei and macronuclei that function in a similar way to these organelles in *Paramecium*. In *Tetrahymena* there is usually one micronucleus and one macronucleus in each cell. Macronuclei contain nucleoli. Chromosomes are visible in dividing micronuclei but not in macronuclei. Since the studies of Andre Lwoff (in 1923) *Tetrahymena* can be cultured in defined media-making this organism is a kind of a model for protozoa research. Much of this research was devoted to the molecular biology of the process in which the non-transcribing micronucleus that results from sexual reproduction is converted into a transcribing macronucleus. *Tetrahynema* also served to study intron splicing (e.g. pre-rRNA intron splicing) as well as the roles of ribozymes and the formation of telomeres.

The conversion into a macronucleus involves major changes in the DNA. During this conversion the five chromosome pairs of the micronucleus that have a mean length of about 2×10^8 nt (about 44 000 kbp) are undergoing conspicuous processings. The processing of the long chromosomes consists of two stages. First, about 6000 internal eliminated sequences (IESs) are removed. This elimination occurs in the non-coding regions of the chromosomes. Then there is a stage of processing in which the chromosomes are cut into about 200 fragments of sub-chromosomal DNA molecules which differ in size (from less than 100 kbp to over 1500 kbp). One molecule is exceptionally small: the 21 kbp fragment encoding the rRNA. In the process of fragmentation, short fragments of about 54 bp are lost at the breakage sites. This stage of elimination has also its own acryonym, BES, based on breakage eliminated sequences, that are removed during this stage. Telomeres are added to the retained fragments and the fragments are multiplied to many copies.

A team of investigators at the University of Rochester, NY, the York University, Canada and the National Institute of Genetics, Michima, Japan, jointly analysed genes that are involved in the processing of the long micronulear chromosomes into the minichromosomes of *Tetrahymena thermophila* (Mochizuki *et al.*, 2002). Actually, these investigators looked into the PPD encoding genes. The PPD stands for PAZ and Piwi domain-containing proteins. We saw that these domains are involved in RNAi and PTGS as well as in quelling in fungal organisms. Several genes that encode PPD such as *AGO1*, required for PTGS in *Arabidopsis*, *qde-2*, required for quelling in *Neurospora*, and *rde-1*, required for RNAi in *C. elegans*, were known. Why did Mochizuki *et al.* (2002) look at the PPD gene? Their rationale was as follows. There were various indications that the DNA molecules of the "old" macronucleus serve as guides to retain the same DNA molecules in the new emerging macronucleus. In other words, whichever sequences that exist in the old macronucleus will not be converted into IES. But this guiding was inferred not to be carried out by the DNA molecules themselves but by RNA sequences as mediators. These small RNAs should be shuttled from the "old" macronucleus into the Anlagen of the emerging macronucleus. An enzyme complex for cleavage-derived small RNA species is therefore required. This complex may contain a PPD protein. The genome of *Tetrahynema thermophila* harbors a sequence that encode a PPD protein, a *piwi* — a related gene termed *TWI1*. Previously, it was revealed that knockout of the *TWI1* gene impaired the IES elimination. Also, a population of small RNAs that is normally formed during conjugation does not accumulate in these knockout mutants.

The investigators employed RT-PCR using RNA from early mating cells of *T. thermophila* and found a single copy gene, *Twi1* which has both PAZ and piwi domains. There are homologs to this gene in all searched metazoa but neither in plants nor yeasts. Because, as noted above, it is assumed that ciliates branched off from the phylogenetic tree before yeasts, the investigators assumed that a *Twi1* gene was latter eliminated from yeasts. Northern blot hybridization showed that the mRNA for *Twi1* was started to be formed only just before conjugation and it was reduced during conjugation. The investigators found a way to eliminate the mRNA from vegetatively growing cells which did not affect the vegetative growth. But such a knockout strongly affected the mating process; although the morphology, right after conjugation, of themicronuclei and the macronuclei appeared normal. When the conjugating pair that had the knockout of *Twi1* was further cultured, there was no further growth and the pair died. When the knockout was only with one of the conjugating pair, there was further normal development. When the two partners were knocked out then the partners that separated after conjugation were arrested at a specific stage: they had two old macronuclei and two micronuclei. Neither of the two micronuclei differentiated into a new macronucleus and there was no further cell division.

The investigators found that two other genes *PDD1* and *PDD2* had a similar affect on the results of conjugation as *Twi1*. Hence, these two genes are also probably involved in IES elimination in the evolving macronucleus. There was also direct evidence that the elimination of specific IESs required the presence of functional *Twi1*, *PDD1* and *PDD2*. The second stage of chromosomal processing of new macronuclei, the chromosome breakage stage, was also affected in *Twi1* knock-outs but not completely blocked.

The investigators also found that during conjugation a considerable level of small RNAs was formed and they had the characteristics of RNase III–derived dsRNA species of less than 28 nt. These small RNAs hybridized much more efficiently to DNA of micronuclei than to DNA of macronuclei. The investigators concluded that these small RNAs were transcribed from micronuclei. Finally, the activities of *Twi1* and *PDD1* was required for the formation of the small RNA species.

These experimental results prompted the investigators to suggest a model for the IES elimination in *Tetrahynema*. In this model both strands of IES sequences are transcribed from the DNA of micronuclei at the appropriate stage of conjugation. This will result in dsRNA representing these IESs. These dsRNAs will be diced, by a dicer-like complex to siRNAs. These siRNAs are transferred from the micronucleus to the cytoplasm and from there to the "old" macronucleus. This occurs in the early stages of conjugation. The siRNA may associate with the PDD1 and the *Twi1* proteins. The siRNA are now serving as scanners to detect which sequences (IESs and possibly also BES) do exist in the micronucleus but are absent from the (old) macronucleus. All the siRNAs that have homologous sequences in the old macronucleus will be degraded. Then those siRNAs that are retained (and in association with Twi1, PDD1 and PDD2) will go to the newly evolving macronucleus and will cause the elimination of homologous DNA sequences which could be either direct or by specific changes in the chromatin that will serve as a signal for the process of elimination. The model suggested by Moshizuki *et al.* (2002) is reasonable and takes into account the findings of these investigators but it requires a lot of verifications. As we shall see, Taverna *et al.* (2002) looked at the programmed

DNA elimination in the formation of macronuclei of *Tetrahymena* from a rather different angle.

The team of David Allis at the University of Virginia, Chalottesville, is engaged in the study of histones and chromatin structure since many years ago. Their early study was actually focused on *Tetrahymena*. They thus looked at the elimination of specific DNA sequences, from the developing macronucleus forming the angle of chromatin remodeling and histone modifications (Taverna *et al.*, 2002). The editor of the *Cell* journal printed the articles of Mochizuki *et al.* (2002) and Taverna *et al.* (2002) sequentially in the same issue. The latter investigators introduced their experimental results by references to histone modifications that serve as platforms for the binding of regulatory proteins and complexes and to the concept of the "histone code" that was elaborated in the past by Allis and associates (see Appendix on Chromatin Remodeling in my book: Galun, 2003). Take, for example, the chromodomain of the protein HP1 binds specifically to methylated lysine 9 at the tail of histone H3, abbreviated as Me (Lys9) H3, but not to the unmodified histone H3, neither to Me (Lys4) H3. Obviously, the chromodomain of certain other proteins may bind specifically to Me (Lys9) H3. By "other" chromodomains. Taverna *et al.* (2002) had in mind the chromodomains of two *Tetrahymena* proteins Pdd1 and Pdd3 (designated above by Mochizuki *et al.*, 2002, as PDD1 and PDD3). These two proteins are expressed only during conjugation and are colocalized with the elimination of DNA from the macronucleus during its maturation (hence their name: **P**rogrammed **DNA D**egradation proteins). PDD1 contains two chromodomains and PDD3 contains one such domain. *In vitro* studies showed that indeed these proteins chromodomains bind to the methylated tail of Lysine 9 of H3. The binding was found to be specific to Me(Lys9)H3. By appropriate synchronization the investigators could show that the methylation of Lys9 in H3 is confined to a specific time during conjugation: it started at 7.5 hours after commencement of conjugation, peaked at 9 hours and then gradually disappeared. This change was traced by binding of specific antibodies. They also found that the PDD1 as well as the methylation was confined to the developing macronucleus and not to the micronucleus. It was further found that Me(Lys9) H3 and PDD1 were physically associated with unique micronuclear-limited sequences before the bulk of DNA elimination; the Me(Lys9)H3 and PDD1 were enriched at the IESs, relative to sequences retained in the maturing macronucleus. With the use of PDD1 knockout strains it was indicated that without this protein the chromatin modification required for DNA elimination was impaired. Elimination of PDD1 also reduced Me(Lys9)H3. The investigators used a procedure to "tether" (bind to a specific location) PDD1 directly to a specific DNA sequence. This procedure confirmed the role of PDD1 in the elimination of IESs.

The authors have concluded that the same epigenetic mark that characterizes transcriptional repression, Me(Lys9)H3, also plays a central role in the programmed DNA elimination in the developing macronucleus. To this mark the PDD1 and PDD3 bind. Still, there is a major question what guides the specific methylation of the Lysine tail of H3 to the IES locations that are to be eliminated. The investigators suggested that this guiding could be performed by small dsRNA and refer to the article of Mochizuki *et al.* (2002). It is possible that a chromodomain also has RNA binding activity. Could it thus be that a specific siRNA binds to a defined location on the long chromosome of an emerging macronucleus and also binds to itself the PDD1 protein, leading to a site-specific methylation of the Lys9 in H3 and finally to the elimination of specific IES? If this is the case we witness a new role for the RNAi system in a very "ancient" eukaryotic organism.

Chapter 7

Examples of RNA Silencing in Lower Metazoa

This chapter will serve as a "drawer" for RNA silencing in several metazoan organisms that are not included in one of the other chapters (e.g. Chaps. 4, 5, 8, 9 and 10) of this book. Thus, there is no logical common denominator for the organisms that will be included in this chapter. Instead, this chapter shall deal with cellular slime molds, (*Dictyostelium*), *Hydra*, and *Planaria*.

RNA Silencing in Cellular Slime Molds

The species of *Dictyostelium discoideum* is a member of a group of organisms termed cellular slime molds. It attracted the attention of biologists due to its rather uncommon lifestyle. In the past it was thought to be related to fungal organisms thus its name cellular slime *mold. Dictyostelium discoideum* is one of about 70 species that are now placed in the phylum Acrasiomycota and not considered to be closely related to fungi. They are probably more related to amoebas (phylum Rhizopoda, see: Chap. 6). Reviews on the life cycle and pattern formation were provided by Firtel (1995) and Loomis (1993) where ample references are provided. In brief, spores of *Dictyostelium* germinate on a substrate with bacteria and amoeba-like cells develop. The slime-mold cells start to feed on the bacteria. A mass of such amoebas is formed and as long as the amoebas are feeding happily on the bacterial lawn they divide, multiply and the cells lead an independent life. But once the feed runs out there is a drastic change. Cell division ceases and the previously solitary cells become responsive to one another. They aggregate and stick to one another in response to the signal of pulses of cyclic adenosine monphosphate (cAMP). The aggregating cells form streams that flow towards a center. Then a differentiation starts. Part of the cells synthesize pre-spore proteins whereas the others start to synthesize prestalk-specific proteins. Then after about 12 hours, a tip is formed at the top of the aggregate. This tip elongates and may fall-over; a *slug* is formed and it starts an extensive migration. The migration may last a few days until certain conditions induce a change in pattern. During migration the anterior part of the slug is composed of future stalk cells and the posterior part of the slug (i.e. most of the slug) is composed of future spore cells. Then the slug culminates into a "Mexican hat" pattern and the tip (prestalk) cells send a stream of cells down, through the prespore cells, until they reach the substrate. These cells also stream up forming the stalk and lifting with them the prespore cells upwards. A ball of prespore cells is thus formed; it is termed *sorus*. When the sorus is situated at the top of the stalk almost all the prespore cells attain their final spore differentiation. This differentiation is manifested by shrinking the prespore cells, expelling water and being encapsulated in a rigid extracellular coat. The height of the stalk is then several millimeters. Though the morphological pattern of *Dictyostelium* development was already clarified in 1940, the intensive molecular genetic study of this morphological differentiation started only in the early 1980s as reviewed by Nellen *et al.* (1987), Loomis (1993) and Firtel (1995). Thus, Firtel (1995) already listed numerous cloned genes that are involved in regulatory functions: aggregation (14 genes), mount/tip (10 genes), prespore cells (three genes), prestalk (two genes), slug (four genes) and culmination (four genes). Although several genes are involved in more than one stage of differentiation, the total number of different genes that were included in the review of Firtel (1995) is less than 37. The molecular-genetic studies in *Dictyostelium* was prompted by the notion that this is a proper model organism to study regulatory networks in multicellular organisms. Among the components of such networks that were revealed in *Dictyostelium* are the stage and cell-type-specific G proteins, serpentine (G protein coupled) receptors, the transcription factor GBF (G-box-binding-factor), MAP (mutagen-activated protein) kinases, PKA (protein kinase A) and GSK-3 (glycogen synthase kinase-3). The two last components, PKA and GSK-3, were also found to control cell-fate decisions in the fly, *Drosophila*, and other components also exist in other metazoa. Obviously, *Dictyostelium* is especially favorable to study the transition from a mass of individual cells to a multicellular organism. This organism "solved" the mean of aggregating by a chemical clue: the cAMP. The cells react to pulses of cAMP coming from the center of aggregation. The full aggregation stage requires about 20–30 such pulses. The surplus extra-cellular cAMP is removed by an extracellular phosphodiesterase (PDE). The level of PDE activity is regulated by transcriptional control and by PD1, an inhibitor of PDE. But this is only the "tip of the iceberg". A series of integrated signaling pathways are involved in the induction and perception factors of aggregation as well as of other pattern shifts in *Dictyostelium*. But these are beyond the scope of this book.

The awareness of mRNA degradation in *Dictyostelium* by a PTGS mechanism existed already for many years as reviewed by Novotny *et al.* (2001) who provided details on the subject of double-stranded ribosnuclease in this organism. This ribonuclease (DdsRNase) is non-sequence specific and digests homopolymer duplexes as well as random sequence dsRNA targets that are generated from *Dictyostelium* genes. It will not degrade ssRNA nor ssDNA. The products of digestion are fragments of 24 or 25 nt. The DdsRNase of *Dictyostelium* appears to be a large 450 kDa multicomponent complex. While functionally similar to the *Dicer* the *Dictyostelium* enzyme is distinct from the RNase III family members that are active in digesting dsRNAs in other organisms.

Wolfgang Netten, of the University of Kassel, Germany, and associates from Kassel and from the University of Chicago, are veterans in molecular-genetic studies with *Dictyostelium discoideum*. They thus looked at the RNAi of this organism (Martens *et al.*, 2002) and focused on two components of this RNAi: RNA-directed RNA polymerases (RdRP) and dsRNase.

The ability of *D. discoideum* to perform PTGS was first tested with a transgene, *β-gal*. Martens *et al.* (2002) used a strain that expresses this gene and introduced various constructs into the cells that will generate, *in vivo*, several types of transcripts that are homologous to about 800 nt of the transgene (*β-gal*). These include sense RNA, antisense RNA and dsRNA (sense and antisense). None of these silenced the transgene. But if an inverted repeat of these 800 nt with a spacer between the repeats was expressed in the cells (a structure which is expected to form, *in vivo*, a stem loop with homology to the transgene) there would be complete silencing of the transgene. It should be noted that feeding the *D. discoideum* cells with *E. coli* that express the dsRNA with homology to *β-gal* did not cause silencing of this gene.

The investigators then asked if endogenous genes can be silenced in a similar manner. For that the gene-family encoding discoidin was chosen and DNA constructs that would produce the respective stem loop RNAs were introduced into the cells. In five cases there was a complete silencing and in four cases there was partially discoidin silencing. In the silenced cells there was no detectable discoidin mRNA. The results clearly indicated that both transgenes and endogenous genes can be silenced by the RNAi/PTGS system provided a construct is utilized that will generate, *in vivo*, the appropriate stem loop RNA.

The investigators turned to the RdRP that is an essential component for quelling in the fungus *Neurospora* and PTGS in the angiosperm plant *Arabidopsis*. They revealed three RdRP-related gene sequences in *D. discoideum*. The proteins encoded by two of these genes, *rrpA* and *rrpB*, differ only by 49 amino acids (i.e. by less than 3 per cent) from the RdRP of other organisms. The third gene *DosA* is less conserved. The similarity is especially great in the N-terminal extension of the proteins. The three RdRP-like genes of *D. discoideum* were disrupted (by homologous recombination) and the respective knockout mutants were obtained. Then when the w.t. and the mutants were tested for their RNAi capability (as described above), it was found the RrpA⁻ mutant completely lost its RNAi capability while the stains with the RrpB⁻ and DosA⁻ mutations retained the RNA silencing capability. In the strains that retained the RNAi capability the investigators found short RNA fragments of about 23 nt with homology to the silenced gene. On the other hand, in the RrpA⁻ mutant, there were no such short RNA fragments. While the RrpA was found to be essential for RNAi capability, the RNase of RrpA⁻ mutants was capable of degrading dsRNA, *in vitro*. We should recall that the homologues for RrpA in other organisms (as *C. elegans*), the RdRP is involved in the upstream movement of the RNAi; this means that in *C. elegans*, a dsRNA with homology to a certain region of a mRNA will not only cause cleavage at the homologous region but also cleave at sites that are upstream of the homologous region. The explanation for this phenomenon in *C. elegans* is that one strand of the dsRNA serves as primer to generate upstream dsRNA by the activity of RdRP. No such activity was yet proven for the RrpA in *Dictyostelium discoideum*, but Martens *et al.* (2002) eluded in their discussion that an amplification of 23 nt (dsRNA) fragment is possible. The investigations of Martens *et al.* (2002) clearly showed that RrpA is essential for RNAi in *D. discoideum* and that RNAi is functional in this organism. Why the introduction of *in vitro* synthesized dsRNA does not impose specific PTGS in *D. discoideum* is not clear yet.

RNA Silencing in *Hydra*

The genus *Hydra* belongs to the phylum Cnidaria which includes polyps (class Hydrozoa), jellyfish and soft corals. All these animals have a radial symmetry. The animals of this phylum were widespread and diverse already in the Precambrian era, meaning 630 or more million years ago. We shall discuss the genus *Hydra* that is a freshwater polyp of relatively simple general morphology. The hydras differ from other metazoa that are also considered primitive (e.g. sponges) by having an extracellular, rather than an intracellular food digestion. The simple general-morphology does not mean that hydras lack sophistication. Hydras are equipped with 20 different cell types. Among these are nerve cells, muscle cells, sensory cells as well as very special stinging cells with nematocysts. The latter cells, termed cnidocytes, exist only in the phylum Cnidaria. The nemotocyst is a powerful "harpoon". It is an elaborated device and when discharged the "harpoon" is powered by an osmotic pressure of about 140 atmosphere. The velocity of the discharge is so high that the "harpoon" can penetrate the hard shell of a crab and release into it a toxic protein. The hydras grow normally as solitary polyps in fresh water. Each polyp sits on a hard substrate by its basal disk ("foot"). The hydra can glide around and bend over so that it is not permanently bound to one spot. Above the foot there is a cylindrical tube (stalk) which ends with a mouth. Around the mouth are the tentacles. The mouth and the tentacles are considered as head. Below the mouth is the gastric region. The body (stalk and tentacles) is made of only two layers of cells — internal and external epithelial cells, also termed ectoderm and endoderm, respectively. Between these two cell layers there is an intervening *extracellular matrix* (ECM). The length of the tube is approximately 5 mm but it is flexible to increase and decrease its length. The total number of cells in a polyp is about 100 000. The hydras can be cultured easily in the laboratory and are fed with freshly-hatched brine shrimps. The possibility of culturing hydras and performing all kinds of graftings with them, as well as their capability to regenerate body parts (as the head with its mouth and tentacles), probably attracted biologists 270 years ago: the Swiss scholar Trembley conducted experimental-development work with hydras around the year 1735. The *Hydra* (together with angiosperm phylotaxis) was also the subject for the mathematical model of morphogenesis, through reaction-diffusion systems, that was developed by Alan M. Turing (Turing, 1952), the mathematician who is known for his pioneer work in programmed computers and artificial intelligence and especially for his dominant role in Bleachley Park in England during the World War II, where the German Enigma code was broken. It appears paradoxal that an apparently simple biological model can trigger 35 pages of high-level mathematical formulations. Furthermore, at the time Turing developed his mathematical elaborations for *Hydra*, the channel for the flow of morphogens in this animal was not revealed yet. This channel, the ECM, is a continuous layer of fibrillar collogen in which morphogens can stream (by diffusion) along the whole polyp. The composition of ECM was analyzed only in recent years. The ECM is a highly elastic matrix. There is a bizarre aspect of the ECM of *Hydra*: the structural features of this interstitial collogen in *Hydra*, mimic in part, what is seen in a human condition known as Ehlers–Danlos syndrome. The latter is an inherited disorder associated with abnormalities in the structure of ECM that results in a matrix that is significantly more flexible than found under normal conditions. The clinical manifestations of this syndrome vary from minor impairment of function to severely debilitation in humans. There is an historical twist in this. There is an assumption that the Ehlers–Danlos syndrome was a contributing factor in Nicole Paganini's (1782–1840) extraordinary ability as a violin virtuoso. Could it be that Paganini's talent to play the violin was due to hypermobility of his joints, which greatly facilitated his ability to perform the remarkable double stoppings and roulades? Obviously, *Hydra* links a mathematical genius (Turing) with a performing virtuouso!

Let us return to biological realities. As detailed by Bosch (1998) numerous developmental genes have been isolated from *Hydra* by homologous cloning, based on detailed studies in other animals (e.g. *Drosophila*). Among these are Hox genes, winged helix genes, T-box genes and genes encoding protein-tyrosine kinases. The sequences in these genes are very conserved in metazoa, but it is not at all clear whether these genes have the same function in *Hydra* as in other more advanced metazoa. Bosch and associates, of the University of Jena, Germany (Lohmann *et al.*, 1999), argued that RNAi should be instrumental in analyzing the function of developmental genes in *Hydra* that have known DNA sequences. Silencing specific genes by the appropriate dsRNAs should reveal the phenotypes that result from these silencings. These investigators (Lohmann *et al.*, 1999) focused on the gene *ks1* which is involved in patterning signals along the apical-basal body axis and is regulated by a complex interaction of inhibitory factors. In other words, *ks1* may be involved in the formation of heads in *Hydra*. For that purpose dsRNA corresponding to the coding sequence of *ks1* was synthesized and this dsRNA was introduced by electroporation up to 60 *Hydra magnipapillata* animals. When analyzed by northern blot hybridization it was found that the transcript of *ks1* was significantly (and specifically) reduced in the treated animals. The same was revealed by *in situ* hybridization of a label that is specific for *ks1*. In the treated animals, the label that is especially strong at the heads of untreated animals disappeared from the heads of treated animals. In a further experiment the heads (mouth and tentacles) were removed from either control polyps or polyps that were previously electroporazed with a dsRNA (with homology to *ks1*). The decapitation was performed 6 days after dsRNA treatment. The results were clear: the heads regenerate 36 hours after decapitation in control polyps but did not regenerate in the treated polyps although the effect of treatment was not permanent. After additional time (i.e. after 72 hours) the treated polyps also regenerated heads. The treatment had no effect on foot regeneration.

A more recent study on RNA silencing in *Hydra* was performed by Cardenas and Salgado of CINVESTAV–IPN in Mexico. The

Mexican investigators (Cardenas and Salgado, 2003) had basically the same approach as Lohmann *et al.* (1999). Only the former investigators focused on STK. The *STK* gene is a hydra homologue of *Src*. The Src family of protein tyrosine kinases, in several metazoa, functions in the regulation of cell adhesion, signaling from growth-factor receptors, cell cycle, development and pattern formation. The *Src* gene is one of a 4-gene family members that are expressed very early on the embryo-development of metazoa. The activity of *src* is especially required to stimulate mesoderm formation by its inducer (fibroblast growth factor). *Src* was found to have additional roles such as a role in the control of cell fate and division-orientation in early embryos of nematodes (*C. elegans*). However, it has been difficult to reveal the precise function of *Src* because of the complicated interactions of signal pathways. The investigators chose *Hydra* because Cnidaria is the lowest phylum where receptor- and non-receptor-tyrosin kinases have been identified. Thus, STK could have important functions in regulating cell-cell communication and/or pattern formation. Actually in a previous study these investigators found, by the use of a *Src* inhibitor that STK is a key component of the signal transduction system that is necessary for cell differentiation in *Hydra*. Now Cardenas and Salgado (2003) intended to explore the function of STK in the regulation of the cell's initial commitment to differentiate into head structures.

Under normal conditions, decapitation caused STK activity at the area below the decapitation a few hours after this decapitation. This kinase activity was prevented by an inhibitor of the STK activity. The investigators used a similar procedure of culture and the introduction of dsRNA as employed by Lohmann *et al.* (1999) but introduced dsRNA with homology to the *STK* gene. By *in situ* hybridization the investigators found that dsRNA-treated animals did not express the *STK* gene as it was expressed in non-treated animals. But the dsRNA-treated animals seemed quite normal. Their tentacles were slim and inefficient, so these animals were fed manually. Though with bud development, the normal vegetative reproduction in the hydras was impaired. Most of the dsRNA (STK) animals lost gradually their capability to bud. The investigators introduced the dsRNA and after the animals had recovered, they removed the head and the foot. Then about 80 per cent of the animal regenerated heads and feet while 20 per cent formed one or two ectopic heads. However, the heads were not normal: they had only one or two "unskillful" tentacles and had to be fed manually. But foot development was normal. The investigators assumed that most STK animals were able to regenerate heads because they retained enough STK before the dsRNA silencing. There were indication that this was the case. A second decapitation of dsRNA-treated animals reduced the percentage of the animals which were able to regenerate heads to 20 per cent. But ectopic heads in the first third of the body were formed and these were functional. The investigators assumed that STK is essential for the formation of heads at the correct location. Finally, the investigators looked at the expression of other *Hydra* genes that are apparently involved in differentiation, in normal and dsRNA-treated animals (see: Technau and Bode, 1999; Martinez *et al.*, 1997 and Hobmayer *et al.*, 2000). They found that the forkhead-type gene *Budhead* was not affected by the dsRNA for STK. The expression of the *Hybral* gene that is analogous to a *brachyary* gene was blocked in STK$^-$ animals but only in the non-regenerated tip. It was expressed in the ectopic heads. Another gene, *HyTcf*, with a probable differentiation role was silenced in the whole STK polyps. Thus, *HyTcf* is apparently a direct target of the STK-regulated differentiation pathway. The expression of *Ks1*, an early head specific gene (see: Lohmann *et al.*, 1999, mentioned above), was completely silenced by the dsRNA for STK.

On the basis of their RNA silencing and other information, the investigators postulated that STK is required for the development of the head organizer, possibly in the signal pathway that transduces the differentiation signal from an unknown activated receptor. What is indeed clear from the study of Cardenas and Salgado (2003) is that the RNA silencing by the appropriate dsRNA is useful to analyse the impact of genes with a known sequence but yet uncertain function. This RNA silencing is especially useful in *Hydra* where sexual reproduction is rare and thus regular genetic procedures are problematic.

RNA Silencing in Planaria

The planarians are flatworms (phylum Platyhelminthes); they are ribbon-shaped and soft-bodied animals with a bilateral symmetry and are flattened dorsoventrally. It is still not clear where, in the phylogenetic tree, Platyhelminthes should be placed. Kiyokazu Agata, Kenji Watanabe and associates (Tazaki *et al.*, 2002) of the Himeji Institute of Technology, Japan, tend to place them close to Mollusca because of the similarity in the coding of a gene for an intermediate filament protein (IF). Did Platyhelminthes branch off from primitive Mollusca before the latter entered their shells or possibly after some lost the shell? Planarians are included in the class Turbellaria. A related class of the Turbellaria is the class Trematoda, to which the parasitic flatworms belong. Under the name planaria there are several free-living animals that are carnivores and scavengers. These range in size from a few millimeters to a few centimeters. The overall body structure is simple though more elaborate than the body of *Hydra*. They have a gut with only one opening. There are muscular contractions in the upper end of the gut where the worm ingests its food and tears it into small bits. The gut is branched and extended throughout the body. The cells that line the gut engulf most of the food (after ingestion) by phagocytosis but the food is also digested extracellularily. There are special "flame cells" that can transfer water in and out of the body and push wastes to the outside. These cells serve as an excretory system and balance the water content of the worms. But much of the waste (of food) goes out through the mouth (pharynx) of planaria. The planaria have a network of nerves in which two nerve cords run in parallel along the body. There is a primitive central nerve system (cephalic ganglia) at the anterior tip of the body where the two nerve cords are connected. There are two eyespots in the head that sense dark and light. The eyespots are connected to the nerve system in the cephalic ganglia, forming a kind of a primitive brain. The planaria can sense food, chemicals and movement of fluid around them and can move towards food. Normally planaria move away from a source of light. Flatworms lack respiratory and circulatory systems and rely on diffusion for oxygen. Planaria may reproduce sexually or asexually. Some species as the species *Schmidtea mediterranea* that is used for molecular-genetic and regeneration experiments (e.g. Newmark and Alvarado, 2002) contain strains that are either sexual or asexual. Cytologically, the asexual strain has a chromosomal translocation that is clearly visible in metaphase chromosomes. This species has four pairs of chromosomes. Other planaria may be polyploid or mixoploid. Take, for example, the species *Dugesia japonica* that is used as laboratory animal by Japanese investigators (Agata and Watanabe, 1999) is mixoploid. The asexual reproduction is by transverse fission: the tail adheres to the substrate, the worm elongates and the fission is at the posterior third of the body. After fission each part regenerates the missing part. The planaria are hermaphrodites and for sexual reproduction two animals mate and deposit egg capsules in which embryos develop. The most significant feature of flatworms, a feature that attracted investigators since many years ago, is their capacity to regenerate. Not only will two halves, after transversal or longitudinal cuts, regenerate to intact and functional animals; flatworm can do much better ... very small pieces can regenerate whole animals. This regeneration capability was the subject of investigations already by Peter Simon Pallas in the middle of the 18th century. After more than 100 years regeneration experiments were conducted by Harriet Randolph in the Bryn Mawr College (Pennsylvania) and this work was followed by Thomas Hunt Morgan who was engaged with planaria studies during the early years of the 20th century. He fed these animals with *Drosophila* eyes. Only in later years he used this fly for studies of inheritance. During experimenting with *Drosophila* in his "fly room" at Columbia University, a white-eyed fly was spotted. The inheritance of this white-eye trait converted Morgan from a skeptic of Mendelian laws of inheritance to a strong advocate of Mendelian genetics and from there to a giant in the emerging field of genetics. A schematic presentation of planaria is provided in Fig. 19. One important component of planaria is not shown in the figure; these are the *Neoblasts*, a term coined by Randolph. These are the only proliferating cells in planaria. They are a kind of stem cells that become active when a damaged or missing tissue requires regeneration. Initially, these neoblasts are located in a dormant form all over the animal but they go into action when required. They then move to the required location, multiply and replace the missing or damaged tissue. When the neoblasts migrate toward the wound epithelium they give rise to the regeneration *blastema* — the structure in which the missing parts will be regenerated. Neoblasts are sensitive to X-rays and thus X-irradiated planaria lose their capability to regenerate and gradually

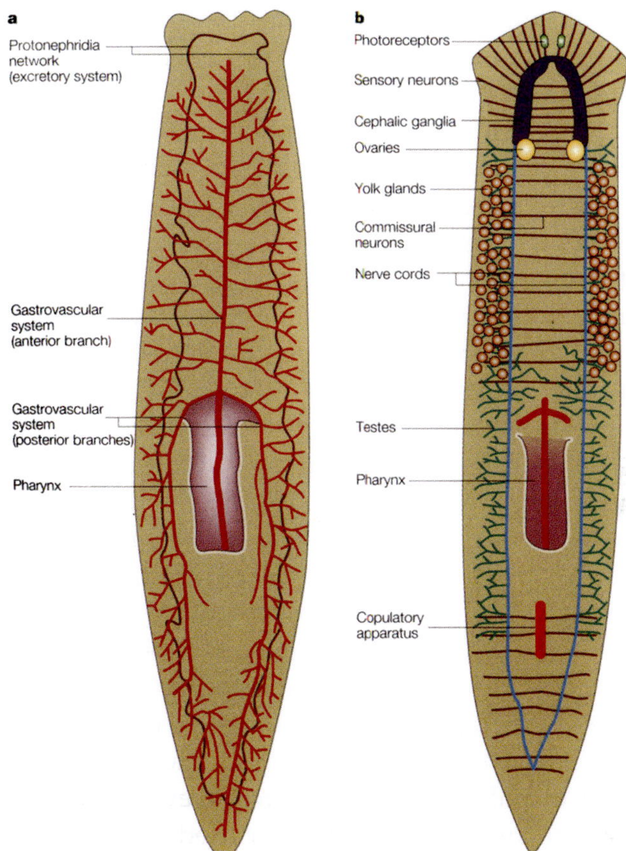

Fig. 19. Diagram of the major organ systems in freshwater planarians. The figure illustrates two of the many morphologies found in the triclades. (a) *Dendrocoelum lacteum* depicts the gastrovascular and excretory systems. (b) A representative of the genus *Schmidtea* in which the reproductive and nervous systems are shown. (From Newmark and Alvarado, 2002.)

degenerate. But when viable neoblasts are transferred to previously X-irradiated animals, the latter animal fully recover. The animals into which the neoblasts were transferred to will then have the characteristics of the introduced neoblasts. More details on the structure and function of planaria were furnished by Agata and Watanabe (1999) and Newmark and Alvarado (2002).

The first report on the use of dsRNA to silence specific gene expression in planaria came from the same Department of the Carnegie

Institution of Washington, from where the first report on RNAi by dsRNA was reported on nematodes (Fire *et al.*, 1998). This report on planarian RNAi was submitted by Sanchez Alavardo and Newmark (1999), about 1 year after the publication of the article on nematode RNAi. Indeed, Sanchez Alavardo and Newmark intended to use the dsRNA procedure to selectively abrogate genes that function during regeneration in planaria.

One of the targets of these investigators was the regeneration of body-wall musculature that consists of four subepidermal muscle-fiber layers and a network of dorsoventral fibers that connect the dorsal and ventral surfaces. Normally, about 3–5 days after amputation this muscular complex regenerates to its original structure. The investigators made cuts in 6–7 mm animals so that the animals were divided into five slices. They then injected dsRNA (or water) into the slices. The dsRNA had a sequence that was homologous to the gene for myosin. After 3 days the animals were analyzed. The dsRNA injected animals showed very little, if any, regeneration of body-wall musculature within their blastemas. Control regeneration was normal. Interestingly, the injection affected not only the differentiating myogenic lineage in the regenerated blastema but also the pre-existing terminally differentiated muscle cells. The effect was restricted to the body-wall musculature and did not appear to affect the musculature of the pharynx (mouth). In the latter there is probably a different myosin encoded in a gene that is not homologous to the body-wall myosin gene. By appropriate *in situ* labeling (by immunofluorescence) the investigators also looked at the effect of myosin dsRNA on another tissue: the ventral epithelium. They found that the ventral epithelium was not affected by the myosin dsRNA; it was regenerating normally. On the other hand, injection of dsRNA of the α-tubulin gene did not affect body-wall musculature regeneration, while the regeneration of cilia in the ventral surface were affected. Finally, they turned to the eyespots. In these there is normally an expression of the opsin gene. The investigators injected opsin dsRNA into the heads of planaria and evaluated the levels of opsin mRNA after injection. At about 12 hours after injection there was a drastic reduction of opsin mRNA and by 24 hours this mRNA was completely eliminated. The study of Sanchez Alvarado and Newmark (1999) thus established that dsRNA can silence specific genes in flatworms and that this RNAi procedure can serve as an efficient tool to study the molecular biology of regeneration in these animals.

The *Drosophila* gene *sine oculis* (*so*) is one of the genes that are required for the development of the visual system. The *so* is a Hox-containing gene. There are mammalian homologues of *so* such as the murine *Six3* gene. The planarian eyespots are considered a very primitive visual system. A group of investigators from Barcelona, Spain; Houston, Texas, Mishima, Japan and Basel, Switzerland (Pineda *et al.*, 2000) explored the possibility that a homologue of the *so* of *Drosophila* is involved in eye spot differentiation in the planarian *Girardia tigrina*. Since *so*-like genes are conserved evolutionarily the authors could isolate the *G. tigrina* orthologue of *so* and termed it *Gtso*. In *Drosophila so* is regulated directly by the master controller *Pax6* and a *Pax6*-like gene was revealed also in the plarian *G. tigrina*. The investigators therefore asked if the role of a *so*-like gene is conserved in the primitive visual system of *G. tigrina*. They used the RNAi system to obtain an answer to their question. *In situ* hybridization detected the transcript of *Gtso* in the photoreceptor cell bodies of the eyespots in adult worms. During the early stages of head regeneration *Gtso* transcripts were also detected in a group of photoreceptor cells close to the dorsal epidermis that constitute an early visible sign of eyespot regeneration. After testing the RNAi procedure and finding that the dsRNA of *G. tigrina opsin* eliminated the *opsin* transcript, the investigators turned to dsRNA of *Gtso*. They found that injection of dsRNA of *Gtso* into regenerating adult flatworms prevented eyespot differentiation for three weeks after a single injection: neither pigment cells nor photoreceptor cells were regenerated. The injected worms did differentiate dorsal blastema so that the dsRNA *Gtso* did not cause a gross inhibition of regeneration. The dsRNA *Gtso* effect was transient: 4–5 weeks after injection, the regenerated heads started to differentiate normal eyespots. To maintain the inhibition of eyespot regeneration the dsRNA had to be injected every 3 weeks.

K. Agata, K. Watanabe and associates in Japan are engaged in the study of molecular aspects of regeneration in flatworms for several years. In one of their more recent studies (Ogawa *et al.*, 2002) they focused on fibroblast growth factor (FGF) receptors (FGFRs). They isolated from their flatworm, *Dugesia japonica*, FGFR genes and sequenced these genes. Two such genes were termed *DjFGFR1* and *DjFGFR2*. Both genes had similarity to FGFR of higher animals. The proteins encoded in these two genes contain two or three immunoglobin-like domains in the extracellulase region and a split tyrosine kinase domain in the intracellular region. In intact planaria *DjFGFR1* and *DjFGFR2* are expressed in the cephalic ganglion. In regenerating planaria (after removing the head and other parts) the accumulation of *DjFGFR1*-expressing cells was observed in the blastema and in fragments that regenerated either the pharynx or the brain. The investigators injected into regenerating flatworms, dsRNAs derived from *DjFGFR1* and *DjFGFR2*. These injections did not prevent the normal regeneration of brain or pharynx. This lack of dsRNA effect did not cause desperation and the same Japanese team, now in collaboration with Sanchez Alvarado (Cebria *et al.*, 2002) launched another study with dsRNA injection. They focused on another growth-factor-like molecule that is specifically expressed in the head region of regenerating *D. japonica*. This protein was found to be encoded by a gene that the investigators termed *ndk* that stands for the Japanese expression: *nou-darake* or "brain everywhere". Actually, the phenotype of "brain everywhere" emerged when planarian fragments were injected with dsRNA representing the *ndk* gene. *In situ* hybridization indicated that (without injections) the gene *ndk* is expressed in the head region, in both the inverted U shaped brain as well as in non-brain cells. After amputation, during regeneration, the expression of *ndk* is first detected at 24 hours post-amputation, only in the anterior blastemal cells, including the new brain primordium. The investigators first injected *ndk* dsRNA into intact *D. japonica* animals and then amputated these animals. Whole mount, *in situ* hybridization indicated that indeed the injection silenced the expression of *ndk*. Seven days after injection the amputation extopic eyes begun to differentiate in dorso-posterior regions of the flatworms body and by 15 days, 94 per cent of the injected animals formed ectopic brain tissue. Such expression of regenerating brain tissue in posterior regions was also found in trunk pieces that were allowed to regenerate. Furthermore, the formation of ectopic brain tissue and extra eyes in injected animals did not require amputation and regeneration. Also, in intact animals that were injected with dsRNA of *ndk* such ectopic brain tissues and extra eyes were observed. These ectopic brain tissues did not develop in X-irradiated worms. Since neoblasts are especially sensitive to X-irradiation the investigators concluded that neoblasts are involved in the ectopic brain formation. If one distances oneself from the biological phenomena we may reach a paradox: in these animals, especially after the amputation of the head, the "sword" of RNA-silencing can cause the regeneration of many head structures! The second labor of Hercules comes to mind, only the Greek mythology refers to Hydra rather than to Platyhelminthes. Will an *ndk* gene be revealed also in Hydra; possibly it will. As for the *ndk* gene, it appears that its normal function is to *prevent* brain and eye differentiation in any part of the worms body, but in the head.

In a later work by the Japanese team (Orii *et al.*, 2003) these investigators found that the application of dsRNA can be simplified considerably. Fragments of *D. japonica* can be soaked for about 5 hours in a solution of dsRNA and then allowed to regenerate in water. This procedure was previously found to be effective in nematodes. Another technical innovation was reported by the team of Sanchez Alvarado and Newmark (Newmark *et al.*, 2003): like in nematodes, planaria can be fed by bacteria that express dsRNA, triggering the respective RNAi.

Whatever the method of dsRNA application, RNA silencing in planaria seems to be a very effective tool for molecular dissection of organ differentiation in these animals and more specifically for a better understanding of regeneration. The utilization of dsRNA in planarium research was discussed in detail by Alvarado *et al.* (2002) and by Newmark and Alvarado (2002).

 Gene Silencing in Non-Mammalian Vertebrates

Model animals are commonly not chosen to study a specific species but rather because they have advantages in experimental procedures and thus serve to gain knowledge on a wide range of other organisms. In this chapter I shall deal with a frog, a zebrafish and chicken embryos. Each of these models serves investigators for several purposes and each of them has its specific advantages and drawbacks as model systems representing vertebrate animals.

RNA Silencing in the Frog *Xenopus laevis*

The frog *Xenopus laevis* (South African clawed toad) is a tetraploid amphibian with a relatively long generation time. Hence, it was not a favorite model for genetic manipulation. But this frog has features that render it an excellent object of embryological studies. Its eggs can be manipulated and methods were developed to use such eggs for the synthesis of macromolecules. DNA can be injected into the nuclei of the eggs and the respective RNAs and proteins will be synthesized in the eggs within a few hours.

Masanori Taira and associates (Nakano *et al.*, 2000) at the University of Tokyo investigated the RNA interference in *Xenopus laevis*. They first attempted to suppress the expression of an exogenous reporter gene and then the inhibition of expression of a gene that is essential for embryo development. First, the expression of a gene for luciferase was studied. For that the investigators injected a luciferase expression-plasmid pCS2Luc. Fertilized eggs were used and the injection was performed into two blastoderms at the 4-cell stage in the dorsal equatorial region.

For analysis of the RNAi effect, the dsRNA representing the whole or part of the coding region of the luciferase genes were co-injected with the luciferase expression-plasmid. The dsRNA representing a part of the luciferase gene reduced the expression to about 50 per cent. The dsRNA representing the whole coding sequence was less effective. These results encouraged the investigators to attempt the silencing of an endogenous gene. They chose the gene *Xlim-1* that is a Hox gene that was known to be involved in head development. The injection of the dsRNA for *Xlim-1* caused a reduction of the respective mRNA. The morphological manifestation of injecting the dsRNA for *Xlim-1* were variable. In some of the injected embryos there was normal head development. In other injected embryos (about 60 per cent) there were anterior defects of various kinds (small eyes, one eye, no eyes or no head). But there were also other development defects. The levels of mRNA for *Xlin-1* were reduced by 30–40 per cent, meaning that there was no complete silencing of the *Xlin-1* gene. It should be noted that injection of the antisense RNA for the *Xlim-1* gene also elicited morphologies that were similar to those of injecting the dsRNA for *Xlim-1*.

We shall see below, while dealing with RNA silencing in mammals, that using long sequences of dsRNA can cause non-specific effects. The dsRNAs used by Nakano *et al.* (2000) were about 1000 nt long. A subsequent study of RNAi in *Xenopus* embryos was performed with the awareness that long dsRNA may elicit non-specific destruction of mRNA. Zhou *et al.* (2002) of the University of Hong Kong thus launched a study in which they searched the inhibition of a transgene as well as of endogenous genes. Their approach was similar to that of Nakano *et al.* (2000). Only the former used synthetic dsRNA of 21 nt. The investigators from Hong Kong also used a transgene for luciferase expression. They injected a plasmid of this gene into the 2-cell stage *Xenopus* embryos. Then the luciferase transgene was expressed from the mid-blastula transition (MBT). The injection of the luciferase gene was done either without other injections or with the injection of other dsRNAs. When the 21 nt dsRNA with homology to the transgene was co-injected it reduced the expression of the luciferase gene by about 60 per cent. The inhibition was specific to the luciferase gene. Turning to endogeneous genes the investigators aimed at the cyclin B1 and cyclin B2 genes. They synthesized the respective 21 nt dsRNAs and found, again, a strong suppression of these endogenous genes. The suppressions were specific, meaning that the cyclin B1 dsRNA affected only cyclin B1 and not cyclin B2 and the cyclin B2 dsRNA inhibited only cyclin B2. Both the levels of the protein and the respective mRNAs were reduced. Albeit, contrary to the RNAi in nematodes, where inhibition by dsRNA is complete (or almost complete), the inhibition in *Xenopus* embryos was about 60–70 per cent, but not 100 per cent.

Gene Silencing in Zebrafish

The zebrafish (*Danio rerio* or *Brachydanio rerio*) is a tropical freshwater fish of small size (about 3 cm long) with a generation time of only 3–4 months. Although tiny, the zebrafish was instrumental in bestowing its investigator, Dr. Christiana Nüsslein–Volhard, with the Nobel Prize. The female of zebrafish lays a few hundred eggs at weekly intervals and the sperm of the male can be kept frozen for extended periods. There are efficient methods to obtain mutants and eggs can be activated by impotent sperm so that the haploid chromosome number is allowed to replicate once while cell division is prevented, resulting in diploid cells (hemizygotes) that will lead to maternal embryos. The development of embryos is rather quick and the larvae are free living 5 days after the eggs are fertilized. The embryos and the larvae are transparent so that abnormal organs can be easily visualized. The zebrafish was chosen by George Streizinger and associates of the University of Oregon, Eugene, Oregon (Streisinger *et al.*, 1981), for genetic studies. In later years the zebrafish became a model for studies in the development of vertebrates at the molecular level (Nüsslein–Volhard, 1994). In 1996, the journal *Development* devoted a special volume to the zebrafish that held 37 articles in 481 pages. The first of these articles by Christiane Nüsslein–Volhard and associates (Haffter *et al.*, 1996) reported on the results of a large scale mutagenesis in which males were mutagenized with ethylnitrosourea (ENU), by placing them in an aqueous 3 mM ENU solution for three 1-hour periods within a week. This work resulted in 4264 mutants of which about one quarter were characterized and most of the latter were assigned to specific genes. A similar genetic screen was performed by Driever *et al.* (1996). Well, in the 1960s some bacterial geneticists claimed that what was learned from bacteria can be applied to elephants. This was obviously a bombastic stalement. But now, if there would be a claim that what is learned from molecular-brain development of zebrafish has relevance for the understanding the development of the human brain, such a claim makes sense.

Anders Fjose and associates of the University of Bergen, Norway, were aware of the early studies on RNA silencing by dsRNA in diverse animals such as nematodes (Fire *et al.*, 1998), planaria (Sanchez Alvarado and Newmark, 1999) and *Drosophila* (Kennerdell and Carthew, 1998). They thus investigated the possible silencing of specific genes by dsRNA in zebrafish (Wargelius *et al.*, 1999). Their rationale was that a great number of mutated genes were revealed already in this fish, but the phenotypic expression was not known yet. Hence, disruption of these genes, by specific dsRNA can lead to the detection of specific functions that will reveal the role of these genes in the development of zebrafish embryos. These investigators already knew that large amounts of RNA injected into zebrafish embryos will produce non-specific defects. They focused on two cloned genes that are expressed very early in the zebrafish embryo: *floating head (flh)* and *no tail (ntl)*. The *ntl* mutants lack tail, notochord and have abnormal somite patterning. In *flh* mutants the notochord does not form and somites are fused medially under the neural tube. These two genes encode transcription factors and initiate transcription about 1 hour before the onset of gastralation. Figure 20 shows the normal development of zebrafish embryos. The investigators (Wargelius *et al.*, 1999) injected dsRNA of *fln, ntl* as well as other genes, into the 2-cell stage. This resulted in variable phenotypic changes that could reflect the silencing of the respective genes but the effects were not sufficiently specific and required rather high doses of dsRNA. Still, the investigators were optimistic about this procedure because it was 10-fold more effective than silencing genes with antisense.

Another team of investigators from the Medical College of Georgia, Augusta, Georgia (USA), had the same goal as Wargelius *et al.* (1999). They were probably not aware of the study in Bergen. The experimental procedures of Li *et al.* (2000) were similar to those of Wargelius *et al.* (1999). There was even an overlap of one target gene: *no tail (ntl)*.

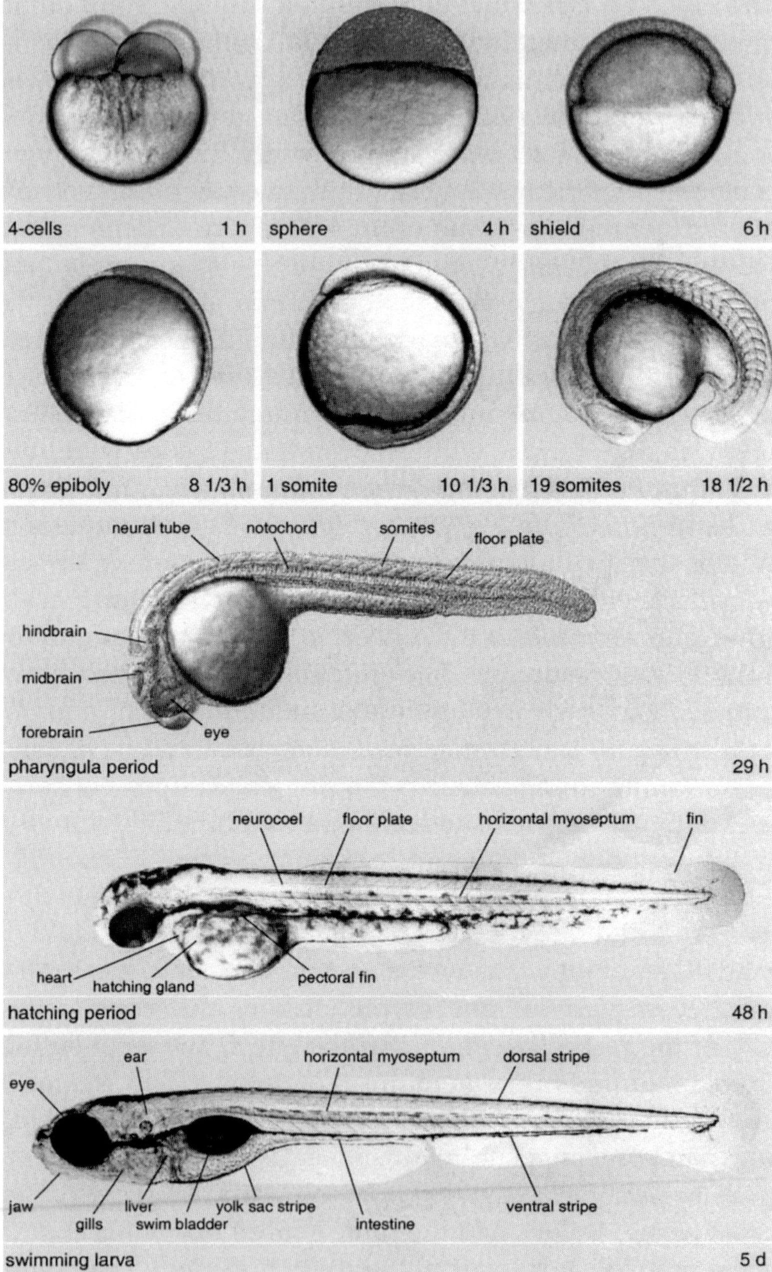

Fig. 20. Living embryos of relevant stages during the differentiation, following the fertilization of the zebra fish egg. (From Haffter *et al.*, 1996.)

The former investigators also used the *Pax6* gene as a target. Mutations in *Pax6* cause defects in eye and brain development. Li *et al.* (2000) injected the dsRNA into eggs of the single-cell (rather than the 2-cell) stage. They first assured that dsRNA can silence the transgene that encodes GFP. As for endogenous genes it appears that when more than 50 000 molecules of dsRNA were injected into an egg — most embryo were defective in the respective silenced genes. In short, Li *et al.* (2000) concluded that dsRNA is a reliable agent to silence specific developmental genes in zebrafish.

The latter conclusion of Li *et al.* (2000) was not shared by a group of investigators (Oates *et al.*, 2000) from Princeton University, headed by Robert Ho, who used similar procedures of dsRNA application as those used by Li *et al.* (2000) including the targeting of dsRNA to the *ntl* gene. But Oates *et al.* (2000) concluded that "the current methodology of double-stranded interference is not a practical technique for investigating zygotic gene function during early zebrafish development". They claimed that the effects of dsRNA were not sufficiently specific. They discussed the possibility that in zebrafish the dsRNA could serve as a "warning sign of viral infection" in a similar way to what was observed in mammalian systems. An RNase L may then cleave both viral and cellular ssRNAs and the synthesis of interferons could be induced.

Investigators from the Tsinghua University in Beijing, China (Zhao *et al.*, 2001) also studied the possibility of specific silencing zebrafish genes, during early embryogenesis, by the respective dsRNA. These investigators came to the same conclusion as Oates *et al.* (2000), namely, that: "It appears that RNAi is not a viable technique for studying gene function in zebrafish embryos". Zhao *et al.* (2001) stressed that the general toxicity of dsRNA is concentration-dependent and that RNAi is incapable of blocking specific gene expression. They also discussed the possibility that injection of dsRNA may induce RNase L activity and thus degrade various ssRNAs.

We shall see that this problem of non-specific degradation of mRNA by dsRNA was also encountered in the early studies on RNAi in mammals. In mammals it was revealed that long dsRNA (a few hundred nt) will elicit non-specific RNAi effects while much shorter dsRNA may cause specific silencing. Until now, very short dsRNA sequences were not used with zebrafish. This possibly will be attempted in the future. The following approach could be applied. A DNA sequence that, after transcription, will cause the formation of an RNA sequence of about 90–100 nucleotides that will fold into a stem-loop structure, may be synthesized. An appropriate strong promoter should be added at the 5′ end of this DNA sequence and the promoter should be inducible at will by a simple effector. This construct could be introduced into the fertilized egg as a plasmid or integrated into the genome of the fish so that a transgenic zebrafish will be established. The transgene will be silent when not induced by intention (with the appropriate effector). When induced, the respective stem-loop will be transcribed. It would then be cleaved into short (∼22 nt) dsRNA sequences. If these sequences have homology to specific genes, the latter could be silenced specifically at the required stage of embryo development. Another way would be to synthesize short (∼22 nt) dsRNA with 1 or 2 nt overhangs on both the 5′ and 3′ ends. These short dsRNA could then be injected into the zebrafish embryos.

The last word on RNA silencing in zebrafish is probably still far away. This assumption is based on a study of Ronald Plasterk and associates (Wienholds *et al.*, 2003) who induced mutations in the *dicer1* gene of zebrafish. Mutated embryos developed normally for about 1 week, during which the maternal (w.t.) *dicer1* was still capable to process pre-miRNA to miRNA; but the accumulation of miRNA stopped after a few days in the young mutated embryos and then there was developmental arrest at about day 10. Hence, *dicer1* seems to be essential for normal zebrafish development, meaning that dsRNA silencing is probably required, at least during early embryogenesis.

RNA Silencing in Avian Embryos

We may wonder about the progression of a given entity toward a specific goal. The wondering can relate to various organisms and specific components in these organisms. We may descent the phylogenetic ladder from man to bacteria. We may thus face extremely different "worlds" from chemotaxis in bacteria that is induced by specific chemical cues and propagated by molecular engines at the base of the flagellum to humans striding toward a desired goal. In each case we may pose "how" and "why" questions. The answer to the "why" question in certain cases is rather obvious and should not worry us; for example, in bacterial chemotaxis: the search for food. For humans strive toward some goals the answers to the "why" are very problematic. Moreover, in humans the *attempts* to reach the goal may be considered as the main purpose of the enterprise (see: *The Myth of Sissyphus* by Albert Camus). There are specific cases in which the "why" becomes obvious because unless the approach to the goal is achieved, nothing is achieved. A good example for the latter cases is the growth of nerve-cell axons. Hence, with axon growth we do not have to ask "why" but only "how". Growing axons may elongate vastly and navigate toward their goal. The navigation is performed by the leading tip of the axon, the growth cone that can detect gradients of attractants and repellents guiding the axon to its target goal. The process of axon guidance is rather elaborate; it requires continuous sampling of the environment for guidance cues. The growth cone has to be equipped with receptors that sense the cues and integrate signals that are derived from multiple molecular interactions at each site along the axons pathway. This sounds complicated and . . . it is. Consequently, finding an appropriate model system in which axon pathfinding can be studied at both the molecular and anatomic levels is a key to solve the "how" for axon guidance. Esther Stoeckli of the University of Zürich focused on axon pathfinding in the chicken embryo, utilizing *in ovo* methodologies (e.g. Stoeckli and Landmesser, 1998). While the zebrafish eggs are laid as signal-celled (followed after fertilization, by quick and frequent cell divisions), the chicken egg (of a fertilized hen) is laid when the embryo contains already 60 000 cells. This is too late for the study of the initial patterning of the embryo but in time for the study of axon pathfinding and the analysis of the genes involved in this process.

Bourikas and Stoeckli (2003) developed the *in ovo* techniques, with chicken eggs to an efficient procedure for large-scale reverse genetic analyses for the identification and characterization of specific genes (Fig. 21). This technique was thus ready for the use of dsRNA interference as an experimental tool to analyze individual genes that are active during chick-embryo development and especially with guidance of axon growth.

 RNA Silencing

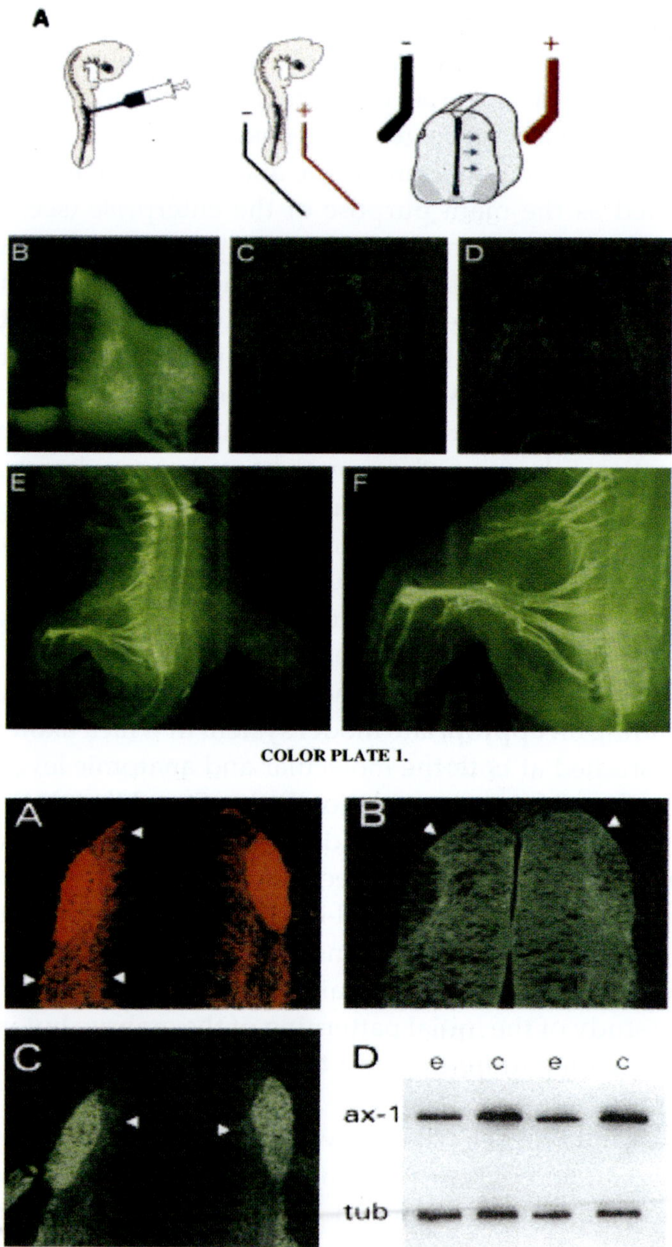

COLOR PLATE 1.

COLOR PLATE 2.

Fig. 21. Plate 1: Delivery of DNA, *in vivo* into chicken eggs. Plate 2: Down-regulation by *in ovo* RNAi of specific target genes. (From Bourikas and Stoeckli, 2003.)

The technique of Stoeckli and associates was intended to replace previous, other "traditional" procedures, to introduce macromolecules (as DNA and RNA) into the chick embryo, by the electroporation method. The "traditional" methods were based on lipofection or transfection with retroviruses. Both these latter methods have drawbacks for *in ovo* studies. The electroporation method consists of opening a "window" in the egg shell, introducing the respective macromolecules by microinjection into specific sites and then causing the penetration of the macromolecules into the embryo cells by micro-electroporation. The application of short electric pulses will create pores in the cell membrane allowing the entrance of the macromolecules.

The actual application of the electroporation technique for delivering dsRNA at specific location in the embryo and at specific stages of chick-embryo development started only recently (Pekarik *et al.*, 2003). The Swiss team (Pekarik *et al.*, 2003) initiated a series of experiments to explore the silencing of specific genes in the development of chicken embryo by *in ovo* electroporation. Their experimental work went through three phases. In the first phase they tested the silencing of a transgenic gene. In the second phase they used genes, active in the guidance of axon, of which the mutants had already a known phenotype. In the third phase they used a coding sequence for which the silencing effect was not yet known.

The eggs were "windowed" after a certain number of days of incubation at 39°C. For testing the silencing of a transgene they introduced plasmid DNA that encoded the yellow fluorescent protein (YFP) under the control of the β-actin promoter. The DNA was introduced by microinjection into the central canal of the spinal cord at the leg level. When electroporation was applied after injection (several pulses of 50 ms duration of (26 V). The YFP gene penetrated most of the cells in the injected area, as could be revealed by their fluorescence. Injection without electroporation causes no or very little fluorescence, indicating that the YFP plasmid did not enter the cells of the injected area. When there was a concomitant injection of YFP dsRNA and the YFP plasmid that was followed by electroporation the investigators observed a substantial reduction in YFP expression. This expression was not reduced when the concomitant injection was with an unrelated dsRNA. This indicated that the dsRNA can silence a transgene in the chicken *in ovo* system.

The investigators then turned to axon guidance. They had previous knowledge on the requirement of some proteins for this guidance. The lack of these proteins adversely affected the short-range guidance in commissural axon pathfinding in the embryonal (chicken) spinal cord. Such information was based on the injection of specific antibodies to such proteins. They took the coding sequences of three such genes, transcribed the sense and the antisense RNAs from each of them and annealed them to produce the respective dsRNA. The genes that encode the following guidance-cue proteins were used: axonin-1, NrCAM and NgCAM. When the dsRNA of axonin-1 was injected into the central canal of the spinal cord, followed by electroporation *in ovo*, this procedure resulted in a decrease of axonin-1 expression in the dorsal spinal cord and in pathfinding errors as well as in defasciculation of the commissural axons. When the dsRNA of NrCAM was applied the results were also pathfinding errors, but without affecting fasciculation. The effects of these two dsRNA were thus the same as those observed after interference that led to the reduction of the functions of axonin-1 and NrCAM proteins, respectively. As expected from previous *in vivo* interference studies, the dsRNA of NgCAM induced a defasciculated axon growth pattern but did not impair midline crossing by commissural axons. These results indicated that the combination of *in ovo* electroporation with dsRNA that represent coding sequences of specific genes that serve as pathfinding cues for axons can be silenced specifically by the RNAi procedure.

The investigators thus turned to genes with "unknown" (or "partially unknown") effects on pathfinding of axons. To identify such genes they used a subtractive hybridization screen for guidance-cues involved in commissural axon guidance. One such putative guidance-cue gene was used for the synthesis of the respective dsRNA and the latter was used in injection/electroporation with *in ovo* chick embryo. Nine out of 12 such treated embryos had a peculiar phenotype that was very different from controls and also different from the phenotypes observed after the injection of dsRNAs of axonin-1, NrCAM or NgCAM. In contrast to commissural axons in the control embryo, after injection of this dsRNA, most commissural axons in the experimental embryos stalled in the floor plate and only very few fibers reached the contralateral floor plate border. Most axons stalled after growing about 60 per cent of the width of the floor plate. The cDNA fragment that was represented by this dsRNA identified, in the database, a Rab-GDP dissociation inhibitor. The visual identification of the phenotype caused by this dsRNA was useful in explaining the function of Rab-GDP in guiding commissural axons.

On the basis of these results the investigators had an optimistic vision — that their methodology of using RNAi *in ovo* will "bring the chicken embryo back on stage as an invaluable model system for developmental studies". The chicken-embryo RNA silencing investigations furnished manifestations of the two mottos of this book that described the usefulness of the swords of King Solomon and of Alexander the Great. The chicken system indicated that the wise use of a "sword" (a *dicer*) can be applied in molecular-genetic and developmental studies.

 RNA Silencing in Mammals I: Experimental Induction of Gene Silencing

Due to the vast amount of studies during the last few years on RNA silencing in mammals, this subject will be divided into two chapters and an appendix. This chapter will focus on RNA silencing induced experimentally. Another chapter will be devoted to RNA silencing that was induced endogenously. The utilization of RNA silencing in attempts to cure human diseases is an emerging theme and Eithan Galun will present it in the form of an appendix on Gene Therapy.

This chapter will be arranged into some "drawers". I shall first present the early approaches of gene silencing in mammals by dsRNA. In another "drawer" I shall record a selected number of advanced studies on RNA silencing performed with various mammalian cells, tissues and organisms in which different methodologies were employed. There will be a section on RNA silencing and mammalian diseases and a final section on genome-wide approaches for gene silencing.

Early Approaches of Gene Silencing in Mammals

As noted in Chaps. 1 and 2, the awareness of RNA silencing in eukaryotes has an elaborated history. RNA-induced gene silencing was a recognized phenomenon in plants since many years ago but then came a "discovery" in nematodes by Fire *et al.* (1998) that established the role dsRNA in silencing specific genes. The regulation of gene expression in mammals by endogenously produced RNA (non-coding) species was also considered many years ago but the experimental proof for this regulation lagged behind the proof for such regulation in other eukaryotes.

Before dealing with the early studies on dsRNA induced silencing in mammals we should note two issues. One concerns a rather "old" prophecy while the other deals with non-specific silencing in mammals by long dsRNA. In their review on the regulation of gene expression by repetitive genomic sequences Davidson and Britten (1979) made an interesting statement. They wrote: "These observations (on repetitive DNA sequences) lead to a model for regulation of gene expression in which the formation of repetitive RNA-RNA duplexes controls the production of messenger RNA". The details of their predicted model will not be presented here because a vast amount of additional information on silencing of mammalian genes has accumulated during the 25 years since the statement was made by these authors.

The other issue concerns the question why long dsRNA in mammals causes a general non-specific effect on gene expression. Such an effect was not revealed in lower eukaryotes as plants, nematodes and insects. This issue was discussed in detail by several authors such as Clemens (1997), Williams (1999) and Alexopoulou *et al.* (2001). Only a summary will be presented below. Mammalian cells produce interferons (IFNs) when infected by viruses or other pathogens. When cells are treated by IFNs they exhibit increased sensitivity (of protein synthesis) to inhibition by dsRNA that is longer than 30 nt. Thus, viral replication is impaired but the suppression also affects the mammalian-cells gene expression. The protein kinase PKR has a major role in this suppression. It is a ubiquitously expressed serine/threonine protein kinase. It is induced to several-fold higher levels by IFNs and is activated by dsRNA (as well as by other effectors as cytokine, growth factors and stress signals). The binding of dsRNA to PKR causes conformational changes in the PKR and consequently, to autophosphorylation. The PKR then undergoes dimerization, that is, the active form of PKR. The activated (dimeric) PKR phosphorylates eIF-2α and hence causes inhibition of protein synthesis. There is an additional role of dsRNA. The mammalian Toll-like receptors 3 (TLR3) recognize dsRNA, causing the activation of NF-κB and the production of type I IFNs. These can induce 2′–5′–adenylatesynthase/RNase L, thus degrading RNAs (Kariko *et al.*, 2004). In short, there is a cascade of interactions that are initiated by long dsRNA. These interactions probably evolved as a defence mechanism against pathogenic viruses.

Silencing of genes in oocytes, zygotes and embryos

A convincing report on specific interference with gene expression by experimental treatment with dsRNA was presented early in 2000. This was a pioneering work by Wianny and Zernicka–Goetz (2000) who used mouse embryos in their study. This publication appeared two years after the publication of Fire *et al.* (1998). But soon thereafter there came a "flood" of reports on dsRNA-induced silencing in mammals.

Wianny and Zernicka–Goetz (2000) of the University of Cambridge, UK, intended to cause specific knockouts of genes in order to study embryo development in mice. For that they searched for a procedure that will disturb specific genes during specific developmental stages in the early embryo. These investigators were aware of the success to silence specific genes by the respective homologous dsRNAs in other organisms as plants nematodes and *Drosophila* flies. Moreover, the investigators were also aware of the problem with dsRNA application to mammals: the activation of a dsRNA-responsive PKR that phosphorylates and inactivates the translation factor, eIF-2α, causing a global suppression of translation and ultimately leads to apoptosis. In spite of the latter awareness the investigators tried the dsRNA silencing. Before their trial they mastered the elaborate techniques of injecting synthetic RNAs into mouse oocytes and into pre-implantation mouse embryos. Also, these investigators established transgenic mice that expressed a modified green fluorescent protein (MmGFP) with a mammalian promoter that could serve as an effective marker gene: the silencing of this gene could be swiftly recognized. They first turned to this GFP and asked whether this gene could be silenced by the appropriate dsRNA and if such a silencing would be specific, meaning without impairing the development of the dsRNA-injected embryo. Indeed, they found that the injection of MmGFP dsRNA into single-cell zygotes prevented the onset of green fluorescence at the two- to four-cell stage. After injection, the young embryos were cultured *in vitro* for a few days to the blastocyst stage.

The uninjected embryos expressed the GFP as expected while those injected with MmGFP dsRNA had a markedly reduced fluorescence but without abnormal phenotypes. Moreover, injected embryos could be implanted into appropriate female mice and had the same implantation frequency as uninjected embryos. The normal development of the injected and implanted embryos was observed till 8.5 days postcoitum. The expression of fluorescence was not impaired when an unrelated dsRNA (from a *c-mus* transcript) was injected. This indicated that the dsRNA effect was specific.

The investigators then turned to the knockout of an endogenous gene: *E-cahedrin*. When embryos were injected with *E-cahedrin* dsRNA the embryos behaved as having a null mutation of this gene. There was an initial normal development up to the compaction stage of the morula. However, 70 per cent of the injected embryos did not form a cavity, the remaining 30 per cent formed a cavity but never developed into normal blastocysts. The expression of the E-cahedrin protein was dramatically reduced by the dsRNA injection. The injection of MmGFP dsRNA did not affect E-cahedrin expression. Finally, the investigators turned to oocytes and asked whether dsRNA would interfere with maternally expressed genes. They focused on the *C-mos* gene that expresses an essential component of a cytostatic factor that is responsible for arresting the maturing oocytes at metaphase in the second meiotic division. In the mutant, *c-mos* mice, most of the oocytes have no such meiotic arrest. When oocytes of normal mice were injected with *c-mos* dsRNA the meiotic arrest was eliminated just as in the oocytes of mutant mice. This indicated that dsRNA was capable of blocking the expression of maternally provided gene products. This study of Wianny and Zernicka–Goetz (2000) indicated that indeed it is possible to cause specific gene silencing by the use of appropriate dsRNA in mice embryos.

A group of investigators from the University of Pennsylvania that included Richard Schultz, Paula Stein and Petr Svoboda continued the studies on dsRNA silencing in mice. They introduced their study (Svoboda *et al.*, 2000) with a citation of William Harvey (1578–1657): "Omne vivum ex ovo" (all living things come from eggs). Harvey practiced medicine and as such he followed the footsteps of Roderigo Lopez. Like the latter physician, Harvey also started his career at the St. Bartholomews Hospital and was later appointed physician of the English court. But luckily for Harvey, unlike the fate of Lopez, who was prosecuted and killed for treason (attempt to poison Queen Elizabeth), Harvey had a long and respectful life. Anyway, Schultz and associates started their series of publications with mouse oocytes. In fact, their first report (Svoboda *et al.*, 2000) was submitted a few months after the publication of Wianny and Zernicka–Goetz (2000) and there was an overlap between the two publications. Svoboda *et al.* (2000) focused on mice oocytes and their intention was to search the possibility to inhibit the translation of maternal mRNA in these oocytes. They choose two mRNAs of which one was the *Mos* mRNA (the same mRNA, handled by Wianny and Zernicka–Goetz, 2000, mentioned above). The *Mos* product causes the activation of mitogen-activated protein (MAP) kinase, whose activity is required to maintain arrest at metaphase II of meiosis in the oocytes. The investigators produced the sense and the antisense RNA sequences from the coding region of *Mos* mRNA (length of about 500 nt) that were annealed and purified as dsRNA. Similarly, dsRNA was prepared for the *Plat* gene. *Plat* knockout causes mild perturbations in phenotype, for example, retardation in neural migration. The oocytes were injected with approximately 10^6 molecules of either *Mos* dsRNA or *Plat* dsRNA. In addition, other oocytes were injected with similar amounts of ssRNAs (sense and antisense) representing these two transcripts. The injection of *Mos* dsRNA caused, after incubation, a marked reduction of *Mos* transcript (about 80 per cent reduction), and the injection of *Plat* dsRNA caused an even stronger reduction of the *Plat* transcript. The reduction was specific, meaning that *Mos* dsRNA did not affect the *Plat* transcript. The reduction of the number of dsRNA molecules per injection also reduced the effect of the dsRNA on the reduction of transcripts. The injection of *Mos* dsRNA reduced the MAP kinase in the cultured oocytes and the injection of *Plat* dsRNA strongly reduced the lever of tPA (tissue plasminogen activator) activity. The strong effects of the *Plat* dsRNA was revealed only when the high dose (10^6 molecules) was injected. Lowering the dose reduced the effect but with respect to tPA activity the antisense (ssRNA) of *Plat* mRNA had a similar effect to the injection of dsRNA.

An additional publication of the Pennsylvania team (Svoboda *et al.*, 2001) reported the authors' conclusion that extract from mouse oocytes and preimplantations embryos can process dsRNA into 22 nt fragments. In this later publication the authors used a different procedure to furnish dsRNA. Rather than synthesizing ssRNA of sense and antisense mRNA and then annealing the ssRNAs to dsRNA, they constructed plasmids in which the sequences representing the coding regions were put tail-to-tail with a space between them. Such a construct should produce *in vivo* a stem-loop structure. As they knew that the stem would be cut into ~22 nt dsRNA fragments, the construct, once in the oocytes should result in the required dsRNA. The tail-to-tail in this study represented the *Mos* RNA. They also constructed plasmids with the coding sequence of the reporter gene for GFP. Actually, they constructed several types of plasmids and also prepared *Mos* mRNA dsRNA as mentioned in their previous publication. Initially, the investigators encountered difficulties in the expression of stem-loop or hairpin constructs. But there was reasonable success when they increased the spacer (loop, after *in vivo* transcription) up to about 50 nt. When plasmids with stem-loop producing constructs were injected into oocytes the resulting *Mos* dsRNA were at least as effective as the dsRNA that was synthesized *in vitro*. The same level of interference with the generation of MAP kinase activity was revealed. The sequence coding for the GFP could be added to the plasmid but the authors found that it was important to place an intron in a specific location in order to optimize the results. The authors claimed that they opened the way for the utilization of dsRNA technology to investigate the role of specific mice genes that are vital for normal maturation of oocytes and early embryos. They did admit that in this system dsRNA does not silence endogenous gene completely but partially. They thus suggested a term from the boxing arena to fit their results. Rather than calling this silencing *knockout* they termed it *knockdown* . . . .

In a further study Schultz and associates (Stein *et al.*, 2002) turned to the genomic sequence of the mouse. They looked into this genome for sequences that have similarity to genes in other organisms that are essential for the RNAi process. They focused on the *Mut-7* of *C. elegans* and on *Qde-3* in *Neurospora*. These genes are essential for RNAi in *C. elegans* and for quelling in *Neurospora*, respectively. The closest sequences to these genes in the mouse genome were those coding for *Wrn*, *Blm* and *RecQ1*. They thus took mice with mutations in these genes and repeated their *Mos* dsRNA experiment with the oocytes of the mutated mice. The rationale was that in oocytes from mice having one of these mutations there will be no *Mos* silencing, meaning that these mammalian genes are essential for RNA silencing in mice. In none of these mutants was the silencing by *Mos* dsRNA impaired. The investigators thus decided that these three mammalian genes (e.g. *Wrn*, *Blm* and *RecQ1*) are not involved in the RNAi system of mammals.

In a further study Stein *et al.* (2003a) focused on RNA-dependent RNA polymerase (RdPR). We noted in previous chapters that this enzyme exists in plants and nematodes but was not revealed in *Drosophila*. These investigators found in mice a coding sequence with similarity to the code for RdPR (*Aquarius*) and this coding region is expressed in mice oocytes and in early embryos of mice. On the other hand, they found that RNA synthesis by an RdPR-like enzyme is not required for the manifestation of dsRNA silencing in these oocytes or early embryos: treatment with cordycepin, an inhibitor of RNA synthesis did not prevent *Mos* dsRNA silencing of the *Mos* gene. Furthermore, the inhibition does not *travel* upstream of the site of homology between the gene and the dsRNA. Such a *travel* was revealed in nematodes and in plants and was thus attributed to the activity of RdPR.

In a subsequent publication of this team (Stein *et al.*, 2003b) the investigators turned to transgenic mice. They used the construct shown in Fig. 22. This construct includes a promoter (Zp3) that is active specifically in mice oocytes. It contains a code for the

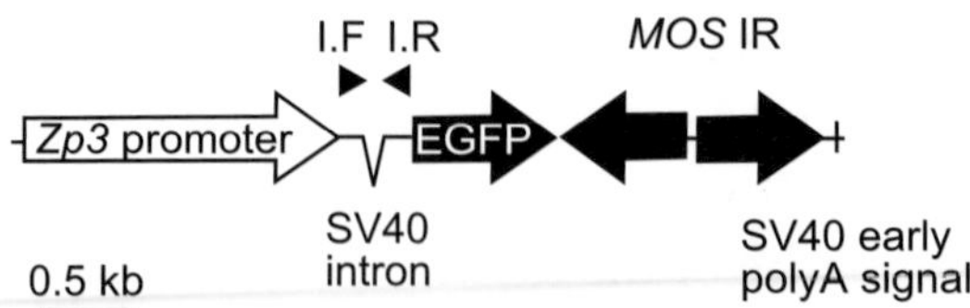

Fig. 22. Scheme of transgenic construct used by Stein *et al.* (2003b) in their study of RNAi in mouse oocytes. Arrowheads depict positions of primers IF and IR used for genotyping and expression analysis. Mos IR is *Mos* inverted repeat; EGFP is enhanced green fluorescent protein. (From Stein *et al.*, 2003b.)

GFP reporter and a tail-to-tail sequence for the *Mos* transcript. The latter will be converted *in-vivo* into a stem-loop structure. After dicing, fragments of dsRNA that have homology to the *Mos* mRNA are expected to be formed. This construct is "swiftly" employed to produce the respective transgenic mice.... The investigators passed this construct to the Transgenic and Chimeric Mouse Facility (TCMF) of the University of Pennsylvania and received from the TCMF B6SJLF1/J mice with this construct inserted in their genome. They found that the male transgenic mice were normal and fertile but the female transgenic mice were sterile. When oocytes were extracted from sterile females and cultured, the oocytes progressed to metaphase II and then underwent parthenogenetic activation. This is consistent with a *Mos* null phenotype. Analyses for *Mos* mRNA in the oocytes from these transgenic female showed a vast decrease of this mRNA, up to a 98 per cent reduction. The mRNA for a non-targeted gene (*Plat*) was not affected. The oocytes also had a reduced level of MAP kinase. Control eggs were arrested in metaphase II. But even among the two transgenic females that were kept by the investigators, there were differences in the level of dsRNA-induced silencing. The transgenic males, while not showing abnormal phenotypes, passed the transgene to their daughters in which the dsRNA was operating.

The studies by the Schultz team clearly paved the way for specific "knockdown" of genes in oocytes and early embryos of mammals. Is this dsRNA silencing restricted to oocytes and young embryos or is it applicable also to mature tissues? An answer to this question was requested by a team of investigators from the Osaka University, Japan. Hasuwa *et al.* (2002) noted that until their investigation it was assumed that dsRNA that is longer than 30 nt will elicit in mammals "antiviral" response that is not specific to the sequence of the dsRNA. To answer the above question Hasuwa *et al.* (2002) also used animals (mice and rats) that were transgenic and expressed the GFP gene in all their tissues. They then synthesized several constructs that should integrate in the animals genomes when the respective plasmids are injected into the pronuclear stage of fertilized eggs. The injected embryos were either cultured up to the blastocyst stage or immediately implanted into pseudopregnant animals to generate transgenic animals. The plasmid that should produce dsRNA with homology to the coding sequence of the GFP had a tail-to-tail configuration with about 30 nt sense and 30 nt antisense separated by a few nucleotide (to form *in vivo* the stem-loop dsRNA). The investigators also added a gene for red fluorescence to mark the genomic integration and expression of the plasmid. As a control the investigators also injected into eggs synthetic dsRNA with the GFP sequence. The results indicated that while the injection of synthetic dsRNA caused a suppression of the GFP that was transient, the suppression caused by integration of the tail-to-tail construct was long lasting and could be observed in the newborns resulting from the implanted embryos. The suppression was especially dramatic in some organs as the heart where in embryos that developed from non-injected eggs there was strong green fluorescence — after injection with the appropriate plasmid the green fluorescence (expressed by the GFP gene) disappeared while the red fluorescence was apparent. This system of gene silencing was successful in both mice and rats.

RNA silencing in mammalian cells

Thomas Tuschl and associates (Elbashir *et al.*, 2001b) of the Max-Planck Institute in Göttingen, Germany, took the RNAi in mammals one step further. They asked whether the RNAi system is applicable to cells in various mammal species. Knowing that long dsRNAs of 30 nt and beyond can trigger non-specific gene silencing they did their experiments with dsRNA of 21 nt. They were not deterred by the labor investment and synthesized these dsRNAs so that they would have homology to target mRNA and have 2 nt overhangs at their 3′ ends. They also put two T or two U bases at these overhangs. To test the silencing capability of their dsRNA they produced plasmids with reporter genes (for GFP) that were co-transfected into mammalian cells with the relevant dsRNA. They investigated the RNAi effect in (mouse) NIH/3T3 cells in monkey COS-7 cells and in (human) Hela S3 cells. In one of these cell lines the co-transfected GFP gene was suppressed by the respective dsRNA 3–12 fold relative to the control. In another cell line, the GFP gene was suppressed even more — to 25 fold relative to the control. A less dramatic reduction in the expression of the target gene was observed with human embryonic kidney cells (293). The standard concentration of the transfected dsRNA was

25 nM but lower concentrations were also effective. The investigators also tested longer dsRNAs and found that in some mammalian cell types there can still be specific silencing. To investigate the silencing of an endogenous gene the investigators choose human Hela cells and the target was the gene that encodes *lamin A/C* (for which antibodies were available). The answer was clear: the *lamin A/C* gene could be silenced specifically by the appropriate dsRNA. Cells with silenced *lamin A/C* had no other abnormalities.

It should be noted that Chinese hamster ovary cells (CHO-K1) were already previously used for RNAi tests by Ui-Tei *et al.* (2000) but these Japanese investigators employed long dsRNA, usually of several hundred nt; their shortest dsRNA was that of 38 nt. The efficiency of their silencing of genes in the mammalian cells was about 2500 fold lower than in *Drosophila* cells.

It should not surprise us that when a new and interesting field of investigation emerges, more than one team of investigators have similar ideas and put them to test at about the same time. This actually happened with the study of RNAi by short (synthetic) dsRNA in mammalian cell lines. Hence, a team of five investigators from four research laboratories in Maryland (Bethesda and Baltimore) had a rather similar approach as that of the investigators from Göttingen (Elbashir *et al.*, 2001b). The investigators from Maryland (Caplen *et al.*, 2001) also decided to synthesize short (21–25 nt) dsRNAs to silence specific genes in mammalian cell lines. The lines used by the Maryland investigators were primary mouse embryonic fibroblasts (MEFs), human embryonic kidney cells (293) and human HeLa cells. Two of these were also used by Elbashir *et al.* (2001b). The dsRNAs were targeted to co-transfected reporter genes: a gene for GFP or a gene for CAT (chloramphenicol acetyl transferase). The results of Caplen *et al.* (2001) were rather similar to the results of Elbashir *et al.* (2001b): the dsRNAs were capable to silence the reporter genes that were co-transfected into the various mammalian cells. The level of inhibition did vary but in all cases the silencing was specific. Moreover, the Caplen *et al.*'s (2001) investigation showed that a dsRNA for the *C. elegans unc-22* gene did not interfere with the expression of the co-transfected reporter gene. The Maryland team also tried longer dsRNAs and, as expected, found non-specific cell deterioration.

DsRNAs of 78 or 81 nt already caused a substantial degree of cell-death in the MEF cells. The Elbashir *et al.*'s paper was submitted on February 20th, 2001 and appeared in a *Nature* issue of May 24th, 2001. The Caplen *et al.*'s paper was submitted on May 18th, 2001 and appeared in the *PNAS* issue of August 14th, 2001.

Toward the end of 2001 there appeared another "pair" of publications on RNA silencing in mammalian cells that reported on rather similar experimental work. The first publication was by Yang *et al.* (2001) from Philadelphia, Pennsylvania. These investigators used (as others before them) the GFP gene as target for dsRNA. Their dsRNAs were all rather long, about 500 nt. The dsRNA was furnished by various means but also by purified dsRNA (i.e. not transcribed *in vivo* from a transfected plasmid but rather annealed from synthesized ssRNAs). The dsRNAs were also furnished by a plasmid with a T7 promoter and inverted repeats of 547 nt from the GFP coding region. It was found that undifferentiated mouse embryonic stem cells (ES) showed sequence specific silencing that was also dose-dependent. On the other hand, in differentiated cells there was no sequence-specific gene silencing. The investigators also found that long dsRNA is cleaved to short dsRNAs of 21–22 nt (i.e. siRNAs) in cytoplasmic extracts of mammalian cells. The readers should note that the GFP protein has a long half-life in mammalian cells, therefore the cleavage of its mRNA will be observed, by the reduction of fluorescence, only after several days.

Between the submission and the publication of the paper by Yang *et al.* (2001), another team from the Friedrich Miescher Institute in Basel, Switzerland, submitted a rather similar report (Billy *et al.*, 2001). These investigators first tested the silencing capability of long (about 700 nt) dsRNA in embryonal teratocarcinoma (EC) cell lines (F9 or P19). When transfected with either plasmids with the reporter genes for GFP or for β-galactosidase (β-gal) these cells expressed the respective genes. But when cells were co-transfected with the GFP gene and dsRNA for GFP most of the cells did not express the GFP gene and even in those cells that did express it, the level of fluorescence was lowered. The dsRNA for GFP did not affect the expression from a co-transfected β-gal gene. Likewise, the dsRNA for the β-gal gene did not affect the co-transfected GFP gene. These results were also verified by northern and western blot hybridization, showing that the mRNA and the protein of the GFP gene were strongly reduced by the dsRNA for GFP. There was no effect by ssRNAs, either sense or antisense. The F9 cells were used in a similar experiment to follow the reduction of expression of endogenous genes for integrins. Again, the investigators found that the dsRNA effect was specific and observed a considerable reduction of about 80 per cent of the integrins by the respective dsRNAs. Mouse ES cells were also examined in a similar manner. The specific silencing by the respective dsRNAs was also recorded in the ES cells. The investigators also found that the long dsRNAs are cut in the cytoplasm into fragments of ~23 nt and identified the *Dicer* enzyme complex in the cytoplasm. The two publications of Yang *et al.* (2001) and of Billy *et al.* (2001) clearly indicated that the non-specific silencing by long dsRNA is not operating in non-differentiated mammalian cells.

The findings of Yang *et al.* (2001) and Billy *et al.* (2001) were further verified and extended by the team of Gregory Hannon of the Cold Spring Harbor Laboratory (Paddison *et al.*, 2002a) although the Cold Spring Harbor team submitted their paper before they saw the publications of Yang *et al.* (2001) and Billy *et al.* (2001). Paddison *et al.* (2002a) also found that long dsRNAs had a specific silencing effect in undifferentiated mice cells (e.g. P19, embryonal carcinoma cells). But there was a non-specific silencing by dsRNA in human embryonic kidney cells and mouse embryo fibroblasts. In HeLa cells the combination of a target reporter and its homologous long dsRNA caused a non-specific silencing.

Paddison *et al.* (2002a, 2002b) also found that the non-specific silencing by long dsRNA can be suppressed by adding a gene that suppresses the PKR-induced reaction. When this was performed in combination of a certain dsRNA, the silencing became specific. The investigators also found that a persistent silencing can be induced by an expression vector that causes the expression of a hairpin RNA that will fold *in vivo* and then be sliced into short dsRNAs. Indeed, these investigators verified the activity of *Dicer* in mammalian cells; they also found that if *Dicer* was silenced by the respective dsRNA, the silencing was repressed. They went further and developed a cell-free system from mammalian cells and found that this system reacts in a similar way as the cells from which the extract was isolated. In short, the RNA silencing mechanism in undifferentiated mammalian cells seems to be via the conventional RNAi mechanism (i.e. as in *Drosophila*).

A further refinement of the RNA silencing in mammalian cells was achieved by Agami and associates of the Netherlands Cancer Institute in Amsterdam (Brummelkamp *et al.*, 2002). These investigators constructed plasmids that were termed by them pSUPER vectors. In these plasmids a polymerase-III H1-RNA promoter drives the transcription of an RNA that starts and ends with five thymidines. After transcription it folds to a stem-loop configuration and results in a stem of 19 nt with 2 nt overhangs, similar to the *Let-7* of nematodes. When the folded construct has the appropriate loop (of 9 nt) it will efficiently and specifically silence genes in mammalian (MCF-7) cells. Well, again a matter of publication priority: Brummelkamp *et al.* (2002) submitted their paper in December 2001 and it appeared in *Nature* in April 2002. Then Hannon and associates (Paddison *et al.*, 2002b) submitted a paper in January 2002 and the latter paper was published in *Genes and Development* before the authors could see the *Nature* paper, but the results of Paddison *et al.* (2002b) with mammalian cells were the same as those of Brummelkamp *et al.* (2002).

Then there appeared four publications that dealt basically with the same approach as the approaches of Brummelkamp *et al.* (2002) and of Paddison *et al.* (2002b). These publications were submitted on December 7th, 2001 (Miyagishi and Taira, 2002), on January 29th, 2002 (Yu *et al.*, 2002), on February 28th, 2002 (Donze and Picard, 2002), and ... on February 28th, 2002 (Sui *et al.*, 2002). The Japanese investigators (Miyagishi and Taira, 2002) intended to overcome the general inhibitory effect of long dsRNAs in mammalian cells by the introduction of a specially designed expression vector. This vector was aimed to transcribe in the cells, short ssRNAs of the sense and of the antisense sequences of a target gene. By using a U6 promoter these short transcripts should have poly U tails so that after *in vivo* annealing a 19 nt dsRNA with two or more U at the 3′ will be produced. The target of their dsRNA was a luciferase gene. The investigators indeed succeeded to reduce luciferase activity in HeLa cells to 10–20 per cent of the control level.

Yu *et al.* (2002) produced the desired hairpin short RNA *in vitro* and found that these hairpin structures, having a sequence that is homologous to the coding region of an endogenous gene in P19 cells will silence this gene (β-tubulin) during neuronal differentiation.

The third team, mentioned above (Donze and Picard, 2002) produced the desired short dsRNA *in vitro*. By using a T7 promoter and a DNA template that is double-stranded at the promoter sequence but single-stranded downstream of the promoter, they synthesized the sense and the antisense ssRNAs. These were annealed and introduced into HeLa cells by lipo-transfection. The targets were (again) the GFP transgene or the human PKR gene. The authors reported efficient silencing and recommended that short dsRNAs of 22 nt rather than of 21 nt should be constructed. The fourth publication was from the Harvard Medical School (Sui *et al.*, 2002).

Looking back at the several publications on RNA silencing in mammalian cells that appeared during the end of 2001 and in early 2002, I remember the motto of the book that I wrote with my son (Galun and Galun, 2001). This motto, from the Jewish Talmud, says: "The Envy of Scholars will Increase Wisdom". Well, this envy is beneficial for the increase of knowledge and wisdom but it is not beneficial to the feeling of the scholars.

For following a bold and direct approach the expression: "To hold the bull by its horns" may be used. It appears that you may also exercise a bold approach by "Holding the mouse by its tail". A team composed of investigators from the Stanford University School of Medicine and the Cold Spring Harbor Laboratory (McCaffrey *et al.*, 2002) went one step further, from mammal cells to adult mice. They devised means to introduce dsRNAs into mice and followed the impacts of these dsRNAs on co-introduced targets. The dsRNA and the targets were injected into the tails of the mice in order to reach the liver tissue. They found that the GFP gene could be silenced by the respective dsRNA to about 20 per cent of the unsilenced levels. When a chimeric target coding for an hepatitis C virus (HCV) gene and the GFP gene were silenced by dsRNA the dsRNA also silenced the chimeric gene (but no specific data were provided on the reduction of expression of the HCV construct).

Phillip Sharp who established his laboratory at the Massachusetts Institute of Technology, and his team (McManus *et al.*, 2002) aimed their RNA silencing research at cells of the immune system of mammals. They choose E10 mouse cells, a thymoma-derived cell line. These cells exhibit the surface expression of the receptors CD4 and CD8α. The introduction of the dsRNAs and vectors for the expression of a reporter gene (for luciferase) into the cells was by electroporation. In summary the investigators found that the silencing was specific, meaning that the dsRNA for CD8α silenced the expression of CD8α but not of CD4. Silencing by the CD8α dsRNA was more effective than silencing by the CD4 dsRNA. The highest level of silencing was to about 20 per cent of control. There was a difference between specific sequences of the dsRNAs for CD8α. The best silencing was achieved when the dsRNA was homologous to the 3'–UTR, 15 nt downstream of the stop codon. In the CD4 mRNA, only a dsRNA that "covered" the stop codon of the coding sequence was affective in silencing. The investigators also performed dsRNA silencing in primary mouse T cells taken from the spleen. When these cells were induced to divide, a similar silencing was achieved as with the E10 thymoma cells. Interestingly, when dsRNA for both CD8α and CD4 were applied simultaneously, the silencing was reduced, hinting for a kind of competition between the two silencings.

David Lewis and associates (Lewis *et al.*, 2002) returned to tail-injection of mice. They injected the plasmids for two genes causing fluorescence and the (synthetic) dsRNA for one of these genes. One day after injection several organs of the mice were analyzed: liver, kidney, spleen, lung and pancreas. They found that in all these organs there was a specific silencing by the dsRNA.

RNA silencing of mammalian brain tissue

From tail to head ... we shall summarize the results of several studies concerned with dsRNA silencing in brain tissue.

Krichevsky and Kosik (2002) of the Brigham and Women's Hospital of the Harvard Medical School used tissues from embryos of pregnant rats (embryonic day 18). They removed the cerebral cortices and the hippocampi and digested (trypsin) the tissues into cells that were cultured. The experiments were conducted 5–8 days after plating the cells. Target genes and dsRNAs were introduced into the plated cells by lipotransfection. It was found that the two reporter genes (for GFP and DsRed2) could be silenced specifically by the respective (21 nt) dsRNAs. Although the silencing was not dramatic: the GFP could be reduced by 42 per cent. Neither the sense nor the antisense RNA for GFP caused such a silencing. A more effective silencing was recorded when a dsRNA for the endogenous MAP2 (microtubule-associated protein 2) was introduced into the plated cells. After 48–62 hours, there was a reduction of 70–80 per cent of the expression of MAP2. The respective dsRNA did not silence another gene (YB-1).

A team of investigators (Calegari *et al.*, 2002) from the Max Planck Institute in Dresden, Germany, and the University of California in San Francisco, used the neuroepithelium of embryonic day 10 mice as their model-target tissue. The investigators studied two kinds of silencing. In one approach they used normal embryos. They injected the target genes and the respective dsRNAs into the lumen of the neural tube and delivered them into neuroepithelial cells by electroporation. The targets were genes expressing either GFP or β-gal. The results were recorded 1 day after injection/electroporation. The silencing was specific: dsRNA for β-gal silenced only β-gal but not GFP. In the other approach embryos of transgenic mice were used. In these mice there was a coding region for GFP inserted in the genome so that it was activated by the Tis21 locus. The Tis21 gene is turned on in intraepithelial cells that switch from proliferation to neurogenesis. When the dsRNA for GFP was directed by electroporation into the neuroepithelial cells the GFP expression was blocked. But one should remember that the bottle-neck of RNAi in mammals is the level of *Dicer* in the cells.

Another research team from the Harvard Medical School (Gaudilliere *et al.*, 2002) focused on a specific gene in the nervous system. They searched the silencing of the expression of a transcription factor in primary granule neurons cultured from the developing rat cerebellum. The target for silencing was MEF2A, a member of the **m**yocyte **e**nhancer **f**actor 2 (MEF2) family. Members of this family (e.g. MEF2A, MEF2B, MEF2C and MEF2D) are expressed in the central nervous system of mammals. The specific gene for MEF2A is highly expressed in cerebellar granule neurons and is regulated posttranscriptionally in these neurons upon neuronal activity. The authors intended to determine the role of MEF2A in activity-dependent granule neuron survival using the RNAi technique. To silence MEF2A the investigators constructed a plasmid that should cause the formation (*in vivo*) of a stem-loop in which the stem had homology to the MEF2A code. They also constructed a plasmid for a stem-loop with homology to another gene: NeuroD. They first tested the silencing of exogenic supplied genes (MEF2A and NeuroD) by transfecting a neuronal cell line (Neuro2A). Several controls were added and the results clearly showed that the stem-loop for MEF2A silenced only this gene and not NeuroD. The investigators then tested the knockdown of endogenously expressed genes (MEF2A and NeuroD) in primary cultures of cerebellar granule neurons from rat pups. The transfection of these cells was done using the calcium phosphate method in which not all the cells are actually transfected. While 100 per cent of HeLa cells can be transfected the transfection of primary nerve cells is problematic. In order to identify the transfected cells an expression plasmid for β-gal was added so that the results could be recorded in individually transfected cells. Medium manipulation could activate the endogenous cells and thus the silencing of this activity could be recorded. Moreover, the investigators could record the expression of genes that are normally transcribed by the induction of MEF2A, the MEF2 response elements (MREs). Indeed, the stem-loop derived dsRNA for MEF2A specifically prevented the expression of MREs. The authors thus claimed that their findings indicated that the vector-based RNAi method should provide a rapid means of analysis of gene function in primary cultures of mammalian neurons. Consequently, one should be able to assess the biological role of a particular member of a given protein family. Furthermore, because RNAi operates in primary postmitotic mammalian neurons, RNAi is likely to be of utility in the study of the nervous system in the intact mammalian organism.

Overview of early approaches of gene silencing in mammals

McManus and Sharp (2002) of the Massachusetts Institute of Technology provided a good review on the early approaches to silence mammalian genes by small interfering RNAs. This review cited many publications that were not mentioned in this section. While basically dsRNAs can silence specifically genes in mammalian cells, there are also numerous problems. As mentioned above cells in mature tissues cannot be silenced specifically by dsRNAs that are longer than 30 nt. While in established cell lines genes can be silenced easily (in HeLa cells silencing can reach 100 per cent), primary cells are notorious for having low transfection efficiencies when DNA plasmids are to be introduced into them. But several methods to improve transfection were developed and short dsRNAs are easier to transfect than DNA plasmids.

It seems that unlike plants and nematodes, mammalian cells lack the dsRNA amplification mechanism. It appears that genes with a low level of transcription are easier to silence than genes with a very high level of transcript.

As for silencing more than one gene by the simultaneous transfection of two silencing plasmids (or dsRNAs), the situation is not clear yet. Such a double silencing seems to be possible but the overall silencing may be reduced.

The reduction of mRNA by the respective silencing can take place within 18 hours but as indicated above many mammalian proteins (as well as GFP) have long half-lives (slow turnover) and there it may take 3–5 days (several doublings of cells) until the impact of gene silencing is manifested.

The question of the region of a target transcript that should be homologous to the engineered dsRNA is still open (but see a discussion at the end of the next section of this chapter). In some studies it became evident that the first 50–100 nt of a transcript (or cDNA) downstream of the translation start site as well as regions around this start site should be avoided. We mentioned above a case in which the 3′ UTR region was a proper one for engineering a homologous dsRNA. But till more information becomes available the construction of dsRNA (or stem-loop RNA that will be processed *in vivo* to dsRNA) shall be determined in each case by trial and error.

Another review on the early approaches of gene silencing in mammals was provided by Shi (2003) of the Harvard Medical School in Boston. Shi discussed several aspects of RNA silencing in mammals such as the desired plasmid vectors, the target selection and the means of delivering the silencing agents (e.g. by viral vectors). As befitting an author from the Department of Pathology of a medical school, Shi also discussed the progress, till the end of 2002, in using RNA silencing in gene therapy.

Advanced Analysis and Application of dsRNA Silencing of Mammalian Genes

A great deal of talent and effort resulted in the establishment of RNA silencing in mammals within about 2 years. This was summarized in the previous section on further studies on this silencing ramified in several directions. The endogeneous gene-silencing in mammalian cells, as during differentiation, by miRNA, shall be discussed in Chap. 10. Below I shall review studies that are intended to provide a better understanding of the mechanism of dsRNA silencing in mammals to improve the delivery of dsRNA into mammalian cells, to handle possible interference with mammalian diseases and to upgrade the silencing by small interfering RNAs as a tool for genome wide genetic-analysis in mammals.

Further insight into the short-interfering–RNA mediated gene silencing in mammals

The fundamental components of the RNA silencing in mammals were found to be similar to those of other eukaryotic organism, indicating a conservation from unicellular plants, fungi and protozoa to evolutionarily advanced organisms. But what about the details? We have already seen that while in plants, fungi and nematodes the RdRP plays an important role in RNA silencing, the RdRP activity was not revealed in flies and mammals. The lack of this enzymatic activity has important theoretical and practical consequences. Investigators thus started to handle specific components of the RNA silencing mechanisms and to compare mammals with other organisms. One such group of investigators was Kaoru Saigo and associates (Doi *et al.*, 2003) of the University of Tokyo, Japan. As a first step toward clarification of the molecular mechanisms of mammalian RNA silencing, these investigators looked for a possible human homologue of the nematode RDE-1 protein and the *Drosophila* Ago2 protein. The exact role of these proteins is not yet known but they are considered to be translation-initiation factors and important components of the RNA silencing in nematodes and *Drosophila*, respectively. The investigators looked at the amino acid sequences of 18 PIWI family members (Fig. 23). Members of the PIWI family have all a PIWI domain and a PAZ domain. The mammalian PIWI family members (elF2C1, elF2C2, elF2C3 and elF2C4) all have another motif, termed PRP motif. This PRP motif also exists in PIWI family members of the *Drosophila* AGO1 and the nematode ALG-1 and ALG-2 but not in the nematode RDE-1 and the *Drosophila* AGO2. The mammalian elF2C proteins thus are not considered by Doi *et al.* (2003) as orthologs of RDE-1 and

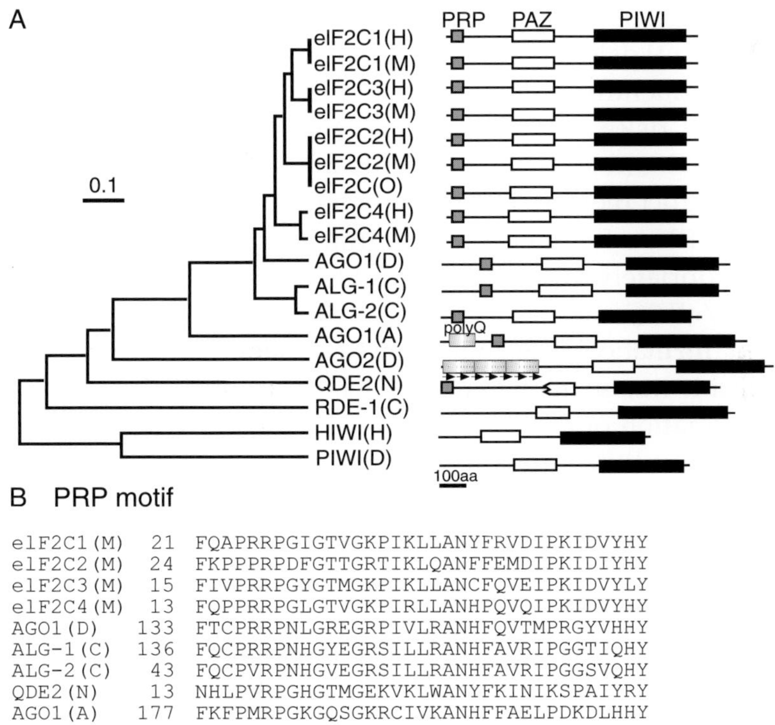

B PRP motif

```
elF2C1(M)    21   FQAPRRPGIGTVGKPIKLLANYFRVDIPKIDVYHY
elF2C2(M)    24   FKPPPRPDFGTTGRTIKLQANFFEMDIPKIDIYHY
elF2C3(M)    15   FIVPRRPGYGTMGKPIKLLANCFQVEIPKIDVYLY
elF2C4(M)    13   FQPPRRPGLGTVGKPIRLLANHPQVQIPKIDVYHY
AGO1(D)     133   FTCPRRPNLGREGRPIVLRANHFQVTMPRGYVHHY
ALG-1(C)    136   FQCPRRPNHGYEGRSILLRANHFAVRIPGGTIQHY
ALG-2(C)     43   FQCPVRPNHGVEGRSILLRANHFAVRIPGGSVQHY
QDE2(N)      13   NHLPVRPGHGTMGEKVKLWANYFKINIKSPAIYRY
AGO1(A)     177   FKFPMRPGKGQSGKRCIVKANHFFAELPDKDLHHY
```

Fig. 23. Phylogenetic tree and domain structure of the PIWI family. Abbreviations are as follows: elF2C(O), rabbit; *D, Drosophila melanogaster; A, Arabidopsis thaliana; N, Neurospora crassa; H, Homo sapiens; C, Caenorhabditis elegans;* and *M, Mus musculus.* PAZ domains are shown as open boxes and filled boxes represent PIWI domains. PolyQ are glutamine-rich sequences. The polyQ domain of *D. melanogaster* AG02 includes eight repeats of 20–21 amino acids. (From Doi *et al.,* 2003.)

AGO2. The mammalian genome lacks orthologs that encode these latter proteins. The investigators constructed a considerable number of specific dsRNAs (RNAis). All these were designed for various targets such as *Dicer*, elF2C1-4, FFP and luciferase. Several human and mice cell lines were employed and the target sequences as well as the dsRNA were transfected into the cultured cells. Appropriate controls were employed. Take, for example, for evaluating the silencing of firefly *luc*, the *Renilla* luciferase served as reference for the transfection. The results indicated that *Dicer*, elF2C1 and probably other elF2Cs are components that are essential for siRNA-mediated gene silencing in mammalian cells (i.e. when these were silenced the dsRNA for GFP or for firefly *luc* could not silence the respective trangenes). The relevance of *Dicer* was not trivial because the silencing was induced by the short (22 nt) dsRNA. Thus, if *Dicer* in mammals would be required only for fragmenting dsRNA it should not be essential for the silencing by short dsRNA. This indicated that *Dicer* may have an additional role. Could it be that there is a kind of RdRP in mammals and the product requires *Dicer* for rendering the polymerized product into short RNA fragments? By immunoprecipitation experiments, with appropriately tagged components, Doi *et al.* (2003) could analyse *Dicer*/elF2C complexes. Indeed, such complexes were revealed and their formation was independent of externally induced silencing. When it was searched as to which domains of elF2C is essential for the formation of the *Dicer*/elF2C complex, it was revealed that it is probably the PIWI domain rather than the PAZ domain. Whatever, the investigators concluded that *Dicer* and elF2C1-4 are essential for siRNA-based gene silencing in mammals.

Improvements in delivery

Several studies were intended to improve the delivery of short dsRNA into mammalian cells. One approach was to utilize the HIV as a Trojan Horse. Inder Verma and associates of the Salk Institute in La Jolla, California (Pfeifer *et al.*, 2002) developed the lentiviral vectors for the expression of transgenes in mammalian cells and carried the system up to the expression of transgenes in mice. The rationale of these investigators was as follows. Mammalian embryonic stem cells (ES) are totipotent; they can be transferred to suitable *"in vivo* environmen*t''* and contribute to various tissues and organs. It is possible first to manipulate ES *in vitro* and then to include them in a developing animal that will beomce chimeric. When cells that were derived from genetically manipulated ES cells reach the germ line, their genome can be transmitted to the sexual progeny. Among the retroviruses the lentiviruses have several advantages as vectors. They can also infect non-dividing mammalian cells, enter the nucleus and integrate into the hosts genome. Moreover, unlike some other retroviruses the transcription of lentiviruses is not silenced after integration. The investigators constructed appropriate lentivirus vectors such as vectors expressing reporter genes (for the expression of GFP and LacZ) so that they could follow the fate of the expression from the modified lentivirus after infecting mammalian cells. It was thus found that when ES cells were infected the transgene was expressed during differentiation of the ES cells *in vitro* and also *in vivo*. The transfer of lentivirus-transduced ES cells into blastocysts resulted in chimeric animals that expressed a transgene (from the lentivirus vector) in several tissues. Moreover, embryos were derived from crossing of chimeric mice expressed the transgene. Also, infection of murine pre-implantation embryos at the morula stage with lentivirus vectors resulted in stable transduction and expression of the transgene over several passages. By these studies of Verma and associates, the lentivirus vectors were established as potential Trojan Horses for the introduction of transgenes into mammalian genomes. Hence, the lentivirus-based delivery system could be further developed to serve the RNA silencing approach. The HIV genome could be manipulated in various ways to remove motifs that are not required for silencing. Sequences could be introduced in a "tail-to-tail" configuration with a spacer between the two tails in order to form *in vivo* stem-loop or hairpin structures. It is assumed that the endogenous *Dicer* (or *Dicer*-like enzyme) of the mammalian cell will cut the hairpin into short ($\sim$22 nt) fragments. This system was developed for RNA silencing by two research teams. One publication on the use of lentivirus as Trojan Horse for RNA silencing was by Parijs and associates (most probably no relation to Paris, King of Troy, whose town was infiltrated by the mythological Trojan Horse). The approach by this team (Rubinson *et al.*, 2003) was thus similar to that by Brummelkamp *et al.*

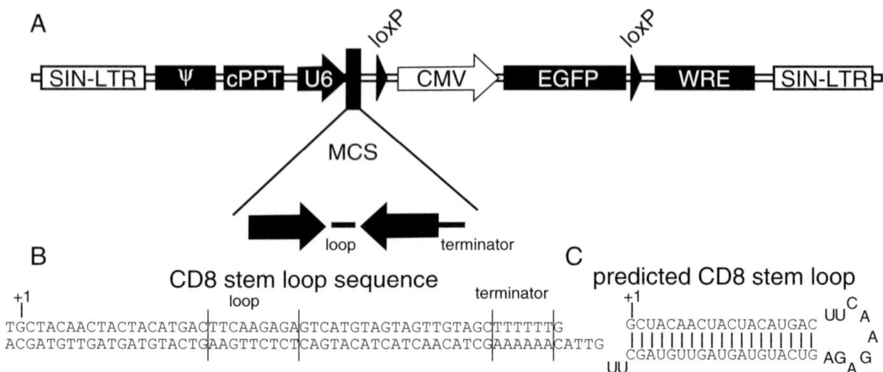

Fig. 24. Stable gene silencing and production of processed shRNAs in a T-cell line by a lentiviral vector. (A) Scheme of the lentivirus vector. Abbreviations: SIN-LTR, self-inactivating long terminal repeat; ψ HIV packaging signal; cPPT, central polypurine track; MCS, multiple cloning site; CMV, cytomegalovirus promoter; and WRE, woodchuck hepatitis virus response element. (B) Sequence of the CD8 stem loop used in the vector. (C) Predicted CD8 stem loop. (From Rubinson *et al.*, 2003.)

(2002) that was mentioned in the previous section, in which a specially constructed vector (pSUPER) was used. Rubinson *et al.* (2003) intended to integrate the (rather elaborate) transgene into the mammalian genome. For that the lentiviral system was rather useful. The construct strategy is provided in Fig. 24. There is a U6 promoter followed by a multiple cloning site (MCS); further downstream there is a code for a reporter (GFP) driven by a cytomegalovirus (CMV) promoter. As mentioned above there are advantages of the lentiviruses. They will also infect non-cycling and post mitotic cells. Also, the transgenes expressed from them are not silenced during development and can be used to generate transgenic animals by infection of embryonic stem cells or embryos. The expression of GFP can be monitored in cells by flow cytometry. The vector constructed by Rubinson *et al.* (2003), termed *pLL3.7* is very versatile. Various stem-loop forming sequences of about 50 nt can be plugged into the MCS. After the investigators had found that pLL3.7 is indeed operating in mammalian cells they added a hairpin-forming sequence, from the coding sequence, into the MCS and obtained the pLL3.7 CD8α vector. The vector was then used with specific CD8α-producing T cells to test whether the expression of CD8α would be silenced. This resulted in 68–82 per cent infectivity and the infected cells had about 93 per cent lower expression of CD8α. The suppression of CD8α was specific. The investigators then turned to dendritic cells and by infection with the pLL3.7 *Bim* vector silenced the expression of the pro-apoptotic Bim molecule. Similar experiments were performed with T cells that were infected with a pLL3.7 CD25 vector and again the infected cells reacted as expected: a marked reduction of the IL-2 receptor α-chain. T cells devoid of CD25 do not proliferate *in vitro* so that this infection caused 75–80 per cent reduction in proliferation after induction with IL-2. In a subsequent experiment the investigators purified hematopoietic stem cells (HSCs) from whole bone marrow cells and infected them with pLL3.7 CD8α. About 30–60 per cent of these cells were infected (evaluated by GFP). Only the infected cells (after being sorted by flow cytometry) were injected into lethally irradiated mice. After 8 weeks the injected HSCs cells contributed to all blood cell lineages in the reconstituted mice. But only about 20–40 per cent of HSC-derived lymphocytes were GFP-positive. When the initial infection was with pLL3.7 CD8α there was a reduction of 90 per cent in the frequency of CD8α T cells (compared to the initial infection of HSCs with base pLL3.7). All these effects were specific and did not affect other surface components.

The investigators then turned to mice embryos. They infected single-cell embryos with pLL3.7 that contained a variety of stem-loops resulting sequences such as those with homology to the codes of CD8α and of p53 tumor suppressor proteins. The suppresson of the respective genes in mature animals was apparent in all tested cases but the levels of the suppression varied. It should be noted that the numbers of lentivirus integrants in the cells that showed gene suppression varied (between 2 and 6 integrants per cell).

Between the submission and the publication of the Rubinson *et al.*'s (2003) paper, the team of Inder Verma and associates (Tiscornia *et al.*, 2003) submitted theirs on gene-knockdown in mice by the lentiviral vector system that was previously described by them (e.g. Pfeifer *et al.*, 2002). The Verma team (Tiscornia *et al.*, 2003) also carried the silencing system into mice pups from the F1 progeny of silenced mice.

The strategy of Tiscornia *et al.* (2003) was similar to that of Rubinson *et al.* (2003). The former investigators also synthesized a short cDNA sequence in which the silencing sequence was put tail-to-tail with a nine nucleotide spacer, in a way that the transcript will fold into a 19 or 20 nt stem and a loop of nine nucleotide. Their promoter was the human H1-RNA promoter. The recombinant lentiviruses were produced by transient transfection in 293T cells. In some of their experiments these investigators used transgenic mice that expressed GFP. These mice could be used in experiments in which the expression of GFP should be silenced by the respective stem-loop forming transgene. Indeed, when eggs from the latter mice were infected with the respective lentivirus vector the fluorescence of the blastocysts was reduced and this reduction was maintained even in the pups that were derived from the F1 generation. The investigators therefore suggested that the system developed by them can be used successfully to generate a large number of mice in which the expression of a specific gene(s) can be down-regulated substantially. The authors believed that their approach of generating "knockdown" mice will aid in functional genomics. So here we had *aid* by *Aids* and we also have three levels of silencing: "knockout", "knockdown" and "down-regulation".

To overcome the non-specific gene silencing by long dsRNA investigators attempted several approaches. One of these was to cut *in vitro* the long RNA into fragments of about 21 nt and then introduce these fragments into the mammalian cells. This approach has a drawback that could also be an advantage.... When a long transcript (usually the mRNA) is used to produce the respective dsRNA one cannot know, before testing, which region of the long dsRNA is the most efficient in gene silencing. It may depend on the "topographic" location, meaning the distance relative to the 5′ or the 3′ end of the transcript. It can also depend on the first few nucleotides of the cut dsRNA. When the long dsRNA is fragmented into many short dsRNA, at least some of these fragments will be effective silencers. Thus, by introducing a bulk of diced dsRNA there is a good chance of silencing, but which of the fragments actually perform the silencing will remain unknown.

A team of Michael Bishop (who received the 1989 Nobel Prize with Harold Vermus) and associates from the Department of Medicine of the University of California, at San Francisco (Yang *et al.*, 2002), decided to try to dice the long dsRNAs *in vitro* and thereafter use the resulting short dsRNA for silencing of genes in cultured mammalian cells. For the dicing of long dsRNA they used the *Escherichia coli* RNase III. Extensive cleavage of long dsRNA by the *E. coli* enzyme cut the dsRNA into small dsRNAs of about 2–15 nt. These short dsRNAs were not effective in silencing. But limited RNase III digestion of dsRNA provided the right size of small dsRNA. Thus, the long dsRNA representing the transcripts of the *Renilla* or the firefly luciferases were digested by a purified *E. coli* RNase III. A few minutes of digestion yielded the short dsRNAs. Similarly, the long dsRNAs of other genes were digested into the proper size of short dsRNAs. The different sizes could be separated on a polyacrylamide gel. The efficiency of the 21–23 nt dsRNAs was first tested with *Drosophila* cells and after these dsRNAs were found effective they were also used to silence transgenes (for luciferase) in mammalian cells (HeLa and C33A cells). Again, the size range of 21–23 nt dsRNAs were effective in silencing the transgenes. The tests were repeated and verified in several other mammalian cell lines. The silencing was effective and specific. The LCa is a mammalian protein that is expressed in HeLa cells. Its half life is about 1 day. The short dsRNAs derived from *in vitro* RNase III digestion of the long dsRNA representing the LCa mRNA were applied to HeLa cells. Again, these short dsRNAs caused specific silencing.

A team from the Stanford University School of Medicine (Myers *et al.*, 2003) looked for a similar way to produce specific and short dsRNAs that will not require the laborious *in vitro* synthesis. Like the other teams of investigators before them (Provost *et al.*, 2002; Zhang *et al.*, 2002), Myers *et al.* (2003) purified a recombinant human *Dicer* (r-*Dicer*) but the latter investigators went further to test the applicability of this r-*Dicer* for cutting long dsRNA into siRNA that will be active in specific silencing of mammalian genes. These investigators found that low levels of r-*Dicer* (of about 0.5 pmole) will cut *in vitro* considerable amounts of 500 nt dsRNA, within 24 hours, yielding 50 pmole of siRNA that is active in gene silencing. This human r-*Dicer* had a molecular mass of 225 kDa and required Mg^{2+}, but not ATP, for its dicing activity. The products of the 500 nt dsRNAs were all of the 20–21 nt size.

The investigators first tested the effect of r-*Dicer*-derived siRNA for reporter genes (two luciferases) that were co-transfected with the reporter genes into HEK293 cells (Human Embryonic kidney cells, also termed 293T). The reporter genes were specifically silenced. But high levels of r-*Dicers*-derived dsRNAs did cause non-specific silencing. The investigators then turned to two endogenous mammalian genes. First, they focused on a gene encoding cyclin E1 (the regulatory subunit of the CdK2-cyclin E complex) and then the gene encoding Cdc25C (i.e. one of three closely related phosphatases that dephosphorylate and activate cyclin-dependent kinases). The cyclin E1 d-siRNA (obtained by cutting the respective dsRNA by r-*Dicer*) caused a significant reduction of cyclin E1 levels in transfected HEK 293 cells. A similar effect was revealed by transfecting such cells with d-siRNA of Cdc25C. The Cdc25C was reduced by 90 per cent without affecting another protein of this family (Cdc25A). The authors found that the d-siRNA is as efficient in silencing genes as synthetic dsRNA and is similar to the latter with respect to the absence of non-specific toxicity, while the former can be obtained by simpler procedures.

Shinagawa and Ishii (2003) of Tsukuba, Japan, developed a different procedure to avoid the non-specific silencing that is commonly induced by long dsRNA in mammalian cells. Their method was based on the knowledge that the triggering of the interferon-response takes place when the dsRNA is in the cytoplasm, but long dsRNA in the nucleus does not induce this response. For that Shinagawa and Ishii developed the pDECAP (**D**eletion of **C**ap and poly-A) plasmid. In this plasmid mRNA can be transcribed from a cytomegalovirus (CMV) promoter but no 7-methylguanosine (m^7G) cap structure nor poly(A) tail (at the 3′ end) are produced. Certain inclusions in the plasmid could prevent the formation of the cap and the poly(A). Thus, the cis-acting hammerhead ribozyme will prevent the formation of the cap. Without the cap and the poly(A) the double-stranded RNA transcribed from pDECAP will stay in the nucleus. Since there is also *Dicer* activity in mammalian nuclei the long dsRNA will be diced while still in the nucleus. As in many other studies luciferase genes served in the experiments of the investigators from Tsukuba. The target cells were mouse embryonic fibroblast cells. Such cells were co-transfected with a plasmid that could express both the *Photinus* (firefly) and the *Renilla* (sea pansy) luciferases and the pDECAP plasmid for long dsRNA of the firefly luciferase. This caused a strong silencing of the firefly luciferase, while the expression of the sea pansy luciferase was not affected. The investrigators then turned to an endogenous gene, *Ski.* The protein *Ski* forms in mammalian cells a complex with histone deacetylases and thus can cause transcriptional repressions of several specific genes. They constructed a pDECAP-*Ski* and transfected it to 293T cells together with a *Ski*-expression vector. The pDECAP-*Ski* reduced the levels of *Ski* and of mRNA for *Ski* in the co-transfected cells. The DNA fragment containing the expression unit of the 540 nt *Ski* dsRNA from the pDECAP-*Ski* vector were injected into fertilized oocytes. This led to transgenic mice with 2–12 copies of the transgene in the genome. Small dsRNAs of 21–22 nt representing the *Ski* gene were then detected in the mouse embryos. No such dsRNAs were found in normal mouse embryos. The embryos with the pDECAP-*Ski* had various neural abnormalities that indicated silencing of the embryos *Ski* gene. Such defects could be traced in further sexual generation of the mice (the defects did not affect all the embryos). The investigators provided evidence that the defects in the pDECAP-*Ski* containing mice were caused by the reduction of *Ski* mRNA. The investigators also suggested that by including in the pDECAP promoters that will be active only in specific tissues, their procedure could be further refined.

In a collaborative study by investigators from the NIH in Bethesda, Maryland, and the Case Western Reserve University in Cleveland, Ohio, the same basic idea as that of the aforementioned Japanese study was put to test. The former team (Wang *et al.*, 2003) constructed another plasmid, termed pBI-Tet-On that will transcribe the sense and the antisense RNA, annealing *in vivo* to dsRNA. The transcription will happen only by triggering a promoter by tetracycline (or doxycycline). By a specially constructed pBI-Grx the investigators could specifically lower Grx at a desired time: NIH 373 cells were trasnfected with pBI-Grx and at a later specific time, induced to silencing by doxycycline. As Grx catalyses deglutathionation of actin, the silencing of Grx at specific time could be followed by actin polymerization and translocation.

I shall mention two additional methods to facilitate gene silencing in mammals by dsRNA. One method concerns improvement in introducing dsRNA into cells and the other method utilizes a synthetic vertebrate transposon.

Matsuda and Cepko (2004) of the Harvard Medical School were interested to promote their study on retinal development in mammals. They claimed that previous methods to introduce constructs that will interfere with retinal differentiation, as the use of viral vectors had several drawbacks and therefore they developed a method based on electroporation. Note that while dealing with gene silencing by dsRNA in chicken embryos, information was provided on such an electroporation method in fertilized eggs of chicken (e.g. Pekarik *et al.*, 2003). The Harvard investigators used either of two procedures. In one procedure the plasmid DNA was injected into the subretinal space of neonatal pups of mice or rats and then electroproation was initiated by tweezer-type electrodes. The other method was to isolate explants from these animals and to perform the addition of plasmid DNA as well as the electroporation in a micro-chamber (Fig. 25). The plasmids were applied to newborn rodents by a syringe near the lens (3–6 µg/µl). Also, a fast-green tracer was added. Five square pulses of 50 microseconds were applied (with 950 microsecond intervals). Newborn mice and rats received, respectively, 100 and 80 V pulses. In the microchamber five square pulses were also applied but of 30 V. One plasmid was for the expression of GFP. Almost all newborn rodents survived the electroporation and expressed the GFP in a wide area of the retina. During further development the GFP was expressed in specific components as rod PR, bipolar cells and Müller glial cells. The expression of GFP diminished after 3–4 weeks. The investigators could use the GFP expression to select useful promoters. The pattern of GFP expression driven by various tissue-specifying promoters provided important information on retinal differentiation in newborn rodents. In the *in vitro* electroporation the investigators studied gene silencing. They found that the plasmid should be introduced from the scleral side rather than from the vitreal side. The injection/electroporation of P0 rat retinae was performed with a plasmid that should express the *Rax* gene (this gene is coding for a homeobox transcription factor). Two retinal transcription factors Crx and Nrl could also be expressed by electroporating the respective genes.

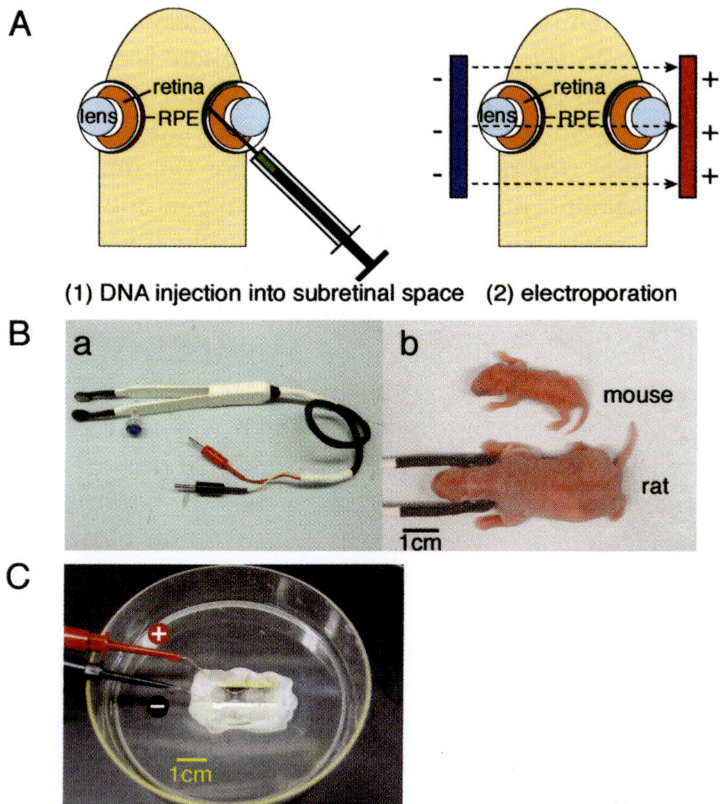

Fig. 25. Procedure and apparatus for *in vivo* and *in vitro* electroporation in rodent retina. (A) Schematic illustration showing the *in vivo* electroporation method. (B) The electrodes used: tweezer-type electrodes (a) are placed to hold the head of newborn (PO) rat or mouse (b). (C) A micro chamber for *in vitro* electroporation. (From the supplement information of Matsuda and Cepko, 2004.)

An interesting approach to introduce dsRNA into mammalian cells is introduced by another Trojan Horse: the Sleeping Beauty (SB) transposable element. There is a bizarre association between SB and the Trojan Horse because originally the SB was developed as an active transposon from the inactive ("skeleton") transposon of a zebrafish (Tdr 1). This synthetic TE is active in fish, rodents and man (see: more information in Galun, 2003 and recent review by Izsvak and Ivics, 2004). The SB-mediated gene silencing by the respective hairpin structure was suggested by a group of investigators (Heggestad *et al.*,

2004) from the University of Florida in Gainesville. They termed their procedure *Maleficent*. In this procedure there is a tail-to-tail coding sequence with a spacer between the two tails. This construct is inserted into the SB element with the appropriate promoters and a selectable marker (for resistance to G418). For the integration of the modified SB (Maleficent) into the mammalian genome a plasmid that produces transposase is required. For testing which hairpin RNA will be the most effective in silencing the GFP gene they first introduced the coding for the expression of GFP into HeLa cells by an appropriate pMaleficent. A cell line that expressed GFP was thus established and then transfected with an SB plasmid that had the code for the hairpin for silencing the GFP, which was thus achieved. The same approach was used to silence the (endogenous) gene for lamin A/C. Silencing by an SB that contained the anti-lamin hairpin code was indeed achieved. Western blot hybridization clearly indicated that both lamin A and lamin C were eliminated by the anti-lamin pMaleficent expressing cells.

For efficient and specific silencing of mammalian genes one should possess a variety of relevant information. Take, for example, an answer to the question as to which region of a gene should be represented in the short dsRNA to cause maximal silencing should be given. It is also important to know whether there are in the genome other coding sequences with identical 19–21 nt as in the chosen target for silencing. Then it is important to know if indeed only a perfect homology between the target gene and the dsRNA will cause silencing or that one or more mismatches will still cause silencing. Also, the minimal quantity of dsRNA that is required for silencing should be known. Overdoses may cause non-specific silencing. A team from the Abbott Laboratories, Abbot Park, Illinois, conducted studies to answer the above and other questions (Semizarov *et al.*, 2003).

Rigorous analyses of the best regions of a gene that should provide the nt sequences for short dsRNAs in gene silencing were performed by Amarzguioui and Prydz (2004) of the University of Oslo, Norway. These investigators tested four target genes: (1) human tissue factor (*hTF*); (2) murine tissue factor (*mTF*); (3) human protein kinase H1 (*PSK*); and (4) human *c-src* tyrosine kinase (*CSK*). They then synthesized dsRNAs that had 19 duplexes and two single hangovers. For each of the four genes they tested which of the various dsRNAs (that had homology to different regions in the target genes) was the most effective in gene silencing. Silencing was defined in at least 70 per cent reduction of gene-expression. The authors also took into account data that were accumulated by other studies and worked out an algorithm that was highly successful in distinguishing between functional and non-functional siRNAs. They added in their calculations the sites of specific base pairs (as a weak pairing A to U, designated W, and a strong pairing G to C, designated S). Then specific bases/location had a strong correction with functionality. Take, for example, the absence of a U in position 1 (out of the 19 dinucleotides) in the sense strand was strongly correlated with functionality. U at position 10 was correlated with lack of functionality while U at position 13 was positive for functionality. Also, S (C/G) at position 1 and W (A/U) at position 19 were of positive functionality. In their conclusion the authors provided several recommendations on how to design the most effective dsRNA for silencing. My recommendation is to consider these "rules" but not to accept them as a rigorous prescription. Somewhat different recommendations were the result of another study by a team of eight investigators from Japan (Ui-Tei *et al.*, 2004). For best dsRNA to silence mammalian genes they recommend; (I) A/U at the 5′ end of the antisense strand; (II) G/C at the 5′ end of the sense strand; (III) at least five A/U in the 5′ terminal one-third of the antisense strand; (IV) absence of any GC stretch of more than 9 nt in length.

After intense investigations efficient methods were devised to stably silence specific mammalian genes. A method of choice was to introduce into the genome, a sequence that, after transcription, will form a hairpin structure. This structure, the pre-miRNA, will be processed *in vitro* to the mature miRNA and silence a specific target transcript. But there is a problem with this method: constitutive silencing can be lethal and is also less informative than a silencing that can be regulated by the investigator. The team of Gregaroy Hannon and Vivek Mittal (Gupta *et al.*, 2004) of the Cold Spring Harbor Laboratory (New York State) found a solution. They ligated the coding sequence for the hairpin structure downstream of an inducible promoter. This promoter could be induced by the application of ecdysone. Thus, silencing will be activated by ecdysone and with the removal of ecdysone the silencing will be reversed to a normal activity of the target gene. This stable but inducible and reversible silencing was applicable in human and murine cells.

About 6 years after the study by Fire *et al.* (1998) a team of seven investigators from Sweden (Xu *et al.*, 2004) returned to the question of whether dsRNA or the antisense of the ssRNA is more effective in gene silencing. But this time the question was asked about mammalian cells. The bottom line of this investigation is that in both mammalian cells and nematodes the dsRNA is by far more efficient in gene silencing. Moreover, it is not because ssRNA is less stable in the cells than dsRNA: when ssRNAs (sense and antisense) are introduced to the cells sequentially, silencing takes place only after both types of ssRNAs are in the cells.

RNA Silencing and Genes Involved in Mammalian Diseases

The silencing of specific mammalian genes is a potential means for the therapy of certain diseases. The applicability of this approach for the treatment of human diseases will be discussed in the appendix on gene therapy. In this section I shall provide a few examples on studies that were intended to explore the feasibility of silencing mammalian genes that are involved in diseases.

The tumor suppressing gene *Trp53* antagonizes *Myc*-induced lymphomagenesis. In null mutants of *Trp53* the Myc induces a highly disseminated disease. A team from the Cold Spring Harbor Laboratory and Memorial Sloan Kettering, New York (Hemann *et al.*, 2003), investigated the impact of silencing the *Trp53* gene on the development of lymphoma in mice. They constructed several hairpin-producing structures with homologies to the *Trp53* gene. All these structures differed by three nucleotides or more from any other gene in the mouse genome. They found that out of three such constructs, one p53-C was the most potent in silencing the *Trp53* gene. In transgenic mice containing the Eμ-Myc lymphomas harboring *Trp53* deletion, the disease arises much earlier and displays a characteristic disseminated pathology. Moreover, the hematopoietic stem cells from Eμ-Myc transgenic mice also cause lymphomas upon adoptive transfer into normal recipient mice The latter lymphomas are greatly accelerated by the loss of *Trp53*. The question asked by Hemann *et al.* (2003) was therefore what would the phenotype of mice be in which Eμ-Myc hematopoietic stem cells were silenced with respect to p53. To answer this question hematopoietic stem cells were used to introduce (by a retroviral vector) the hairpin-producing Trp 53 constructs into these stem cells. Mice were then irradiated with a dose that killed their own hematopoietic stem cells and those later stem cells were replaced with the Eμ-Myc that contained the hairpin-forming construct (shRNA for p53). The mice with the replaced hematopoietic stem cells had palpable lymph nodes 3–5 weeks after stem cell replacement. Thereafter, there were some differences among the mice according to the potency of the shRNA to silence p53. The shRNA with the strongest silencing (p53-C) caused the development of B-cell lymphomas in all mice and the mice reached a terminal stage after about 70 days. This clearly indicated that p53-C strongly accelerated the tumor formation *in vivo*. The severity of the disease was correlated with the level of p53 silencing that was induced by the different hairpin-producing constructs.

Several RNA silencing studies were related to liver diseases. Fas (CD95) mediates cell-death (apoptosis) in liver cells (hepatocytes). The *capsase 8* encodes an enzyme that is involved in the activation of Fas. Therefore, a team of 14 investigators from the School of Medicine at Hannover, Germany (Zender *et al.*, 2003), intended to explore the use of RNA silencing in mice models to silence the expression of *capsase 8* and thus to reduce liver-cell death. The mice model was intended to represent the very serious cases of human acute liver failure (ALF) that has a very high mortality. The details of the various kinds of ALF is beyond the scope of this book. The Hannover team therefore looked for a novel way to negate ALF. In ALF the cell membrane of hapatocytes releases signals that trigger the suicide pathway, leading to the activation of caspase cascades that ultimately cause apoptotic death of the hepatocytes. The approach by Zender *et al.* (2003) was therefore such that therapeutic treatment of ALF could be achieved by the inhibition of death-receptor mediated apoptosis. The investigators first used human HepG2 hepatoma cells. To these cells they introduced siRNAs for *caspase 8* and subsequently treated the cells with AdFasL (adenovirus Fas ligand, known to induce ALF). The siRNAs clearly inhibited *caspase 8* activity and prevented effectively, FasL-mediated apoptosis. To shift the investigation into living mice the investigators first tested the procedure of injection of siRNA into the tail vein of mice. Indeed, by this procedure it was found that the expression of the *lacZ* gene in the liver cells of appropriate transgenic mice could be silencing by siRNA targeted to this gene. Once the applicability of the tail-injection procedure was assured the investigators injected siRNA targeted to the *8 caspase* gene. This caused a reduction of the mRNA for *capsase 8* in the liver of injected mice. The injection of siRNA *caspase 8* also protected the mice during ongoing acute liver failure. The same or even better results were obtained when a more local injection (into the portal vein) was performed. These investigations thus show promise but the application to treatment of human ALF is still in the realm of hope.

Another team of investigators (Song *et al.*, 2003) that included collaborators from the Harvard Medical School and from medical institutions in Guangzhou, China, also handled hepatitis and attempted to protect liver cells by the use of siRNA silencing. The target of these investigators was the *Fas* gene that encodes the *fas* receptor. They used two models of autoimmune hepatitis in mice for their experiments. Fas is highly expressed by hepatocytes and the latter cells are very susceptible to Fas-mediated apoptosis. This apoptosis can result from various insults such as viral diseases, autoimmunity and transplant rejection. Thus, Fas-deficient, *lpr*, mice survive challenge with factors that induce fulminant hepatitis in normal mice. The rationale of the investigators was that by *in vivo* suppression of Fas through specific RNA silencing, one can protect the liver of mice from fulminant hepatitis. This is a similar rationale to that of Zender *et al.* (2003) although the latter investigators intended to silence the *caspase 8* gene rather than the *Fas* gene. The Song *et al.* (2003) team first verified the efficiency of introducing dsRNA into hepatocytes by the "hydrodynamic tail vein injection" of labeled dsRNA (siRNA). They found that when 50 μg were injected almost 90 per cent of the hepatocytes had taken up the siRNA. Several dsRNAs that were homologous to different regions of the *Fas* gene were synthesized. There were great differences with respect to the fas-silencing capability of the various siRNAs but some of them almost completely eliminated the Fas expression. This was a specific effect because other non-homologous siRNAs had no effect on

Fas. Also, by the Fas-siRNA only Fas itself but not *Fas*-related genes (FasL, FADD, FAF, TNF etc.) were suppressed. The effect of injecting Fas-siRNA was relatively long-lasting. The reduction of Fas was maintained for 10 days after the last injection, a relatively long period. But the effect is also not permanent. Fas-mediated fulminant hepatitis is known to be caused by two mechanisms. In one mechanism there is ligation of the Fas receptor on hepatocytes that induces massive apoptosis. This is accompanied by an infiltration of inflammatory cells and secondary necrosis. The other mechanism also provokes hepatic inflammation by inducing the expression of hepatic chemokines that recruit and activate immune cells, leading to hepatocytes death.

To test the efficiency of fas-siRNA, hepatocytes from mice that received the Fas-siRNA injections were challenged *in vitro* with an agonistic Fas-specific antibody (Jo2). The hepatocytes from Fas-siRNA treated mice showed only a 8 per cent death by the Jo2 while control hepatocytes had a 88 per cent mortality. There was a good positive correlation between the capability of a specific Fas-siRNA to silence Fas and its ability to protect hepatocytes from apoptosis. Another approach to test the role of Fas-siRNA was to use hepatic mononuclear cells isolated from Con A treated mice. Hepatocytes from Fas-siRNA treated mice were not lysed by these hepatic mononuclear cells. *In vivo* experiments also showed that after injection with Fas-siRNA the mice were resistant to the subsequent injection of Con A. Likewise, the injection of Fas-siRNA also protected mice from two noxious chemicals. In all these cases, no serious "side-effects" were revealed. It thus appears that the tail-vein injection of Fas-siRNA has the potential therapeutic effect in negating acute and chronic liver injury by viral and autoimmune hepatitis. The tail thus plays a positive role in such therapeutic treatment. Should man regret the absence of a tail? Probably not, hepatologists will surely find another "direct" route to live cells.

The hepatitis delta virus (HDV) is a human virus that is found only in patients that are also infected with hepatitis B virus (HBV). One may define the HDV as a "double parasite". It parasitizes humans and uses human enzymes for the replication of its genome. It also utilizes the HBV because the HDV genomes are assembled by using the envelope proteins of HBV. The HDV has an RNA genome that is single-stranded, circular and contains 1679 nt. There are actually two additional HDV RNA species: the exact complement of the genome (antigenome) and small amounts of an 800 nt polyadenylated RNA, translated to a 195 amino acid protein (small delta antigen-δAg-S). The circular RNA genome of HDV can "fold" into a rod-like double-stranded RNA with 74 per cent of intramolecular base-pairing (Taylor, 2003). Such a long (about 800 nt) rod of dsRNA could be a substrate of *Dicer* but Chang *et al.* (2003) found that it is not fragmented by *Dicer*. The reason is probably that a minimal length of 100 per cent base pairing is required for *Dicer* cleavage. Such a minimal level probably does not exist in the genome of HDV. Its rod-like RNA has "bumps" caused by non-pairing bases after every few nucleotides. Did these evolve to protect the HDV genome? Whatever, in cells infected with HDV there are no short ($\sim$21 nt) dsRNA species that represent fragments of the HDV genome. Also, the apparent dsRNA of the HDV does not induce the interferon defence mechanism.

The Human Immunodeficiency Virus Type I (HIV-1) is a rather "clever" virus and up to now has resisted its eradication by human efforts. A research team (Boden *et al.*, 2003) attempted to silence HIV-1 by dsRNA. This team focused on the transactivator protein gene *tat*. The investigators chose a sequence of 21 nt of the coding of *tat* and constructed the respective dsRNA to match this 21 nt sequence. They also used a AAV DNA vector that also contained a selectable gene (neomycin resistance) and the sense and antisense 21 nt of the *tat* region in a tail-to-tail configuration with 6 nt between the two tails. This tail-to-tail configuration was driven by appropriate promoters so that the respective transcript should fold *in vivo* into a hairpin structure. The idea was that this hairpin structure will form, after being cleaved in the cells, the required *tat* dsRNA. The whole construct was flanked by long terminal repeats to cause the integration of the construct into the cells genome. Because of the selectable marker cells that integrated this construct (and expressed the desired dsRNA) could be selected. Transient reduction of HIV by co-transfecting HIV and synthetic *tat*-dsRNA showed promise of reducing viral replication in 293T cells. The investigators then turned to prolonged suppression of HIV. For that a cell line (H9) from a human cutaneous T-cell lymphoma was transfected with the vector that should express the desired dsRNA (*tat* shRNA). Once established (with due selection) these cells were infected with HIV. Everything went well for about 3 weeks: the viral replication was strongly reduced. But on day 25 the viral replication reached already 25 per cent of infection without *tat*-shRNA and on day 35 the protection by the dsRNA was lost completely. It was found that during these 35 days one nucleotide at base number 9 (out of the 21 nt) was mutated from A to T. Remembering that the viral RNA is replicated by a reverse transcriptase and that the fidelity of this transcriptase is low, such mutations are not surprising. The investigators had suggestions on how to improve the silencing of HIV in the future, such as the use of better designed shRNA that are targeted to highly conserved region of the viral genome (as *gag* and *pol*) or to co-express several dsRNAs against various coding sequences of the viral genome. At the onset of the HIV suppression attempts I have indicated that this is a "clever" virus. By its mutations it evades natural and medical suppressions of its replication. It "mutated" the Latin maxim: *"si vic pacem para bellum"* (the one who wants peace should prepare to war) to "the one who wants victory should prepare to war". This is not a major change of the maxim because the Roman Peace (Pax Romana) actually meant victory to the Romans and defeat to its foes.

Genome Wide Approaches for Gene Silencing in Mammals by Small Interfering RNAs

When the silencing of specific mammalian genes became a reliable procedure, several groups of investigators looked for a genome wide approach rather than dealing with silencing of specific individual genes.

One such group of investigators was a team of Patrick Brown and associates (Chi *et al.*, 2003) of the Stanford School of Medicine in California. They intended to evaluate the changes in the expression of many genes ("genome wide") rather than focusing on one target gene. They also re-evaluated the possibility of "transitive" RNAi activity in human cells. Such 'transitive" silencing was mentioned above in this book because it was revealed in nematodes and in plants. In this RNAi activity a short dsRNA can silence not only a region of the gene that is homologous to the dsRNA but also regions that are upstream of the homologous sequence. With respect to "transitive" gene silencing in human cells the study by Chi *et al.* provided a simple answer: this silencing does not exist in humans. As for the effect of a given dsRNA on genes that were not targets of this dsRNA, the strategy of the investigators was to use human cDNA microarrays. These corresponded to about 36 000 human genes. The mRNA of control cells or of cells expressing GFP was challenged by dsRNA against the GFP genes. This approach was used to evaluate the genome-wide impact of specific silencing dsRNAs. The procedure was rather elaborate in order to provide all the required controls but the bottom-line of the results indicated that a specific dsRNA did not affect significantly any other human gene.

A more recent study (Scacheri *et al.*, 2004) by 11 investigators from the NIH in Bethesda, Harvard Medical School and Agilent Technologies, Andover, MA, provided a different conclusion from that of Chi *et al.* (2003). The approaches of the West Coast (Stanford) and the East Coast (MD & MA) differed considerably. The East Coast investigators choose one target gene, *MEN1* (multiple endocrine neoplasma type 1) and evaluated changes in its expression as well as such changes in two unrelated genes: *TP53* and CDKN1A. The latter two genes are considered functionally unrelated to *MEN1* but they are sensitive markers of the mammalian cell state. For silencing *MEN1*, 10 different dsRNAs were constructed by annealing the respective synthesized ssRNAs that matched different regions of the target gene. None of the 10 siRNAs had exact matches with other genes in the human genome. After these 10 *MEN1* dsRNAs were transfected into HeLa cells the expressions of *TP53*, *CDKN1A* and the gene for actin were evaluated. Significant up-regulations or down-regulations of the genes *TP53* or *CDKN1A* were revealed by the transfection of several of the 10 dsRNAs for *MEN1*. Moreover, certain of the latter dsRNAs that did not change the expression of *MEN1* did change significantly the expressions of *TP53 or* CDKN1A. Interestingly, a change of a single nucleotide in dsRNA could cause a drastic difference in silencing. But none of these 10 dsRNAs affected the level of actin. Even reducing the level of the dsRNA did not abolish the non-specific changes in the levels of the non-target gene-expressions. Notably the data for changes in mRNA levels did not always match the changes in protein levels. The sequences of the dsRNAs for *MEN1* did not match coding sequences in either *TP53* or *CDKN1A* although a miRNA, partially matched could not be excluded. There were some matchings in the siRNAs for *MEN1* with the 3′ UTR of the transcripts of the two former genes. Why these non-target genes were silenced (but the actin gene was not affected) was not unequivocally explained but the authors recommended a careful validation of downstream effects mediated by specific siRNAs. They hoped that in the future information will provide knowledge that is required to construct dsRNA that will have no off-target effects. Thereafter, the way will be open for a large-scale use of RNAi in functional genomic studies and in gene therapy.

A short publication of an extensive team from Rosetta Inpharmatics (Merck) in Kirkland, Washington (Jackson *et al.*, 2003), reported on an experimental approach that was similar to the approach by Chi *et al.* (2003). The Kirkland investigators also used transcript profiles to reveal siRNA-specific changes in target and non-target silencing. The conclusion of this team was that silencing of non-target genes can happen whenever there are 11 (or more) contiguous nucleotides in the dsRNA and the non-target genes.

Mousses *et al.* (2003), a team of 12 investigators from the USA, Brazil and Finland, developed a model for RNAi microarray analysis in cultured mammalian cells. This method should provide information on the response of individual cells that were plated in monolayers. The model was developed for transformed HeLa cells that were expressing the transgene for GFP. Briefly, the method consists of putting *spots* of dsRNAs that had sequences that are homologous to the transcript of the GFP gene on a glass slide. The slides were then overlayered with a monolayer of adherent (HeLa) cells and incubated, and the dsRNAs were allowed to transfect the overlayered cells. The cells that were transfected by the dsRNA (GFP-siRNA) could be recognized by due labeling. The reduction of GFP expression was evaluated by the reduced green fluorescence and the cells could be visualized by DAP1 staining of the nuclei. By appropriate techniques the authors could show that arrays of siRNAs on the surface of glass slides did enter the overlayer of cells via transfection and the siRNAs mediated potent inhibition of the target-protein expression in a spatially confined manner. The procedure required quantitative image analysis with single-cell resolution. To enable the resolution for thousands of array element the authors developed an automated high-resolution microscope-based digital image acquisition system for RNAi micro-arrays as well as custom algorithms for quantitative image analysis. Also, three images for three different fluorescent channels were required from one area within each microarray element/spot: the blue (DAPI) for cells/nuclei; the green for GFP expression; the red for evidence of transfection. The whole procedure required considerable investment in instrumentation and evaluation but the procedure worked for the model and can probably be developed for other transgenes and/or endogenous genes with due amendments in the detailed procedure.

Peter Schulz and collaborators of the Scripps Research Institute in La Jolla and the Genomics Institute of the Novartis Research foundation, San Diego (Zheng *et al.*, 2004), developed efficient procedures for the construction of siRNAs and for the production of many different siRNAs by using special PCR procedures. Basically, they first verified the efficiency and specificity of silencing ectopic and endogenous genes. Thereafter, they applied the PCR-derived siRNAs library to target thousands of genes and finally focused on genes involved in the NF-κB signaling pathway.

In order to construct the dual promoter vector (pDual) for the expression (*in vivo*) of siRNAs, the investigators used the strategy summarized in Fig. 26A. In this plasmid the two siRNAs that will anneal in the cells are transcribed by the mouse U6 and the human H1 promoters, respectively. These are pol III promoters. Special care was used in the construction such as the insertion of five Ts at the 3′ ends of the two apposing sequences to serve as termination signals.

To test for the efficiency and specificity of the pDual system for silencing a transgene, 293T cells were cotransfected with plasmids that express two luciferases (one as reference for transfection efficiency and one for testing the silencing). Indeed, the pDual plasmid for siRNA of firefly luciferase silenced this gene (70–90 per cent reduction) at a similar level as the respective hairpin (pSUPER) expressing plasmid. The (reference) *Renilla* luciferase was not affected by the siRNA-pDual plasmid for firefly luciferase.

The single-step PCR strategy for producing siRNA expression cassettes, based on the above mentioned pDual system, is schematized in

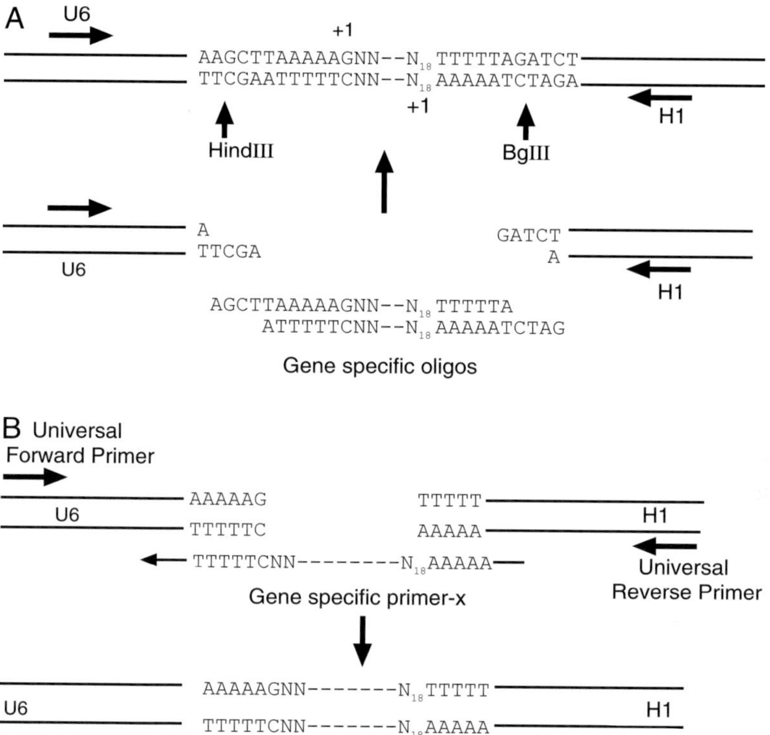

Fig. 26. The pDual siRNA expression system for genome-wide screening of expressed siRNAs in mammalian cells. (A) Strategy for generating siRNA by using two opposing PolIII promoters; the mouse U6 and the human H1 promoter sequences were cloned into pBluescript SK in opposite directions; appropriate mutations were made to define termination signals for siRNA transcription or facilitate inserting siRNA-encoding sequences; to create a gene-specific siRNA expression plasmid, a pair of complementary oligonucleotides (35–37 nt) were annealed and ligated into pDual digested with *Bgl* II and Hind III. (B) A single-step PCR strategy for producing siRNA expression cassettes based on the pDual system. (From Zheng *et al.*, 2004.)

Fig. 26B. Into these cassettes any of thousands of dsRNAs representing mammalian genes could be inserted.

As mentioned above these cassettes were then utilized to screen for genes involved in the NF-κB signaling pathway. Some information on the regulation of NF-κB was already known (see: Trompouki *et al.*, 2003) such as the role for the CYLD protein that negatively regulates the activity of the tumour necrosis factor receptors (TNFRs) and the latter are activators of NF-κB that is required for appropriate cellar homeostasis of the skin appendages. To search for additional genes involved in NF-κB expression Zheng *et al.* (2004) applied their "genome-wide" screening by siRNAs cassettes into which short dsRNAs from many genes were inserted. They transfected mammalian cells with a pNF-κB-Luc reporter plasmid. They co-transfected HEK293T (termed HEK293 or 293T by others) cells with the pDual-based cassettes that contained sequences from 8000 human genes. The target-sequences for each gene were selected according to previously published guidelines for effective siRNAs. About 12–14 hours after this co-transfection TNF-α was applied to activate the reporter plasmid, the investigators revealed 96 genes that differentially modulated (after 4 days) the reporter activity. Of these, 20 genes were tested further by targeting several regions in each gene. By that the investigators detected genes that were previously known to be involved in the NF-κB signaling pathway as well as additional "new" genes. Among the latter were the FAP48 encoding gene (FAP48 is a FK506 binding protein) and a gene encoding the SonA protein that represses the transcision of hepatitis B virus (HBV). In summary Zheng *et al.* (2004) designed and constructed a dual-promoter (pDual) system for the expression of siRNAs in human cells that can silence efficiently gene function. They also developed a "single-step" PCR protocol for the production of effective siRNA expression cassettes in a high-throughput fashion.

While Zheng *et al.* (2004) applied the strategy of siRNAs formation an extensive team of 17 investigators (Paddison *et al.*, 2004) had a similar approach for large-scale RNAi screening in mammals but the latter investigators used plasmids that will produce *in vivo* the respective hairpin-forming transcripts, thus producing an extensive shRNA expression-library. This library was for targeting 9610 human genes and 5563 mouse genes. At the time of publication (March 2004), this library consisted of 29 000 sequence-verified shRNA expression cassettes... and the number of cassettes was still growing. While such extensive "silencing libraries" existed for nematodes, this was the largest of its kind for mammals. Briefly, each of the finalized hairpin-designed constructs of this library consisted of a 27 nt U6 leader sequence followed by 29 base-pairs of dsRNA (with at least 19 bp that are congenial to the target in the silenced gene) and a 4 nt "loop". These shRNAs were expected to be processed by the mammalian cell's Drosha and *Dicer*, resulting in defined siRNAs. The shRNAs were designed to match several regions of 10 000 human and 5000 mouse genes each. Care was taken to ensure that these shRNAs should have more than three mismatches to any of the non-target genes in the human and mouse genomes, respectively. To avoid the use of constructs with mistakes each construct was sequence-verified. The present library covers already 85 per cent of known genes for kinases and phosphatases (with three or more shRNAs for each gene). To evaluate their silencing-library these authors focused on one complex of proteins: the 26S proteosome. They asked which shRNAs that represent specific genes will compromise the function of the proteosome. Defective proteosomes could be evaluated by a Zs-Green-MODC dragon fusion. When this "dragon-fusion" is applied, the fluorescence will be increased by defective proteosomes (but not by intact proteosomes). By the analyses of 6712 shRNAs targeting 4873 genes the authors revealed about 100 RNAi constructs that increased the fluorescence by the ZsGreen-MODC dragon fusion, indicating impairment of the proteosomes. Twenty-two of these RNAi constructs corresponded to 15 known proteosome subunits. Additional shRNAs that affected proteosomes were revealed by subsequent analyses.

The systematic generation of synthetic short siRNAs that will, after introduction into mammalian cells, produce hairpin RNAs (forming alien microRNAs) is plausible for targeting a few genes. But for large-scale silencing of many genes it is not practical. We should recall that several siRNAs (or hairpin RNAs) have to be tested for each mRNA in order to find the most effective one for silencing its target. As was summarized by Inder Verma and associates in their Commentary (Singer *et al.*, 2004), solutions for large-scale production of gene-silencing RNAs were proposed almost simultaneously by three research teams. These teams were located in very different places on the globe but all three proposed very similar methods to enzymatically synthesize libraries of silencing RNA via hairpin-forming cassettes from sources of double-stranded DNA. Such libraries should allow the selection of functional siRNAs for any target-gene, known or unknown, as long as it is represented in the mRNA source of the library. The three teams were as follows: Kenzo Hirose, Masumitsu Iino and associates (Shirane *et al.*, 2004) of the University of Tokyo, Japan; Helen Blau and associates (Sen *et al.*, 2004) of the Stanford University, California, and Harvey Lodish and associates (Luo *et al.*, 2004) of the MIT in Cambridge, MA. As I shall briefly describe below, the outlines of generating siRNAs from dsDNA are similar in the research of Shirane *et al.* (2004), Sen *et al.* (2004) and Luo *et al.* (2004). Each of these teams was unique in giving its procedure a different acronym. Shirane *et al.* (2004) termed their technology *enzymatic production of RNAi libraries*, hence: EPRIL. Sen *et al.* (2004) called their procedure *restriction enzyme-generated siRNA*, hence: REGS. Luo *et al.* (2004) named their procedure *interfering RNA production by enzymatic engineering of DNA*, hence: SPEED. The starting point of all the three teams was to synthesize dsDNA from a mRNA source. The derived dsDNA was then fragmented by either (well controlled) DNase I digestion (Luo *et al.*, 2004) or digested with a mixture of restriction endonucleases (Shirane *et al.*, 2004; Sen *et al.*, 2004). The fragments of DNA were ligated to a hairpin-linker that contained a site for *Mme*I that cleaves 18–20 nt away from its recognition site. By that a mixture of 20 bp fragments that were all attached to a hairpin linker was generated. Then a second linker was ligated to the *Mme*I-generated termini and a DNA polymerase converted the short-stranded hairpin DNA to linear dsDNAs. These dsDNAs were ligated to an inducible promoter. Further digestion with a unique restriction enzyme and self-ligation eliminated most of the first linker sequences, retaining a short sequence between the sense and the antisense sequences in the final siRNA transcripts that will fold *in vivo* to hairpin miRNAs. The three procedures differed with respect to the specific promoter used during the engineering (U6 or H1 promoters). The three teams used an integrating retrovirus for delivery into tissue-culture cell lines. Singer *et al.* (2004) recommended that the lentiviral delivery system that was detailed previously by the Verma Laboratory (Tiscornia *et al.*, 2003) be used instead of the retrovirus.

The three teams of investigators actually tested their procedure. Shirane *et al.* (2004) obtained the expected results when a cDNA library from mRNA expressed in mouse-myeloid-precursor cells was the source of their siRNA. Luo *et al.* (2004) validated the effectivity of their technique by using a mouse embryo cDNA library as their source for siRNAs. Sen *et al.* (2004) tested their technique on a transgene (encoding GFP) and also after successful silencing on two endogenous genes (Oct 3/4 and MyoD) both of which were silenced, as expected. It is plausible that the abovementioned technique will be further improved, possibly by the addition of a promoter that can be activated by an external stimulation (e.g. a heat-shock promoter) so that vital genes can be silenced at a specific time.

The Gregory Hannon Laboratory came up with an advanced RNA-interference microarray procedure for high-throughput loss-of-function genetics in mammalian cells (Silva *et al.*, 2004). These investigators based their procedure on a cell-microarray technique that was developed by Ziauddin and Sabatini (2001) by which cells grown on a glass substrate can take up DNA-lipid complexes that have been deposited on a glass-slide before the cells were plated. This causes the cells to be transfected *in situ.* The authors amended this technique so that the mammalian cells can be transfected with either siRNAs or with DNA constructs that will direct the expression of short hair-pin RNAs (synthetic miRNAs) that may cause the silencing of specific transcripts of the cells. Two previous publications of the Hannon Laboratory that were mentioned above (Mousses *et al.*, 2003 and Paddison *et al.*, 2004) provided a useful background for the development of the Silva *et al.*'s (2004) procedure. Briefly, the protocol of the latter investigators for reverse-transfection was as follows. Transfection mixtures (that included controls and transfection markers) were spotted on CAPII glass slides by a robotic devise that delivered ("printed") small dots of lipids containing the transfection mix. Then each square of nine dots was combined into one spot of 400–500 µm in diameter. Each slide was thus covered with an array of combined spots and then dried. For transfection the slides were placed in 10 cm tissue-culture dishes that were filled with 15 ml of medium that contained 10^6 cells per ml. The cells were then incubated for 60 hours. A slide could be printed with about 160 (combined) spots. Four groups of 40 spots each could thus be included in a single slide. The procedure permitted to include three controls in the test-transfection. One of the controls was intended to evaluate the efficiency of the transfection and the silencing of a reporter gene (e.g. a gene expressing firefly luciferase). In assays in which the hairpin-producing constructs were included in the trans-fection mix, these constructs were from the previous study (Paddison *et al.*, 2004) by the Hannon Laboratory. Several mammalian cell lines were subject to this procedure (e.g. NIH3T3, HeLa and HEK293T). The investigators first verified the efficiency and specificity of the procedure. For that, they included in the transfection mix an ectopic marker gene for GFP as well as a siRNA that should silence this marker. To assure specificity and trace the transfection of silenced cells another gene for red fluorescence was also added. The procedure was found satisfactory. Both siRNAs and RNA-hairpin-producing DNA were effective in this *in situ* procedure to silence target mRNAs in a specific and efficient manner. Of the various cell lines tested HEK 293T cells showed the highest efficiency. These cells therefore served in fur-ther experiments. "Printed" slides were useful even after 2 months of storage at 4°C when the spots contained DNA for hairpin RNA sequences (shRNA) but the siRNA printed slides were not useful beyond 2 weeks of storage. The investigators then choose mRNAs encoding subunits of a protein-degrading particle — the proteosome as targets for shRNA induced silencing. For that, part of a large shRNA library was used for *in situ* silencing. Indeed, silencing of genes for specific subunits was achieved. Finally, the investigators turned to genes essential for cell division. One of these encodes the protein Eg5. When Eg5 is lacking defects in spindle formation ("rosettes" of micro-tubules) can be observed. For tracing the transfected cells a plasmid that encoded a chimeric protein of α-tubulin and GFP was added. By the analysis of the array the investigators found two hairpin RNAs that caused the "rosette" abnormality. These results were verified by more direct silencing of the Eg5 encoding gene. The study of Silva *et al.* (2004) thus not only extended the results obtained by Mousses *et al.* (2003) and by Kumar *et al.* (2003) but also showed that this procedure can be used to silence *endogenous* genes. Endogenous silencing will be detailed in the next chapter.

Before rounding up this chapter I would like to note that there are constant efforts to improve the procedures of gene silencing in mammals by the application of small dsRNAs. These were either pro-vided by an *in vitro* synthesis of dsRNA, or by a plasmid that encodes a sequence that will produce a hairpin-RNA forming transcript. In the cells the hairpin will be processed to dsRNA. The hairpin can be expressed transiently or the respective sequence will be integrated in the genome of the host. Host-genome integrated sequences can be transcribed constitutively or after due induction. I shall only mention some of the past efforts without summarizing them. It is probable that additional efforts of this kind are underway when this book goes to print and more reports on these efforts are expected. So here is a short list of the past efforts: Kawasaki *et al.* (2003); Kumar *et al.* (2003); Aza-Blanc *et al.* (2003); Schwarz *et al.* (2003); Fritsch *et al.* (2004); Wu *et al.* (2004); Gupta *et al.* (2004); Ui-Tei *et al.* (2004) and Khvorova *et al.* (2003).

Chapter 10

RNA Silencing in Mammals II: Silencing by MicroRNAs

In Chap. 9, I have already discussed miRNAs in mammals but these were mentioned as tools produced experimentally rather than being transcribed from the natural genomes of mammals. We now turn to the miRNAs that are regular components of mammalian cells. To avoid possible confusion, especially of those who are new in this field, I shall first introduce some terms. As will be briefly reiterated below, the field of miRNAs was opened in nematodes by Lee, Feinbaum and Ambros (1993) and by Wightman, Ha and Ruvkun (1993) who revealed simultaneously that *lin-4* miRNA inhibited a gene involved in heterochronic lineages (*lin-14*). After 7 more years the Ruvkun team (Reinhart *et al.*, 2000) discovered another miRNA in nematodes, *let-7*, that inhibited the expression of other genes (e.g. *lin-41*) involved in heterochronic lineages. Thus, in the early literature the term *small-temporal RNA* (stRNA) was used. Only later, when it was revealed that additional miRNAs were not involved in heterochronic lineages the term miRNA replaced stRNA. But for brevity microRNA is sometimes shortened to miRNA or even miR. The elimination of the term stRNA avoided another possible confusion. In nematodes a *small transitive RNA* (see: Hannon, 2002) that silences a mRNA that is upstream of its binding to the transcript was revealed. The latter small RNA could be also abbreviated to stRNA. The genomic sequence that encodes a miRNA generates a linear transcript as shown schematically in Fig. 27. This transcript folds in the nucleus into a stem-loop or a hairpin structure (the terms stem-loop and hairpin are interchangeable; when the "loop" is small the term hairpin is more appropriate) that was coined *pre-miRNA*. After

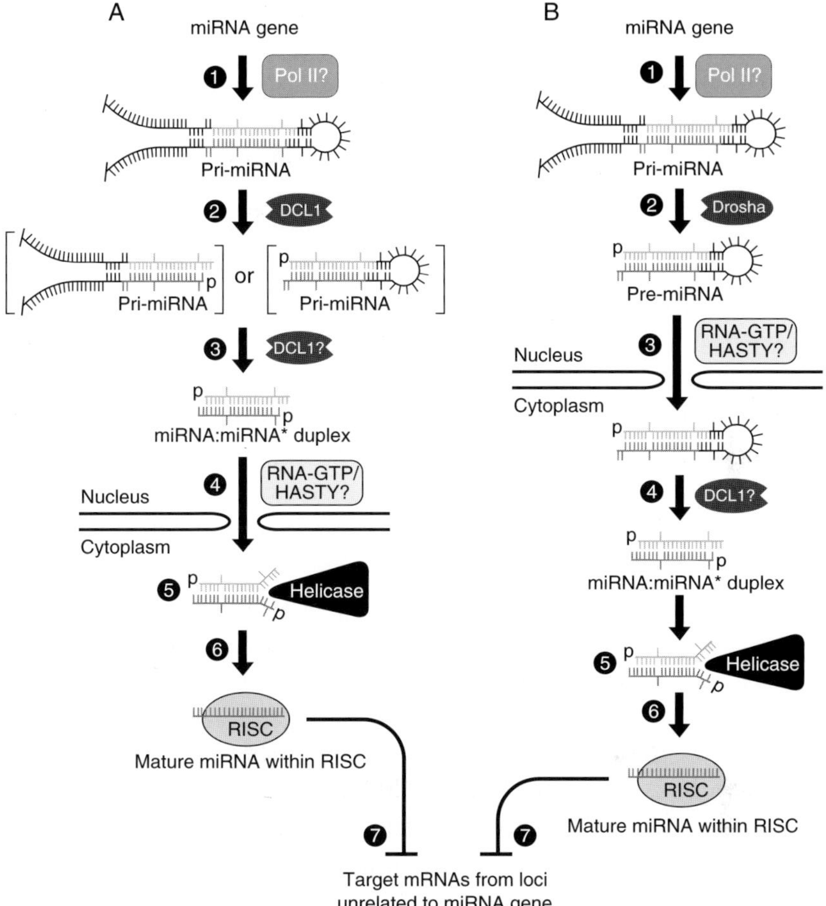

Fig. 27.   The biogenesis of miRNAs. (A) The biogenesis of plant miRNA (steps 1–6) and its hetero-silencing of loci unrelated to that from which it originated (step 7). The pre-miRNA intermediates (bracketed), thought to be very short-lived, have been isolated in plants. The miRNA is incorporated into the RISC (step 6) whereas the miRNA* is degraded (not shown). A monophosphate (P) marks the 5′ terminus of each fragment. (B) The biogenesis of metazoan miRNA (steps 1–6) and its hetero-silencing of loci unrelated to that from which it originated (step 7). (From Bartel, 2004.)

cleavage the *pre-miRNAs* are established. According to the model of Bartel (2004), there are two kinds of *pre-miRNAs*: one without the "loop" and one without the "swallow tail". Either of these may be cleaved again in the cytoplasm to result in a ~22 nt dsRNA termed miRNA:miRNA* duplex. The duplex is then separated (by a helicase) to yield a mature single-stranded miRNA that joins the RISC, and a single-stranded miRNA*. This miRNA is commonly termed *mature microRNA*, *mature miRNA* or mature *miR*. All this can be confusing because in the literature the term *miR* is used frequently for the gene that codes for the mature miRNA. This confusion will hopefully be resolved below when I discuss the processing of miRNAs in more detail.

In this chapter I shall first provide a short history of miRNAs. Thereafter, information about the features and the roles of miRNAs in mammals will be provided. A further section shall be devoted to the identification and "fishing" of miRNAs in the mammalian genomes. Toward the end of this chapter information on miRNAs in mammals will be updated.

The Emergence of MicroRNAs

Just as the discovery of the impact of dsRNA on gene silencing, the first miRNA was also discovered in the nematode *Caenorhabditis elegans*. This discovery of the role of *lin-4* in *C. elegans* was reported by the Victor Ambros team (Lee *et al.*, 1993) and the Gary Ruvkun team (Wightman *et al.*, 1993). The history of miRNA started in the laboratory of Sydney Brenner who chose *C. elegans* as a model animal for studies of the genetics of differentiation. There, one gene, then termed e912, was identified because its mutations caused remarkable developmental defects. These mutants, later termed *lin-4* mutants, failed to stop molting and underwent extra-larval stages. These were thus considered as mutations of the *heterochronic lineage* of the developing larvae, causing a failure of temporal developmental switches. It was then considered that *lin-4* encodes a master regulator of developmental timing. Another nematode gene entered the picture: *lin-14*. Mutation of *lin-14* could suppress the effect of *lin-4* mutation. The *lin-14* gene encoded a protein, LIN-14, that was formed in different levels during the development of the worms and it was subsequently found that the level of LIN-14 is regulated by *lin-4* and that the 3′ untranslated region (3′ UTR) of *lin-14* is involved in this regulation. After several years of intensive investigations the *lin-4* was found to code for a short transcript that did not encode a protein but rather could fold into a hairpin with a dsRNA stem of about 20 nt. The laboratory of Gary Ruvkun that identified the role of the 3′ UTR in *lin-14* mRNA received the sequence of the short dsRNA of *lin-4* from the laboratory of Victor Ambros and sent their 3′ UTR sequence of *lin-14* to the laboratory of Victor Ambrose. Both laboratories simultaneously found that there were homologies between the short dsRNA of *lin-4* and the 3′ UTR of *lin-14.*

Ruvkun *et al.* (2004) highlighted an earlier publication of Ambros (Ambros and Hurvitz, 1984) and thus entitled their historical account as "The 20 Years it Took to Recognize the Importance of Tiny RNAs". Whatever the "real" birthdate of miRNA in animals is, it took 7 more years of "pregnancy" from the discovery of the *lin-4/lin-14* interaction until a second nematode gene (the *lethal-7* or *let-7*) was identified as generating a miRNA that down-regulates a protein-encoding gene, *lin-41*, that is also (like *lin-14*) involved in temporal patterning. This later discovery was made by the Ruvkun team (e.g. Reinhart *et al.*, 2000). From the commentary of Ruvkun *et al.* (2004) it is not clear what was the real trigger that led to the conclusion that the *let-7/lin-41* in nematodes is an indication of a general trend in the animals. On the one hand, Ruvkun noted the connection between *let-7* and siRNA by the number of 22 nt — a magical number in the Jewish mystical Kabbalah that was elaborated by Jewish scholars in the late medieval years, in Spain (and southern France). There, in the Kabbalah, a special value was given to the 22 letters of the Hebrew alphabet. This was the irrational connection — although it should be noted that Kabbalists still exist today and are firm believers of their mysticism and do not consider it as irrational. The other connection between *let-7* and a wide-range silencing of protein-encoding genes emerged with the availability of the full sequences of the human and the *Drosophila* genomes. It was then revealed that *let-7* has homologues in these genomes. Analyses of RNAs from a wide range of animals showed that the *let-7* sequence is conserved among these animals (Pasquinelli *et al.*, 2000). The *let-7* thus sparked the interest in miRNAs in many animals including mammals and involving other miRNAs. Relevant websites were established to assist investigators in the field of miRNA studies. One of these is: http://www.sanger.ac.uk/Software/Rfam/mirna/index.shtml

But the collection of miRNAs (miRs) by computer-assisted screening of animals genomes followed by due verification preceded the establishment of websites that are accessible to the public. The amiable relationship between the Ambros and the Ruvkun laboratories was noted above but the exchange of information between many other investigators of animal miRNAs (e.g. D. Bartel, R.W. Carthew, G. Dreyfuss, S.M. Hammond, G.J. Hannon, R.H.A. Plasterk, P.A. Sharp, P. Svoboda, T. Tuschl, P.D. Zamore) was practiced throughout the years. This helped cause a quick development of this field. One recalls the vast progress in maize genetics that followed the "open" tendency of R.A. Emerson (of Cornell University) in the early years of the 20th century to share information and genetic lines with other maize geneticists. True enough, "The envy of scholars will increase wisdom." (Talmud Bavli, Baba Batra p. 21/1) but sharing information and material among scholars is of even greater benefit to science. Thomas Tuschl, who in earlier years, collaborated with several RNAi investigators (e.g. Bartel, Sharp, Zamore) while staying in the USA (Tuschl *et al.*, 1999) moved to the MPI in Göttingen, Germany, and thereafter moved to the Rockefeller University in New York. While in Göttingen Tuschl's team came up with pioneering reports on the identification of many miRs in mouse and man (Lagos–Quintana *et al.*, 2001, 2002, 2003). The term miRNA (or miRNA) was coined in these publications (to replace the term stRNA) by agreement with RNAi investigators in *C. elegans* (Ambros and Bartel). Lagos–Quintana *et al.* (2001) used total HeLa cell RNA and ligated 5′ and 3′ adopter molecules to the ends of a size-fractionated RNA extract. They performed a reverse transcription polymerase chain reaction (PCR) amplification with concatamerization, cloning and sequencing. This procedure led to 21 novel human miRNAs (in addition to the *let-7* like miRNAs). The predicted precursor of these human miRs could be folded by an appropriate computer program into stem-loop structures with bulges in the "stem" and various "loop" sizes. The finding of Lagos–Quintano *et al.* (2001) that the *let-7* sequence discovered initially in worms also exists as a miRNA gene in man, clearly indicated the conservation of miRNAs among metazoa. This indication was amply verified by additional studies. Actually, when you detect small RNAs with 5′ phosphate and 3′ hydroxyl groups in the cells and a length of

21–22 nt, you know that *Dicer*, rather than an uncharacterized break-down, was active. When northern blot hybridizations were performed it became evident that the accumulation of the miRNAs is rather tissue-specific. Moreover, the mature miRNAs tended to associate specifically with a Germin-containing protein complex that included Germin3, Germin4 and human eIF2C2. This 15S particle was revealed by the G. Dryfuss team (Mourelatos *et al.*, 2002) and termed *miRNP complex*. The main findings of the teams of Ambros, Bartel, Dreyfuss and Tuschl (e.g. Lee and Ambros, 2001; Lau *et al.*, 2001; Mourelatos *et al.*, 2002; Lagos–Quintana *et al.*, 2001) on mammal miRs will be reiterated in the next section on features of mammalian miRNAs.

In the early months of 2002 the basic features of miRNA in mammals became evident. Toward the end of 2002, 13 investigators, from laboratories in the USA, United Kingdom, Australia and Germany, submitted a collaborative article (Ambros *et al.*, 2003) to the journal *RNA*. In this article the investigators suggested a uniform system for miRNA annotation that included criteria for the expression and biogenesis of miRNAs in eukaryotic organisms. While the exact biochemistry of the involvement of miRNAs in regulating the levels of specific proteins required further clarification, the regulatory role of these small RNAs became evident. The miRNA clearly opened a new field in the molecular genetics of mammals and thus attracted the attention of numerous investigators.

Features of Mammalian MicroRNAs

The awareness of the existence of microRNAs in mammals was derived from the discovery of miRNAs in nematodes. As indicated above, after investigators had revealed the first miRNAs in these worms, *lin-4* and *let-7*, they searched for similar sequences in other organisms. Indeed, a thorough search for homologs of *let-7* was performed by a team of 19 investigators, headed by Gary Ruvkun, in which researchers from various laboratories around the globe took part (Pasquinelli *et al.*, 2000). Consequently, sequences were found in the genomes of the fruit fly (*D. melanogaster*) and man (*H. sapiens*) that represented inverted repeats of the *C. elegans, let-7* sequence. The transcripts of these sequences in the fly and in man, could be

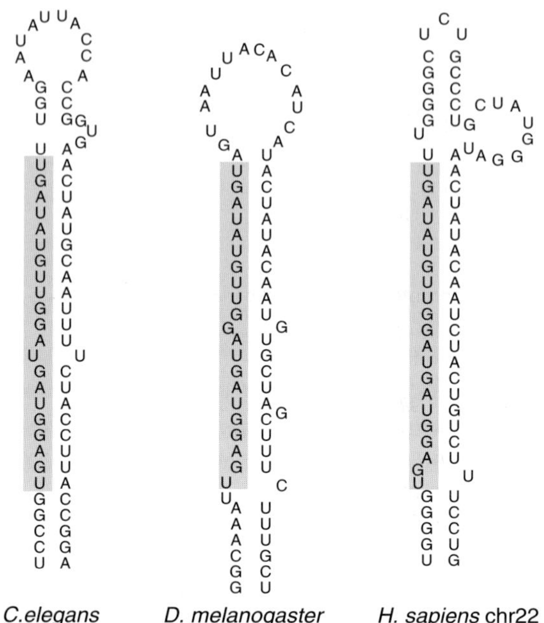

Fig. 28. The *let-7* sequences: the presumed longer transcripts of the precursors of *let-7* in *C. elegans, D. melanogaster* and *H. sapiens*; the "mature" 21 nt *let-7* are shaded. (From Pasquinelli *et al.*, 2000.)

computer-folded into the respective stem-loop structures. One of the stems had almost perfect homology to one of the stems of the folded precursor of *let-7* (Fig. 28). By looking at Fig. 28 it is evident that while the stems of the *let-7* in *C. elegans, D. melanogaster* and *H. sapiens* are homologous, the three *let-7* precursors differ considerably in their respective loops. Moreover, the three precursors (pre-miRNAs) had UG in their 5′ ends. The same team (Pasquinelli *et al.*, 2000) screened the numerous organisms and found that *let-7* like expression exists in most metazoan phyla (it was not found in some Cnidarians and Poriferans). The phylogenetic scheme of *let-7* RNA expression is shown in Fig. 29. Based on the information in which organisms the *let-7* is expressed, namely, in all the three main clades of bilateral metazoa, the authors suggested that the gene for *let-7* evolved after the divergence of diploblastic and bilateral animals. Although it is possible that some metazoa lost this gene during evolution and the "root" of *let-7* is even more ancient. Computer-folding the transcript

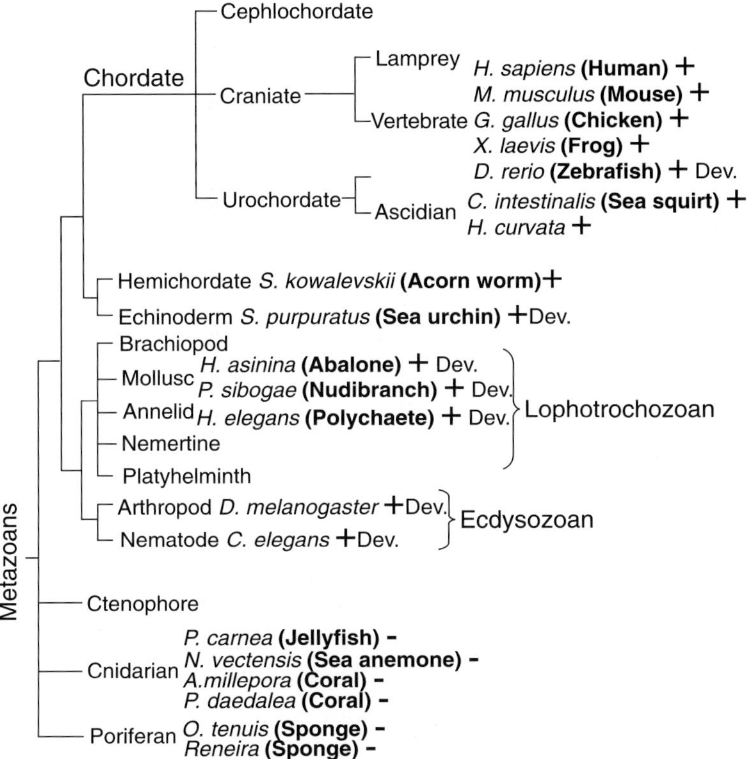

Fig. 29. Phylogenetic comparison of *let-7* RNA. The phylogenetic tree shows species that either express (+) or do not express (−) *let-7* RNA. In some species there is a developmental pattern of *let-7* RNA (no let-7 RNA in early stages but *let-7* RNA expression by adulthood; these were marked by Dev). (From Pasquinelli *et al.*, 2000.)

of the *let-7* gene resulted in a stem-loop folded-structure, with stems that are about 21 nt (or more) long. This led to the first characterized feature of a mammalian miRNA.

As detailed in the review of Bartel (2004) numerous other characteristics of mammalian miRNAs and their processing were gradually defined during the years 2001–2004 where more than 200 miR genes were found in mammals so that the common denominator of these genes could be formulated. We shall see below that it is also common to mammalian miRs that they are cleaved from their respective transcripts (in the nucleus) by *Drosha*, an RNase III endonuclease that cuts the dsRNA of extended stems so that the resulting dsRNA of the pre-miRNA has a 5′ phosphate and a ~2 nt 3′ overhang. An additional cleavage of dsRNA is performed in the cytoplasm by *Dicer* that also has an RNase III activity. Because *Dicer* has a "ruler" that measures ~21 nucleotides between its "cutters" the resulting miRNA:miRNA* duplex, has a dsRNA "stem" of ~19 nt, phosphates on its 5′ end and 2 nt overhangs with hydroxyls at its 3′ end. The miRNA:miRNA* is thus apparently identical in general structure to siRNA that is produced endogenously (e.g. in plants as defense against viral pathogens). There is information that indicates that there is no symmetry in the miRNA:miRNA* duplex that I shall detail later. In one end of dsRNA there are stronger bonds between the bases (due to C and G bases) than in the other end (due to A and U bases). This may cause the miRNA:miRNA* to be separated (like a zipper) from one end (e.g. by a helicase). By this asymmetry the silencing mechanism (RISC) can identify if the strand is miRNA or miRNA*. These basic characteristics enabled three research groups to conduct efficient "fishings" that resulted in a wealth of miRs in mammals (Lagos–Quintana *et al.*, 2001; Lau *et al.*, 2001; Lee and Ambros, 2001). True enough, the "fishing" was primarily intended to reveal new miRs in the namatode *Caenorhabditis elegans*. But as the biblical Saul who searched for his father's donkeys and found the Kingdom, these investigators found mammal miRs. The leaders of these teams though settled in different locations, were obviously familiar with one another due to their former collaboration (Zamore *et al.*, 2000) during their MIT residency. They managed to report their results in three consecutive articles in the same *Science* issue. Thus, the Tuschl team (Lagos–Quintana *et al.*, 2001) identified several dozens of miRs in human HeLa cells and in mouse kidney tissue. The Bartel team (Lau *et al.*, 2001) briefly mentioned that potential orthologs of miRs were identified in the human genome without providing further details. As for the *C. elegans* miRs these investigations provided several characteristics that could also exist in mammalian miRs. They include: most miRs expressed from independent transcription units that do not contain an open reading frame and none of the *C. elegans* miRs matched a transcript that was validated by an annotated expressed sequence tag (EST); most miRs were at least 1 kb away from the nearest annotated sequence. Moreover, even miR genes near predicted protein-coding regions or within predicted introns are probably expressed separately from annotated genes. But since the miRs are expressed in specific tissues and/or in a fixed temporal manner, the question of how the expression of the miRs in mammals is regulated, remains unanswered. The Ambros Laboratory revealed 15 new miRs in *C. elegans* and three of these had orthologs in mammal: *mir-1* that is expressed throughout the development in nematodes has an ortholog in man and *mir-2* that is expressed in the L1 larval stage of nematodes also has orthologs in man. All the revealed miR genes were predicted to produce ~65 nt stem-loop forming transcripts that may be processed to the ~22 nt forms. While in *C. elegans* the expression of *mir-1* is not stage specific the expression of its orthologs in man is tissue specific (heart) and in mice, stage specific (during embryogenesis). This probably means that while the basic role of miRs and their sequences were conserved during evolution the specific tasks of the ortolgos in *C. elegans* and mammals ramified considerably. Parenthetically, some authors use the abbreviation *miR* while others use *mir*. I try to follow the abbreviation of the authors.

A very elegant study was performed by the Gideon Dreyfuss team of the University of Pennsylvania in collaboration with researchers from Denmark and France (Mourelatos *et al.*, 2002). These investigators looked for a ribonucleoprotein (RNP) complex that includes the mature (single-stranded) miRNAs of human HeLa cells. Their study was based on their previous finding that complex-components such as the **S**urvival of **M**otor **N**eurons (SMN) and five Germin proteins interact, resulting in the promotion of assembly and function of ribonucleoprotein (RNP) complexes (e.g. spliceosomal small nuclear RNPs, small nucleolar RNPs, heterogeneous nuclear RNPs and transcriptosomes). They also took note of the eukaryotic initiation factor 2C (eIF2C) that is a member of the family of Argonaute proteins which were known to be important in RNAi and in the developmental regulation imposed by miRNAs (e.g. *lin-4* and *let-7*). In both these pathways a cleavage of dsRNA is required. In *Drosophila* an Argonaute protein was found to be part of the RISC complex that co-purifies with siRNA; the latter is probably leading the RISC to its mRNA target. This and additional information led to the possibility that the Argonaute proteins are also involved in the processing of miRNAs of human cells. The strategy of Mourelatos *et al.* (2002) was therefore to isolate, from HeLa cells, particles that contain Germin proteins (Germin3 and Germin4), eIF2C2 and small ~22 nt RNAs (miRs). Such particles were identified from HeLa lysates by the respective antibody. Further analyses verified the presence of Germin3, Germin4 and eIF2C2 in the HeLa-cell-derived particles. The researchers could co-precipitate Germin3, Germin4 and eIF2C2. These three protein components were also complexed *in vitro.* When a Germin3-eIF2C2 complex was isolated from HeLa cells it also contained 22 nt ssRNAs that were presumed to represent mature miRs. Fractions from a sucrose gradient of HeLa cell lysate were obtained and one of the fractions contained 15S particles. These were identified as miRNP particles that contained the Germin3, Germin4, eIF2C2 and the putative miRs. The miRs were then cloned and sequenced to search for homologues in the human genome. It was found that the miRs were homologous to genomic sequences that had reverse-repeat configurations, suggesting stem-loop forming precursors for these mature miRNAs (miRs). In this way numerous new human miRs were identified. These mature miRs were in the range of 16–24 nt (most of them 20–22 nt). Several of these new miRs were assigned to specific chromosomes. The secondary structure (stem-loop) of each putative pre-microRNA could be predicted. This prediction was formulated by using 70 nt sequences from upstream of each miR homolog as well as 70 nt downstream of it. The whole region was then computer-folded and the resulting stem-loop was "trimmed" to a total of 80 nt. In almost all cases the putative mature miRNA represented one of the two strands of the stem. In one case each of the two strands of the same stem represented a different miR (miR-91 and miR-17).

In summary the study by Mourelatos *et al.* (2002) clearly indicated that the mature miRs of man are bonded to miRNP particles that also contain Germin3. Germin4, eIF2C2 and possibly additional components. These RNP particles can bind a plethora of different miRs. Thus, different specific miRNPs can affect different target transcripts. The inclusion of Germin3 in the miRNP is noteworthy because of its presumed helicase activity. The Germin3 may be instrumental in "opening" the double-stranded precursors of the mature miRs so that one of the two strands can bind to the miRNP and scout the miRNP to its specific mRNA target.

The purpose of the study of P.A. Sharp and associates (McManus *et al.*, 2002) was to design hairpin (stem-loop) RNA species that will be effective in silencing mice genes. As they examined several features of potential hairpin RNA silencers their findings are very relevant to understanding the characteristics of the miRNA that are affecting silencing. But we should note that these investigators followed PTGS rather than the reduction of translation levels. The study was performed with E10 cells of mice and three types of short RNA species were synthesized as "miRNA". One type was actually siRNA (i.e. a double-stranded $\sim$19 nt with 3' and 5' overhangs). Then they produced two kinds of "miRNAs" with either no real loop (a "pin" of 4 nt). These were the class I hairpins, and "miRNAs" with a loop of 12 nt. The latter were termed class II hairpins. The targets for silencing were the HIV co-receptors CD4 and CD8α. We shall not detail the results of this study. There was one main message indicating that even slight changes in the structure of the hairpin RNAs cause vast differences in silencing. Just one example: the "pin" can be at either the 3' end or the 5' of the antisense strand. It was found that silencing was induced only when the "pin" was at the 3' end of the antisense strand.

A detailed study on the process of miRNA "maturation" in mammalian (HeLa) cells was performed by V. Narry Kim and associates (Lee *et al.*, 2002) in Seoul, South Korea. He was previously from the laboratory of Gideon Dreyfuss, Pennsylvania, studying pre-mRNA splicing. Lee *et al.* (2002) followed the biogenesis of miRNAs from the initial transcripts to the mature miRNAs in HeLa cells. They first asked about the length of the transcripts that include miRs in HeLa cells. Two miR clusters were chosen. They were located outside of regular genes (that encode proteins). The investigators found that the miR-containing transcripts are formed as single transcriptional units. These transcripts were always longer than 70 nt, a size attributed to pre-miRNAs (that could form the respective stem-loop structure). The investigators also found that the long primary transcripts could be cleaved in cell extracts to fragments of $\sim$65 nt and even to fragments of $\sim$23 nt. They concluded that the long transcripts, termed pri-miRNA, were processed to pre-miRNA of about 65 nt and subsequently the pre-miRNAs were processed to ~23 nt "mature" miRNAs. The meaning of their "mature" miRNA was probably dsRNA. The first cleavage of pri-miRNA to pre-miNRA apparently takes place in the nucleus. The pre-miRNA is then transported to the cytoplasm where the pre-miRNA is cleaved to "mature" miRNA. The possibility that the initial cleavage, the export from the nucleus and the second cleavage in the cytoplasm are regulated was mentioned by Lee *et al.* (2002). But the authors left the question of how this regulation takes place unanswered. They also did not address the question of what happens to the "mature" (double-stranded) miRNA during the silencing process.

In a subsequent study in which the Korean team of V. Narry Kim collaborated with investigators from Canada and Sweden (Lee *et al.*, 2003) the *Drosha* was revealed in human nuclei. Immuno-purified Drosha was capable to cleave *in vitro* pri-miRNA to release the respective ~70 nt pre-miRNA. Thus, it was suggested that the two RNase III proteins, Drosha (in the nucleus) and *Dicer* (in the cytoplasm) collaborate in a stepwise processing of miRNAs.

Another set of relevant information on miRNA in mammalian cells came from the laboratory of Bryan R. Cullen of the Duke University Medical Centre (Zeng *et al.*, 2002; Zeng and Cullen, 2003; Zeng *et al.*, 2003). In their first publication on human miRNA (Zeng *et al.*, 2002) the investigators of the Cullen Laboratory focused on the human miR-30 miRNA which they detected by cloning the cDNA that contained the sequence of the predicted 71 nt miR-30 precursor. They used a somewhat ambiguous term for the mRNA that called this putative precursor *"irrelevant,* endogenously transcribed mRNA". Probably, the authors meant that this mRNA did not encode a protein. The investigators then asked whether only the precursor, 71 nt, of miR-30 in a transcript would be processed in the human cells to the mature 22 nt miRNA or if the shorter miR-30 itself would be processed likewise. The investigators constructed appropriate plasmids in which a promoter was put upstream of either the precursor or the short miR-30. The plasmids were transfected into human (293T) cells and the investigators then looked for the generation of mature, ~22 nt miR-30. The latter were found only after transfection with plasmids containing the 71 nt precursor, but not when the plasmid with the short miR-30 sequence was used. This was repeated with other cell types (e.g. HeLa, NIH3T3) and the results showed that the mammalian cells can process the precursor to mature miR-30 miRNA. The mature miR-30 was commonly encoded by the 3′ arm of the putative hairpin folded precursor. But the investigators also found mature miRNA derived from the mir-30 precursor's 5′ arm in transfected cells. In another set of experiments the investigators found that transfection of the precursor (~71 nt) can block the translation of a synthetic miR-30 target that has the appropriate 3′ UTR (untranslated region). The investigators went further and synthesized precursor-miRNAs for miRs that do not exist in the human genome and introduced plasmids that should express these precursors *in vivo* as well as synthetic targets that were also capable to be expressed *in vivo*. It was found that the silencing process also takes place with synthetic precursor miRs and synthetic targets. It became clear that the precursor-miRNA (and not the mature miRNA) will cause translational silencing of its target transcript when transfected into human cells. In a further study Zeng and Cullen (2003) studied the miR-30 as well as another miRNA, miR-21. They confirmed that in human cells the long transcripts that contained the mature miRs in a sequence that can fold into a hairpin secondary structure will be processed to the mature ~22 nt miRs. This capability was confirmed for both miR-21 and miR-30. As for the inhibition of translation it was found that single point mutations in miR-21 or in miR-30 rendered these miRNAs inactive in inhibiting the translation of the respective mRNA targets. The maturation of precursors to mature miRs was only slightly affected by mutations. Take, for example, changes that should cause a reduction of the loop size. These had only minor affects on the silencing capability. Also, changes that will cause the elimination of bulges in the stem of the precursor had no effect on silencing while creating extensive bulges was detrimental to silencing. As was previously found by Lee *et al.* (2002) and as mentioned above, Zeng and Cullen (2003) also showed evidence for a step-wise maturation from very long miR precursors (pre-microRNA) to ~65 nt pre-miRs. The "basis" of the stem of the miR precursor (i.e. opposite the loop end) could play a role in the efficiency of processing and the silencing.

Disruption of the base-pairing in this basis was detrimental to silencing capability which could be monitored by a reporter mRNA. The investigators constructed a plasmid that can cause the expression of the firefly luciferase but the sequence was changed so that the 3' UTR had miR-30 or miR-21 target sequences. Clearly, miR-30 precursors silenced only the luciferase with 3' UTR targets of miR-30 but not if the luciferase mRNA had 3' UTR targets for miR-21. The investigators also found that two different precursors can be included in the same transcript and both will be processed.

In a further study (Zeng *et al.*, 2003) the Cullen team investigated the mode of silencing by microRNA versus the silencing by siRNA. The accumulated previous information indicated that miRNA silences by affecting the 3' UTR of the target mRNA, causing a reduction of translation but not the degradation of the mRNA. On the other hand, siRNA causes silencing by the degradation of the mRNA. These investigators used the miR-30 and miR-21 to ask whether sequence changes could also change the silencing mode. When the endogenous miRNA was targeted to a synthetic mRNA that had full homology to the miRNA, the respective mRNA was degraded (rather than the translation being retarded). On the other hand, when a synthetic siRNA that had only partial homology to a mRNA target was used the result was not mRNA degradation, but rather an inhibition of the messengers translation. This seems to indicate that the "origin", whether from a siRNA or a pre-miRNA is not decisive for the mode of silencing. What determines the mode of silencing is the specific sequence (i.e. full homology or partial homology to the 3' UTR of the target mRNA). The author thus suggested that, at least in cultured human cells, siRNAs and miRNAs may be functionally interchangeable. The results of Zeng *et al.* (2003) are actually compatible with the previously reported results of Hutvagner and Zomore (2002a, 2000b) who found that in human cell extracts, the miRNA *let-7* endogenously enters the RNAi pathway, indicating that only the degree of complementarity between a miRNA and its RNA target determines its function (or, mode of silencing). A further contribution to this subject was provided by Sharp and associates (Doench *et al.*, 2003) that I shall discuss below.

Since the discovery of the first two miRNAs in nematodes (*lin-4* and *let-7*) it was assumed that these endogenous and tiny RNAs are primarily involved in temporal and tissue-specific gene regulation. A team of Carlo Croce (of the Thomas Jefferson University in Philadelphia) and 14 associates (Calin *et al.*, 2002) focused on two human miRNAs, miR15 and miR16 genes that are located at chromosome 13q14. Because this region is deleted in more than half of B cell chronic lymphocytic leukemias (B-CLL) these investigators followed the correlation between B-CLL and *miR15/miR16*. The loss of this region is also frequent in other cancer types (e.g. prostate cancers). This hinted to the possibility that this region contains one or more tumor-suppressor genes. But no protein coding genes could be identified in the (1 Mbp) 13q14 region that correlated with the malignancy. On the other hand, these investigators found that the two miR genes, *miR15* and *miR16* are located at 13q14, within a 30 kbp region of loss in CLL and that both these genes are deleted or down-regulated in the majority of CLL cases. Northern blot hybridization indicated that *miR15* and *miR16* are strongly expressed in normal CD+ lymphocytes and therefore presumably play an important role in normal CD5+ B cell homeostasis. In normal tissues *miR16* is expressed at higher levels than *miR15*. The targets (protein encoding genes) of miR15 and miR16 were not yet identified, thus the idea that they are related to tumor formation awaits further research. Even later studies of the same team (Calin *et al.*, 2004a, 2004b) did not identify targets of miR15 and miR16 in humans that are involved in B cell chronic lymphocytic leukemias (B-CLLs) or in any other cancer disease. But this latter study did show a very significant phenomenon — that among the 186 miRs analyzed, about one half are located in fragile chromosomal sites and in other sites that are correlated with cancer. In other words, 98 out of 186 of the miR genes are in cancer-associated genomic region or in fragile sites. Northern blot hybridizations indicated that several miRs that are located in sites that are subject to deletion have low levels of expression in cancer samples. Again, how exactly the expression of miR15 and miR16 is regulated and how exactly these miRs affect cancer formation require further clarification. What could be concluded by the authors with confidence was that the inactivation of miRs can cause overexpression of yet unidentified targets while miR activation may lead to the down-regulation of targets that are supposed to participate in apoptosis, cell cycle and/or angiogenesis.

I have mentioned the study by the Cullen team (Zeng *et al.*, 2003) who concluded that the origin of the silencing RNA (i.e. whether it originated as an siRNA or as microRNA) does not determine the mode of silencing (degradation of mRNA or blocking translation). What seemed to be decisive is the level of homology between the silencing RNA and its target. Phillip Sharp and associates (Doench *et al.*, 2003) studied this subject further. They reviewed the siRNAs and the miRNAs in animals and noted that while no endogeneous (natural) siRNAs were detected in mammals, siRNA could be introduced into mammalian cells and these may then cause mRNA degradation. On the other hand, miRNAs are intrinsic components and commonly silence target genes by blocking the translation. This reduction of translation happens when the miRNA is partially homologous to the 3′ UTR of the target mRNA. But there are exceptions: the *let-7* miR can degrade a mRNA that has full homology to this miR and the team of Cullen (Zeng *et al.*, 2003) found other exceptions as I have mentioned above. Doench *et al.* (2003) thus asked whether a synthetic siRNA that will have partial homology to the 3′ UTR of a specific mRNA will cause silencing by the inhibition of translation in mammalian cells. In other words, can a specially designed siRNA act in the mode of miRNAs? For that they choose a siRNA that was known to cleave the mRNA of the cell-surface receptor CXCR4 due to full homology to the target mRNA. The target mRNA was either with a full homology to the siRNA or the target was constructed so that bulges will form at its 3′ UTR when paired with the siRNA. To visualize the silencing the investigators used the *Renilla reniformis* luciferase as reporter gene. The experiments were performed by transfecting HeLa cells with the respective constructs. It was found that the CXCR4 siRNA caused a drastic (10-fold) silencing but the mode of silencing was different according to the sequence of the target. When there was full homology between siRNA and target, there was degradation of the mRNA. If there were bulges in the 3′ UTR/siRNA the silencing was not related to degradation of the mRNA. However, the siRNA reduced the mRNA only marginally while silencing was about 10-fold. The results clearly showed that a siRNA can silence (reduce the translated product) without causing significant mRNA degradation. Subsequently, the authors constructed different (modified)

CXCR4 siRNAs and obtained detailed information on the correlation between the sequences of the bulges formed between the siRNA and the target and the level of gene silencing.

Search for MicroRNA Targets

While targets were assigned years ago for the first miRNAs found in *C. elegans*, (*lin-4*, and *let-7*), hundreds of miRNAs were identified in mammals before gene targets with known functions were identified in these animals. Then Kawasaki and Taira (2003a, 2003b) of the University of Tokyo looked for the target of the miR-23 in human NT2 cells. When they looked for sequences in the human genome that were fully homologous to miR-23, they did not find any such coding sequence. But they did find a gene that had a 77 per cent homology. This was a gene termed *Hes1*. The HES1 protein is a helix-loop-helix transcriptional repressor that is expressed only in undifferentiated cells. It acts as an antidifferentiation factor. When the human miR-23 was computer-matched to the coding sequence of *Hes1* the 77 per cent homology was located in the coding region near the termination codon. A similar situation was revealed with the mouse miR-23 and the mouse *Hes1*. This suggested that the role of miR-23 has been conserved phylogenetically. Indeed, computer-assisted alignment indicated that the mouse miR-23/*Hes-1* and the human miR-23/*Hes-1* complementarity were conserved phylogenetically. In order to further confirm that the *Hes-1* mRNA is a target for miR-23 in man the investigators used human NT2 cells: human embryonal carinoma (EC) cells which will differentiate into neural cells after treatment with retinoic acid. The change in HES-1 expression was thus evaluated during retinoic-acid-induced differentiation. It was found that before the induction of differentiation the HES-1 was amply expressed but following induction (i.e. during 3 weeks) there was a gradual deminition of HES-1. While the level of HES-1 protein was reduced drastically there was no reduction in the level of HES-1 mRNA. This indicated that the level of HES-1 is reduced during retinoic-induced differentiation due to reduction of translation and not due to reduction of transcription. Furthermore, the retinoic induction caused the appearance of miR-23 (Fig. 30). This

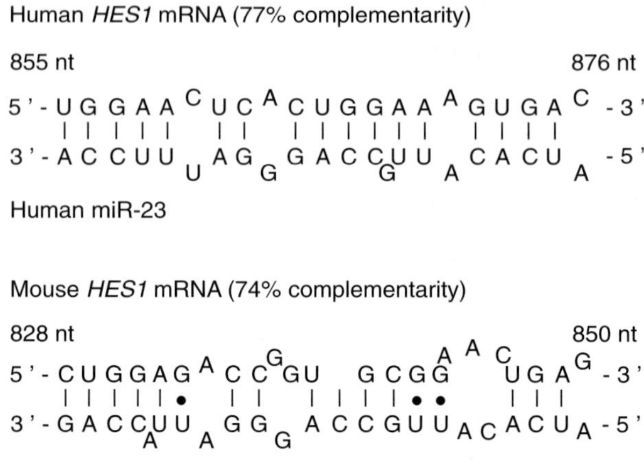

Fig. 30. Expression of Hes1 and miR-23. Complementarity between *Hes1* mRNA and miR-23; the region of nearly complementary to human and mouse miR-23 is located in the coding region, near the termination codon of human and mouse *Hes1* mRNAs. (From Kawasaki and Taira, 2003a.)

suggested that upon retinoic acid induction the miR-23 is strongly increased and this high level of miR-23 inhibits the translation of HES-1, thus causing the differentiation. To test this hypothesis the investigators added miR-23 to undifferentiated NT2 cells. This addition indeed reduced the level of HES-1 in these cells. When instead of an authentic, non-mutated miR-23, a mutated miR-23 was applied — there was no change in the HES-1 level. The mRNA for *Hes-1* was not changed by authentic, neither by mutated miR-23. The reduction of endogenous miR-23 could be performed by RNAi techniques. The interaction between miR-23 and the expression of *Hes-1* was further investigated by a clever approach. An expression plasmid containing a chimeric gene was constructed. In this plasmid the mRNA for a luciferase gene was engineered to contain targets for the miR-23. In the absence of endogenous miR-23 (undifferentiated cells) luciferase was highly active but this activity diminished with the onset of differentiation (that caused in an increase of endogenous miR-23). The results were also verified by following the changes of differentiation markers as MAP-2 or SSEA-3 that increase or decrease, respectively, during differentiation of NT2 cells.

The studies on miRNA that I have reviewed above provided an impression that there is generally a simple siRNA/miRNA relationship, with respect to gene silencing. Apparently, siRNAs with homology to the coding sequence of a target mRNA will cause cleavage in this mRNA and consequently induce transcriptional silencing. On the other hand, miRNAs that have partial homology (with some mismatches) to sequences that are 3' UTR of mRNAs will cause the inhibition of translation without affecting the level of intact mRNAs. Investigators of the University of Virginia in Charlottesville (Saxena *et al.*, 2003) probed further into this siRNA/miRNA/mRNA relationship. They noted the previous information that a single base mismatch between a silencing short-RNA (siRNA) and the coding region of a mRNA target can prevent the degradation of the target mRNA. But some mismatches between a silencing RNA (commonly miRNA) and the 3' untranslated region of the target-mRNA will not prevent the inhibition of translation from this target. These investigators then found that translational inhibition may also result when there is a mismatch between a single site in the coding region of the target mRNA and the silencing short RNA. This may happen when the antisense of the silencing RNA has a 5'-phosphate but not a 3'-hydroxyl group. The investigators used PC3 and HeLa cells in their study and choose two mammalian genes, *p21* and *geminin* as targets. When the target was *p21* mRNA and dsRNAs that had mismatches to the 3' UTR of *p21* mRNA were used, there was a reduction of translation from *p21* even while dsRNA had a few mismatched bases. Increasing the mismatching eliminated the reduction of translation. But a reduction of translation also resulted when there was a mismatch between dsRNA and a coding region. In this case the synthetic dsRNA (siRNA) acted as a miRNA. Moreover, because the investigators observed cases in which the mRNA was only partially degraded (due to homology between dsRNA and the target mRNA), it is plausible that in some cases there is degradation of mRNA and also an inhibition of translation. Similar results were obtained when the target was mRNA from the *geminin* gene and dsRNA had specific mismatches to the target. In summary the evidence that a mismatched dsRNA designed to target the coding sequence of a gene can function as a miRNA clearly shows that not the origin but rather the sequence of the interfering RNA determines the mode of gene silencing. The exact degree of mismatches that still maintains silencing can obviously vary in specific genes. Are synthetic miRNAs that have full homology to the coding region of genes acting as siRNA by causing degradation of the respective mRNAs? They probably are but further experimental evidence will give a decisive answer.

The expression of a protein from its gene is regulated by several control mechanisms. Grossly, these control mechanisms start with controls at the chromatin level (e.g. heterochromatin *versus* euchromatin, methylation, deacetylation, etc.) and then go through the regulation of transcription (promoters, inducers, transcription factors), stability of mRNAs and finally stability of the protein product. I have mentioned above that miRNAs were found in recent years to control the expression by the inhibition of mRNA translation and possibly also by the degradation of mRNA. In order to control the expression of a gene there must be a specific recognition of the mRNA by the respective miRNA. There are many thousands of mRNAs at any time in a mammalian cell. Therefore, the specific recognition should be quick and efficient. How is this performed? Before one can probe into this question we have to be aware that the silencing of mRNA (whether by degradation or by the inhibition of translation) takes place in the cytoplasm. This is the cellular compartment where the mature (i.e. spliced) mRNA is located. Studies published during 2003 and 2004 (e.g. Yi *et al.*, 2003; Lund *et al.*, 2004) reported that in man the long pre-miRNAs are processed in the nucleus to ~65 nt hairpin-folded pre-miRNAs. A specific protein, Exportin-5, then mediates the export of the pre-miRNAs from the nucleus to the cytoplasm where the pre-miRNAs are further processed to the mature miRNAs as noted above (in Fig. 27). The miRNAs are expected to join RNP complexes and will scout this complex to the correct region on specific mRNAs. This scouting is the arena where the specific recognition should take place. Due to the plethora of mRNAs in the cytoplasm, finding the correct mRNA in order to cause the miRNA:mRNA binding is as complicated as looking for a needle in a haystack.

Doench and Sharp (2004) of the MIT in Cambridge, USA, intended to investigate the "rules" of miRNA:mRNA pairing. The results of this study that will be summarized below indicated that nature, due to its accumulated wisdom of many hundreds of million years found a way to locate such "needles" in the cytoplasmic "haystack". Doench and Sharp (2004) first tried to find the part of the 3'-UTR of a specific mRNA which is important for the repression of protein translation by the miRNA. For that they choose the 3'-UTR region of the mRNA for a gene that encodes CXCR4. To simulate the miRNA that represses the translation from this mRNA, the investigators used the antisense of a siRNA (rather than the miRNA) in which the two most 3' nucleotides were deoxythymidines (rather than uridines). They then introduced several mutations in the mRNA to reveal which nucleotides are affecting the translational repression. To monitor the rate of the repression of translation the investigators used a marker, the *Renilla* luciferase. The 3'-UTR of CXCR4 was ligated to the coding region of luciferase mRNA so that an effective binding of the antisense siRNA to the 3'-UTR of CXCRL4 could be monitored by luciferase assays. The results indicated that the ability of a miRNA (represented by antisense siRNA) to translationally repress a target mRNA is largely dictated by the free energy of binding of *the first eight nucleotides in the 5' region of the miRNA*. However, the G:U wobble base-pairing in the binding region interferes with the translation beyond the level predicted on the basis of thermodynamic stability. It thus appears that the miRNA does not search for a homology (or partial homology) of its total ($\sim$19 nt) nucleotide sequence. The search starts for 8 nt and possibly even less of its 5' end nucleotides. Only after matching nucleotides are found among the nucleotides of the respective target mRNA, the search goes on so that the whole length of the miRNA is hybridized to its target region. The step-wise search for homology seems to be the process that helps the miRNA locate its target and bind to it. Further studies may reveal additional factors that influence the binding of miRNA to mRNA. Moreover, although the hybridization starts with only seven or eight base-pairs, thus forming a weak duplex, it could be sufficient for translational repression. This should not surprise us. By analogy, bacterial endonucleases require only short sequences (commonly of 6 nt) on a DNA region for correct recognition and cleavage. These investigators also found that protein translation from a given mRNA can be repressed simultaneously by more than one miRNA. Also, the repression of protein translation can probably be modified, by the relative ratio of mRNA to miRNA and by other cellular conditions.

A study by another team from MIT in Cambridge, USA, that included D. Bartel, Christopher Burge and associates (Lewis *et al.*, 2003) had a different goal from Doench and Sharp (2004). In spite of the formal difference in goals important conclusions of the two teams were very similar. Lewis *et al.* (2003) submitted their manuscript only a few weeks before Doench and Sharp (2004) submitted theirs. Thus, although the two studies were conducted at the MIT they were independent efforts. The approach of Lewis *et al.* (2003) was to predict as many as possible mammalian (human, mouse and rat) targets for miRNAs. They intended to provide computational and experimental evidence for the authentity of these targets. For that they also used the genome sequence of a non-mammalian vertebrate: the pufferfish (*Fugu*).

A first step in the study of Lewis *et al.* (2003) was to develop an algorithm for the prediction of mammalian miRNA targets. They developed an algorithm called Target Scan which takes into account the thermodynamics of RNA:RNA duplex interactions and comparative sequence-analyses to predict targets that are conserved in different organisms. MiRNAs that are conserved among these organisms as well as the 3′ UTR of the respective orthologous transcripts were taken into account. It was found that the complementarity of nucleotides 2 to 8, counted from the 5′ on the miRNA, are the key for the identification of the target. This finding, based on a computational approach, is very similar to the conclusions of Doench and Sharp (2004) that was based on a rather different experimental approach. Lewis *et al.* (2003) introduced useful terms in their publication. The set of seven bases from base 2 to base 8 at the 5′ end of miNRAs was termed "miRNA seed" and the seven bases at the UTR of the mRNA that have perfect matches with the miRNA seed were termed "seed matches". The analyses were started with "seed matches" but then the search for additional base-pairs was performed. I shall not detail the bioinformatics analysis but note that it is essential to take into consideration orthologous pairing among human-mouse, among human-mouse-rat and among human-mouse-rat-*Fugu*. This analysis pointed to an estimate that in mammals each miRNA has about four targets.

But this analysis is prone to errors of different kinds. One possibility is that in some mRNAs the target of the miNRA is outside of the 3′-UTR (as was found for some plant miRNA:mRNA interactions). The investigators accumulated 854 miRNA:UTR pairs that represented UTRs of 442 distinct genes conserved in mammals (human, mouse, rat) and deposited this list in a supplement. Of these, an abbreviated list is provided in Table 2.

Several important conclusions emerged from this study. One of these is that the most conserved miRNAs of mammals also have the greatest number of probable target mRNAs. When the investigators focused on 15 predicted targets and devised experimental procedures to analyze if indeed they were specifically silenced by the respective miRNA, a positive answer was obtained in 11 cases. It was not clear why the remaining four predicted target 3′-UTRs did not respond to the respective miRNAs. As inferred previously the author's experimental evidence suggested that mammalian miRNAs are generally negative regulators of gene expression. While the first two miRNAs revealed in nematodes were both involved in the regulation of development, the target genes identified by Lewis *et al.* (2003) are of diverse types, as regulators of development, transcription factors, and nucleic acid binding proteins, transporters, etc. The relative ratios of the different types of miRNA targets in mammals seem to differ from the ratios found in plants. Take, for example about 70 per cent of the miNRA targets in plants that were analyzed were members of transcription-factor gene families. A much lower ratio of transcription factor genes was found among the miRNA targets in mammals. Moreover, nearly all the transcription-factor targets in plants that were revealed have known or predicted roles in development. In mammals only about 13 per cent of the predicted targets of miRNA may be involved in development. Hence, it seems that mammalian miRNA-mediated control of gene expression has a broad diversity of biological processes.

When an author of plant science sees a publication entitled "Dicer is essential for mouse development", the rhymes of *Mother Goose* come to his mind. Specifically the rhyme on the *Three Blind Mice* whose tails were cut with a carving knife (a *dicer*). But this title was given to a communication submitted by Gregory Hannon and associates

Table 2. Predicted targets of mammalian miRNAs.

Category	Seed	MiRNAs	Ensembl ID	Gene name
Regulation of transcription/ DNA binding	AGUGCAA	miR-130, -130b	169057	Methyl-CPG-binding protein 2(*MECP2*)
	GUGCAAA	miR-19a	169057	Methyl-CPG-binding protein 2(*MECP2*)
	AAAGUGC	miR-20, -106	101412	Transcription factor *E2F1*
	GAGGUAG	let-7(a-g,i), miR-98	100823	DNA-(apurinic or apyrimidinic site) lyase (*APEN*)
	GAAAUGU	miR-203	125347	Interferon regulatory factor 1 (*IRF-1*)
	ACAGUAC	miR-101	134323	N-MYC protooncogene protein
	GAGGUAU	miR-202	134323	N-MYC protooncogene protein
	AAUCUCA	miR-216	065978	Nuclease sensitive element binding protein 1 (*YB-1*)
	UAAGGCA	miR-124a	163403	Microphtalmia-associated transcription factor
	GCUGGUG	miR-138	054598	Forkhead box protein C1 (*FKHL7*)
	AAAGUGC	miR-20, -106	103479	Retinoblastoma-like protein 2 (*RBR-2*)
	UCCAGUU	miR-145	151702	Friend leukemia integration 1 transcription factor (*FLI-1*)
	GCAGCAU	miR-103, -107	137309	High mobility group protein HMG-I/HMG-Y (*HMG-1(Y)*)
	GGAAGAC	miR-7	136826	Kruppel-like factor 4 (*EZF*)
Signal transduction/ cell-cell signaling	UAAGGCA	miR-124a	168610	Signal transducer and act. of transcription 3 (*STAT3*)
	UGGUCCC	miR-133, -133b	010610	T cell surface glycoprotein CD4 precursor
	UCACAUU	miR-23a, -23b	107562	Stromal cell-derived factor 1 precursor (*SDF-1*)
	GCUACAU	miR-221, -222	157404	Mast/stem cell growth factor receptor precursor (*C-KIT*)
	GGAAUGU	miR-1, -206	176697	Brain-derived neurotrophic factor precursor (*BDNF*)
	UAAGGCA	miR-124a	154188	Angiopoietin-1 precursor (*ANG-1*)
	GGCAGUG	miR-34	148400	Notch homolog protein 1 precursor (*HN1*)
	CCCUGAG	miR-125a, -125b	128342	Leukemia inhibitory factor precursor (*LIF*)
	AGUGCAA	miR-130, -130b	184371	Macrophage colony stimulating factor-1 precursor (*MCSF*)
	UCACAGU	miR-27a	184371	Macrophage colony stimulating factor-1 precursor (*MCSF*)
	AAUACUG	miR-200b	008710	Polycystin 1 precursor
	GAAAUGU	miR-203	122641	Inhibin beta A chain precursor (*EDF*)

Table 2.   (*Continued*).

Category	Seed	MiRNAs	Ensembl ID	Gene Name
	AUUGCAC	miR-25, -92	065559	Dual spec. mitogen-activated protein kinase 4
	GCUGGUG	miR-138	070886	Ephrin type-a receptor 8 precursor (*HEK3*)
	GUAAACA	miR-30(a-e)	156052	Guanine nucleotide-binding protein G(1), alpha-2 subunit
	AUUGCAC	miR-25, -92	156052	Guanine nucleotide-binding protein G(1), alpha-2 subunit
	GAGAACU	miR-146	175104	TNF receptor-associated factor 6 (*TRAF6*)
	GGCUCAG	miR-24	166484	Mitogen-activated protein kinase 7 (*ERK4*)
	GAGAUGA	miR-143	166484	Mitogen-activated protein kinase 7 (*ERK4*)
	AGCUGCC	miR-22	166484	Mitogen-activated protein kinase 7 (*ERK4*)
	GCAGCAU	miR-103, -107	141433	Pituitary adenylate cyclase act. polypeptide precursor
Other	GUGCAAA	miR-19a, -19b	171862	Phosphatidylinositol-3,4,5-trisphos. 3-phosphatase (*PTEN*)
	AGUGCAA	miR-130, -130b	130164	Low-density lipoprotein receptor precursor (*LDLR*)
	GGAAUGU	miR-1, -206	160211	Glucose-6 phosphate 1-dehydrogenase (*G6PD*)
	UUGGCAC	miR-96	101986	Adrenoleukodystrophy protein (*ALDP*)
	AGCACCA	miR-29b, -29c	168542	Collagen alpha 1(111) chain precursor
	AGCACCA	miR-29b, -29c	114270	Collagen alpha 1(VII) chain precursor
	AUUGCAC	miR-25, -92	168090	*COP9* subunit 6
	AAGUGCU	miR-93	168090	*COP9* subunit 6
	AAAGUGC	miR-20, -106	168090	*COP9* subunit 6
	CCCUGAG	miR-125a, -125b	160613	Proprotein convertase subtilisin/kexin type 7 precursor

The 442 predicted targets conserved between human, mouse and rat were ranked based on the number of references listed in the RefSeq GenBank flatfiles (11/10/03 download). The top 37 most referenced predicted targets are shown, grouped on the basis of Gene Ontology annotations. The last six digits of the Ensembl ID are shown (ENSG00000#). MiRNAs with different seeds that target the same UTR are listed on separate lines. (From Lewis *et al.*, 2003.)

(Bernstein *et al.*, 2003) of the Cold Spring Harbor Laboratory, New York. This communication reports on a very elegant molecular genetic study that probed into the process of endogenous silencing of mammalian genes (i.e. miRNA-type genes) and their role in mammalian development. These investigators asked what the effect of disrupting the processing of endogenous RNA silencers on mice development would be. More specifically, the question was posed on how mice development would be affected when the *Dicer* was eliminated from the mouse genome. For that, the investigators created a mouse strain with a chromosomal lesion in a *Dicer* gene, *Dicer1*. They prepared a gene-targeting vector for *Dicer1* using an *in vivo* recombination strategy and replaced exon 21 of *Dicer1* with a neomycin-resistant cassette. The replacement caused the splice donor and acceptor sites of the exon to remain intact, thus increasing the chance that the entire neomycin-resistant cassette would be incorporated into the *Dicer1* transcript while disrupting the transcript with respect to *Dicer1* expression. From experiments with mutated human *Dicer* the investigators knew that a defect in *Dicer* causes an inability to cleave long dsRNA into siRNA. The construct with the exchanged exon 21 (having neomycin-resistance) was introduced into ES cells (these could be selected on neomycin-containing medium). The investigators then obtained ES cell lines with disrupted *Dicer1*. Such ES cell clones were used to create chimeric mice. Two of these mice transmitted the disrupted *Dicer* allele through the germ lines. When 62 mice were born from heterozygous intercrosses, none of the viable offsprings was homozygous for the defective *Dicer1* allele. By following embryo development it appeared that in homozygous defective-*Dicer1* embryos the development was disrupted at a rather early stage before the body plan is configured during gastrulation. In conclusion, an intact *Dicer* activity is essential for the very early stages of mice-embryo development, indicating that some miRNAs are active in the very early stages of mammalian embryonal differentiation. In a way the study by Bernstein *et al.* (2003) is an extension of the previous study by Doi *et al.* (2003) of Tokyo and Shizuoka, Japan. The latter investigators already showed that for RNA silencing by an *externally* supplied dsRNA to mammalian cells, the *Dicer*, as well as the PIWI family members, the eIF2C translation initiation factors are required.

An approach to search for miRNA targets in mammals that was very different from those of Lewis *et al.* (2003) and Doench and Sharp (2004) was taken by a team of investigators from Southern Australia (Michael *et al.*, 2003). These investigators looked for such targets by tracing a possible association between the formation of solid tumors and specific miRNAs. For that, they first identified miRNAs that are found in uniquely colorectal tumors. On the other hand, they also looked for miRNA that consistently *fail* to accumulate to normal levels in precancerous and cancer tissues. To identify miRNAs the investigators fractionated the RNA and focused their attention on those sizes that could be miRNA. These latter RNAs were cloned and analyzed. In the tumor tissue 19 putative miRNA clones were found. Some of these were known human miRNAs. Others were known from mice or were previously not identified. More putative miRNAs were cloned from normal tissue. Again, some were new and others known from previous studies. Using northern blot hybridization with the putative miRNA sequences as probes, the normal and the tumor tissues were compared. The levels of two (murine) miRNAs, miRNA-143 and miRNA-145 were significantly lower in tumor tissue. Interestingly, this difference was with respect to the mature miRNA but not with respect to the hairpin precursors. Probing northern blots from different cancer tissues with miRNA-143 (and miRNA-145) indicated similar results, meaning that miRNA-143 (and also miRNA-145) is down-regulated in cells derived from breast, prostate, cervical and lymphoid cancers as well as colorectal tumors. Finally, the investigators compared the sequences of miRNA-143 and miRNAs with known mRNA targets and came up with a list of nine human miRNAs that are putative targets of miRNA-145. In a way the study by Michael *et al.* (2003) is an extension of the previously mentioned study by Calin *et al.* (2002) but the latter focused on B-CLL and revealed a deletion in chromosome 13 that codes for two miRNAs rather than a reduction of the processing from pre-miRNA to mature miRNA, as reported by Michael *et al.* (2003).

Thomas Tuschl and collaborators (Meister *et al.*, 2004), now based at the Rockefeller University, returned to the question of miRNA silencing mediated by either or both repression of translation and mRNA degradation in mammals. To answer this question these investigators considered that the validation of targets for miRNAs is an essential step. They assumed that the inactivation of specific miRNAs at the stage of mature (single-stranded) miRNAs that bind to miRNP complexes will reveal which potential targets are not silenced after the specific inactivation of the miRNA/RNP complexes. Consequently, they developed a method to render the mature miRNA inactive in a specific manner. For that, the investigators focused on miRNA-22. They constructed a 24 nt 2′-O-methyl oligo-ribonucleotide that had homology to the mature miRNA-22 and transfected it into HeLa cells. It was previously known that the highly expressed miRNA-22 will quickly degrade a substrate RNA that has homology to miRNA-22. When the substrate is tagged by ^{32}P-cap-labeling the rate of substrate degradation can be monitored. The level of miRNA-22 in the (S100) cell extract was previously found to be 50 pM. The activity of the substrate degradation by the 2′-O-methyl oligoribonucleotide was completely blocked by 3 nM of this inhibitor and 0.3 nM already reduced the degradation by 60–70 per cent while 0.03 nM had no affect. Further experiments indicated that RISC and miRNP complexes (with specific miRNAs) can be effectively and sequence-specifically inhibited with 2′-O-methyl oligonucleotides that are antisense to the guide (miRNA) in the RNA silencing complex. This technique should therefore be useful in assessing the function of miRNA genes expressed in cultured cells.

The gospel of 2′-O-methyl oligoribonucleotides as a possible specific inhibitor of small RNA-mediated gene silencing reached not only the Rockefeller University (Meister *et al.*, 2004) but also the Medical School of the University of Massachusetts in Worcester. Hence, a team of Phillip Zamore and associates (Hutvagner *et al.*, 2004) reported on a research that was similar to the research of Meister *et al.* (2004). The latter investigators submitted their manuscript on November 18th (2003) and it was accepted on December 11, while the Zamore team submitted theirs on November 4th (2003) and it was accepted on January 30th (2004). The Zamore team used the 2′-O-oligonucleotides to inhibit the cleavage of mRNAs in *Drosophila* and in humans. In *Drosophila* it was found that when an embryo lysate with RISC complexes and siRNA that was homologous to a reporter mRNA (coding for firefly luciferase) was challenged with an appropriate 31 nt

2'-0-methyl oligonucleotide, the degradation of the reporter target-mRNA was prevented. There were several experimental results that suggested that the inhibition of cleavage reflects binding of the oligo to the RISC (rather than to the mRNA). But the exact mechanism may only be revealed by further studies. From flies the investigators turned to man, performing *in vivo* experiments with HeLa cells. Various levels of siRNAs were transfected into the cells, then reporter mRNA was transfected together with the appropriate 2'-O-methyl oligonucleotides. The results were again measured by the luminescence of the reporter (coding for luciferase). Increasing the levels of the oligos gradually eliminated the degradation of the reporter mRNA. From this experimental approach it became clear that the inhibition of silencing could not be a consequence of the oligo displacing the sense strand of the siRNA from the RISC. The experiments were also performed with oligos that are complementary to *let-7* (i.e. a miRNA in many animals including nematodes and humans). Furthermore, *C. elegans* worms that were challenged with oligos complementary to *let-7* miRNAs showed similar defects of larval development as loss of function of *let-7* such as weak cuticles, defects in egg-laying and loss of adult-specific cuticular structures. The assumption that the *let-7* effect was indeed suppressed by the oligos was verified by using worms with mutated *lin-41*, the expression of which is the natural target of *let-7* silencing. Indeed, in such mutants the oligos did not cause the developmental defects. Practically, as in the study by Meister *et al.* (2004), the study by Hutvagner *et al.* (2004) indicated that the 2'-O-oligonucleotides can disrupt the function of specific miRNAs and thus are useful tools for dissecting the function of numerous miRNAs found in animals.

The lethal gene *let-7* brought vitality to the miRNA research.... As narrated already previously in this book, *let-7* was first found in the nematode *C. elegans* as the second miRNA gene and soon after it was found (with very slight changes in nucleotides) in all other metazoan model-organisms. One of its variants, *miR-7a*, exists in the human genome and was successfully used by a team from the School of Medicine of the University of Pennsylvania (Nelson *et al.*, 2004), to further study miRNP:mRNA association. This team (e.g. Mourelatos *et al.*, 2002; Dostie *et al.*, 2003) used the term *miRNP* and identified *miRNP* as an essential particle for miRNA-mediated gene silencing that contained several protein components as the argonaute-family member eIF2C2, Gemin 3 and Gemin 4. It was also previously found that another complex, SMN, that is involved in the spinal muscular atrophy (SMA) disease in humans shares components (Gemin 3 and Gemin 4) with miRNP. Hence, loss-of-function mutations of SMN in SMA may also affect the activity of human miRNP. Is Gemin 3 (a presumed helicase) associated with the miRNPs of motor neuron cells that are specifically affected in SMA? The investigators from Pennsylvania (Nelson *et al.*, 2004) therefore study the characterization of miRNAs associated with miRNPs of neuronal cells of mouse and of man. Analyses (by immuno precipitation and appropriate labeling and gel electrophoresis) indicated that in extracts of mouse motor neuron (MN-1) cells and human retinoblastoma (Weri) cells, Gemin 3 is associated with small RNAs, suggesting that Gemin 3 interacts with small RNAs in the miRNP of Weri cells and probably also in other mammalian cells. Nelson *et al.* (2004) argued that information on the composition of miRNP particles and their assembly will be helpful to understand the role of miRNP particles in the recognition of the target mRNA by the specific miRNAs. An additional term was suggested by these investigators. They referred to the sequences that the miRNAs recognize their mRNA targets as *miRNA recognition elements* or MREs. The investigators used lysates from the human retinoblastoma cell line Weri. They separated the lysate (on sucrose gradients) and identified the various components by immunological methods. These methods and a further analysis of the polysomal fraction clearly showed that the miRNP (15S) particles are associated with polysomes and contained eIF2C2, Gemin 3, Gemin 4 and miRNAs that were individually (e.g. *miR-124a*) identified miRNAs. It was also found that *let-7b* and 28S rRNA were associated with eIF2C2 in the polysomal fraction. The question was then posed if MERs can be recognized in which specific miRNAs will be bound to specific mRNA regions. It should be recalled that while in plants there is commonly an extensive complementarity between miRNAs and their targets on the mRNA, there is only a partial complementarity in animals. The investigators thus had to devise a computational-experimental approach to identify putative MREs, such as those for the *let-7b* in the 3'-UTR of the respective mRNA (*lin-28*). The *lin-28* MRE for *let-7b* should cause miRNA-dependent translational repression and so the investigators constructed a plasmid in which the 3′ end of the Renilla luciferase mRNA was replaced with the MRE of *lin-28* (or its mutant, for control). Hybridization of the MRE with *let-7b* should silence the reporter gene. This construct (and controls) were transfected into Weri cells (that normally express *let-7b* and other *let-7* paralogs). Two days after transfection there was a 5-fold reduction of the reporter's translation without a reduction of the respective mRNA level. Similar results were also obtained with other mammalian cells that normally express *let-7b*. It was thus concluded that human and murine *lin-28* mRNAs contain the required MRE and thus are targets for *let-7b* silencing. Further experiments indicated that the *lin-28* mRNA is associated with polysomes and with eIF2C2, and that probably the miRNPs associate with the mRNAs exclusively in the polysomal fraction. In other words, the miRNAs are present in the form of miRNP in the polysome-containing fraction where they form a stable association with their mRNA targets. These findings require the clarification of several additional issues. One unsolved question was whether there is a direct binding between the miRNA (*let-7a*) and its MRE (of *lin-28*) or the binding is "bridged" by other factors. It could also be helpful if a cell line with a mutated *lin-28* that lost the MRE for *let-7* become available. Such a mutant should have a constantly high level of the Lin-28 protein. It is also not clear how a miRNP with the appropriate miRNA "decides" either to degrade the mRNA or to suppress translation. *In vitro* experiments showed that *let-7b* containing human miRNPs can cleave RNA sequences that have sequences that are homologous to *let-7b*. The authors favored the possibility that the "key" to the "decision" is an Argonaut protein (as eIF2C2) that has endonuclease activity and senses the rate of MRE/miRNA homology. When fully homologous the Argonaut protein will be structurally changed and become an active endonuclease, and the mRNA is cut. But when there is only partial complementarily the Argonaut protein will have no endonuclease activity. Finally, the investigators suggested that the biochemical isolation and cloning of miRNA targets from polyribosome-associated miRNPs should be a plausible procedure. In a way this team used this concept (Dotsie *et al.*, 2003) in their big-scale cloning and characterization of miRNAs that were derived from miRNPs of neural cells of mouse and man.

Further Information on miRNAs, Their Tissue Specificity and Their Targets

Extensive reviews on various aspects of miRNAs, including methodologies used to study the occurrence of miRNAs in mammals and the identification of the miRNA targets, were published by Lai (2003) and by Bartel (2004). These authors, from the West Coast (Berkeley) and the East Coast (Cambridge, MA) of the USA, respectively, covered the subject thoroughly but in a very different manner. This difference is apparent already from the respective titles, *miRNAs*: *Runs of the Genome Assert Themselves* (Lai) and *miRNAs*: *Genomics, Biogenesis, Mechanism and Function* (Bartel). For readers who intend to go more in depth in RNA silencing in mammals, both reviews are recommended.

An effort to obtain comprehensive data on the expression of a great number of specific miRNAs in several mammalian tissues was already made by Lagos–Quintana *et al.* (2002) of the Tuschl Laboratory, when Thomas Tuschl was based in the MPI in Göttingen (Germany). These investigators extracted total RNA from nine tissues of 18.5 week-old mice and from these RNAs they cloned short RNAs that had characteristics of miRNAs. These clones (of ~21 nt) were sequenced and assigned to locations on the mouse genome where the coding sequences were components of inverse repeats. This cloning yielded many miRNAs of which 34 were novel. In some of the extracted tissues (e.g. liver) there was only one miRNA (e.g. miR-122) while in other tissues there were one or two prominent miRNAs but also additional miRNAs. Some of the miRNAs were expressed in several different tissues. Also, an ortholog of the nematode *lin-4* miRNA was revealed in the brain tissue of the mouse (miR-125a, miR-125b).

An extension of this study was performed by the same investigators (Lagos–Quintana *et al.*, 2003). In this study human osteoblast sarcoma cells (Saos-2) were extracted in addition to tissues of adult mice, and the RNA was used to clone a total of about 600 putative miRNAs; 31 of these were revealed as novel miRNAs. The latter study actually confirmed the previous study of the Tuschl team but also added important information. It revealed the existence of certain miRNAs in many mammalian tissues while others, such as miR-208, were found only in one tissue, the heart. Some miRNAs are notably excluded from the neuronal tissues. The human Saos-2 cell line yielded six novel miRNAs. This may indicate that the final number of mammalian miRNA is larger than initially estimated (e.g. 200–250). In several cases, a human miRNA (miR-10) was found in a very large range of metazoa (e.g. mammals, fishes and insects). This wide-range existence of a miRNA in different metazoa was previously found for *let-7*.

The team of Phillip Sharp at MIT (Houbaviy *et al.*, 2003) focused on ES of mice and searched for miRNAs in undifferentiated and differentiated ES cells which are totipotent cell lines that are derived from the inner cell mass of the mammalian blastocyst. They can be induced to differentiate *in vitro*. Houbaviy *et al.* (2003) cloned 20–26 nt RNAs from undifferentiated and differentiated cultured ES cells. To identify the onset of differentiation, the levels of Oct4 mRNA was monitored since this mRNA is high in undifferentiated ES and decreases during differentiation. A total of 681 short RNA clones were sequenced and of these, 53 were identified as putative miRNAs (they constituted regions in genomic sequences that could transcribe RNA that will fold to hairpins). Of these, 32 miRNAs were identical to previously identified miRs. Additional five clones were homologous to known miRs and 15 clones were unrelated to previously identified miRs. Some novel miRNAs were encoded by genomic loci that were clustered. Take, for example, miR290, miR291 and miR292 were in one cluster while miR293, miR294 and miR295 were at a nearby locus. These clustered miRNAs were expressed in undifferentiated ES cells but repressed during differentiation. They are undetected in mature mice. It thus seemed that some miRNA are (causally?) related to the maintenance of the undifferentiated, pluripotency of cells.

Gary Ruvkun, Kenneth Kosik and other investigators from the Harvard School of Medicine (Kim *et al.*, 2004) focused on the mammalian brain and their earlier work was mentioned above. They

searched for endogenous miRNAs in mammalian brain preparation and explored the function of these miRNA and their regulated
expression. By these studies these investigators intended to get information on the temporal sequences in brain development. They considered the accumulated information on the impact of translational
control on neural development and on the maintenance and plasticity of neural connections. It has previously been found that some
mRNAs and their translational machinery, including ribosomes and
other non-coding RNAs, are localized to dendritic regions of neurons
and axons. Also, it was previously found that synaptic activity activates translation of certain mRNAs. In all these translational controls
there is a possibility that specific miRNAs play an important role.
These considerations tempted Kim *et al.* (2004) to search for miRNAs
in the brain of mammals. Their "lampposts" were primary cell cultures (E18) of cerebral cortices derived from rat embryos. The miRNAs
were cloned from extracts of cells grown in culture for 1.5, 7 or 14 days.
Cloning, sequencing and verification (e.g. by northern blot hybridizations) were performed by procedures that I have mentioned above.
The predicted hairpin structures and respective codes in the genome
were derived and these coding sequences were identified in the mouse
and rat genomes. Then it was determined which of the identified miRs
are novel. In total this effort yielded 86 miRs, 40 of which were
novel.

The number of times a specific miR was cloned varied greatly.
There was only a single clone for each of 32 miRs while miR-125 was
found in 71 clones. Thirty two out of the 40 new rat miRs had codes
in both the rat and the mouse genome. Probably, some of the rat miRs
are not coded by the mouse genome but the lack of a code in the existing sequence of the rat genome could result from the incompleteness
of the rat genome at the time of this investigation. Some novel miRs
could not be detected by northern blot hybridization to total RNA of
adult rat cortex. This could result from either very low expression of
these miRs so that they evaded detection or that indeed these miRs
were expressed in the cultured cells but not in the cortex of adult rats.
An examination of a sample of 12 new miRs indicated that eight of
the miRs found in rat neurons were "endemic" to neurons while four
are also expressed in other tissues. The former eight miRs were found

in association with polysomes, suggesting a role in the regulation of translation.

The Harvard Medical School team (Krichevsky *et al.*, 2003) that was also mentioned above, utilized the novel rat miRs identified by Kim *et al.* (2004) as well as miRs from the mouse brain identified by Lagos–Quintana *et al.* (2002) to obtain more detailed information on the regulation of miRNA expression during mammalian brain development. To seek this information Krichevsky *et al.* (2003) developed an oligonucleotide DNA array. For that oligonucleotides (antisense to the miRs) were spotted (at 7 μM) on appropriate membranes. RNA samples enriched in short RNAs were radiolabled and hybridized to the spotted membranes. The procedure included replications and due controls.

Forty-four miRNAs were included in this array. The radio-labeled RNAs were derived from the forebrain of prenatal, juvenile and adults rats. The array assay identified nine miRNA candidates that had differential expression during brain development. Some miRNAs (e.g. miR-131) change about 4-fold from one developmental stage to another but in other miRNAs (e.g. miR-124a) the signal increased 13-fold from stage to stage. Also, specific miRNAs were increased only in later corticogenesis (e.g. miR-128). Generally, the miRNAs were up- or down-regulated during brain development. The phylogenetic conservation of miRNA regulation was rather amazing. Take, for example, miR-9 that was down-regulated during rat brain development was previously found to be down-regulated during the development of *Drosophila*. The miR-9 is also conserved in man (with codes in chromosomes 1, 5 and 15) and in mice.

Presenilin-1 (PS1) is essential for the normal development of the central nervous system (CNS): PS1 deficient mice show lethal defects during CNS development. When RNA from the forebrain of PS1-deficient mice was analyzed the two miRNAs, miR-9 and miR-131 were found to be reduced. In conformation with results narrated by Kim *et al.* (2004), miRs were especially detected in the polyribosome fractions of neural tissue, further strengthening the assumption that the miRNAs in this tissue have a role in the regulation of translation.

David Bartel, Harvey Lodish and associates (Chen *et al.*, 2004a) chose the hematopoietic-lineage differentiation for investigating the possible role of specific miRNAs in modulating this differentiation in mice. The differentiation of the hematopoietic lineage in mammals is a continuous process from pluripotent hematopoietic stem cells in the bone marrow to lymphoid and myeloid progenitors, and from these to at least eight different blood lineages (for more details, the readers can refer to Fig. 7 in the book of Galun and Galun, 2001).

As a first step toward testing the notion that miRNAs may play a role in the regulation of mammalian hematopoiesis, Chen *et al.* (2004a) cloned about 100 miRNAs from mouse-bone marrow and sequenced them. The sequences indicated that most of these miRNAs were already identified in vertebrates. The investigators choose four miRNAs according to their expression in mice organs. The miR-16 was found in diverse mice organs and was therefore further used as a control/reference miRNA. The miRNAs miR-181, miR-223 and miR-142s were differentially or preferentially expressed in hematopoietic tissues. MiR-181 was strongly expressed in the thymus (the primary lymphoid organ that "educates" T-lymphocytes to react with antigens) but there was also expression of miR-181 in other mice organs (e.g. brain, lung, bone marrow, spleen). MiR-223 was nearly exclusively expressed in the bone marrow where hematopoietic stem cells, as well as their derivatives of various differentiation stages, reside. MiR-142s was highly expressed in all tested hematopoietic tissues (e.g. liver, bone marrow, spleen, thymus) but with little or no expression in non-hematopoietic tissues (e.g. heart, kidney, lung, muscles). The investigators therefore isolated cells from within the bone marrow and sorted them according to lineage markers. Then the RNAs from sorted cells were analyzed by northern blot hybridization to each of the four miRs: miR-16, miR-181, miR-223 and miR-142s. Indeed, cells with the different lineage markers expressed the miRNAs differentially, with the exception of miR-16 that was expressed in all of the cell types. The expression of miR-142s was lowest in erythroid and T-lymphoid lineages and highest in B-lymphoid and myeloid lineages. The expression of miR-223 was confined to myeloid lineages. MiR-181 was very low in undifferentiated bone marrow cells (Lin$^-$ cells) but then increased only in B-lymphocytes. These tests clearly indicated a correlation between the differentiation in the various lineages and the three miRNAs but the results did not assure that these three miRs actually affect the direction of differentiation. To seek indications for the miRNAs role in differentiation the investigators introduced the codings for the miRNAs into hematopoietic progenitor cells. They thus constructed expression cassettes for retrovirus-mediated introduction of the codes. It was first found that a short coding sequence for the hairpin precursor will not cause the expression of the mature miRNA. The sequence with the hairpin code should be $\sim$270 nt or longer in order to be transcribed and processed correctly; the hairpin-forming region should be flanked by about 125 nt genomic sequences. Vectors with the appropriate codes for the miRs could then cause the expression of the respective miRs in 293T cells and in bone marrow cells. Lin$^-$ hematopoietic progenitor cells from mouse bone marrow were then transformed with viral vectors that expressed miR-181, miR-223, miR-142s or miR-30. The onset of differentiation in the transformed cells could be traced by the surface markers. Each of the ectopic miRs, but the miR-30, had a specific effect on the differentiation.

The investigators then turned to *in vivo* experiments with ectopic miR-181. For that mouse Lin$^-$ bone marrow cells were transformed with the miR-181 containing vector and the transformed cells were then transplanted into lethality irradiated mice where they reconstituted all the blood lineages. After 4.5 weeks the lineage composition of peripheral blood cells was found to have a substantial increase in B-lymphoid cells. In parallel the T-lymphoid cells were decreased especially in the CD8$^+$ T cells from 16 to 1.2 per cent. The *in vivo* results were in accord with the *in vitro* experiments. In the case of miR-181 the results suggested that this miRNA is a regulator of B-cell differentiation. But which, if any, regulatory protein is directly affected by each of the three miRs is still an open question. Obviously, other mechanisms of gene modulation can play roles in hematopoietic lineage differentiation.

Several reviews, such as those of Kaempfer (2003), Kawasaki *et al.* (2004), Dorsett and Tuschl (2004) and Mittal (2004) should be useful for readers who wish to obtain more detailed information on miRNAs in mammals.

Chapter 11

RNA Silencing in Angiosperm Plants I: Externally Induced Silencing

I shall begin this chapter with a tribute to my late friend Professor Raphael Frankel (with whom I wrote my first book: Frankel and Galun, 1977). It is probable that Frankel was actually the first who observed RNA silencing in plants (Frankel, 1956) while he was a graduate student at UC–Davis. He reported in the *Science* magazine on a transgraft transmission of male sterility (ms) from an ms stock to a fertile scion in Petunia. He found that a fraction of the sexual progeny of the grafted scion was ms. A certain percentage of the latter ms plants further transmitted the ms to the next sexual generations. These results were, at the time, considered "bizarre" but they were independently repeated by others and even served commercial Petunia breeders to convert fertile Petunia lines to ms and thus provided seed-parents for the production of hybrid Petunia cultivars. The mechanism of this transfer of ms remained a riddle. As mentioned briefly in Chap. 2, after more than 40 years, the team of Harve Vaucheret and associates (Palanqui *et al.*, 1997) of the INRA in Versailles, France, reported on transgraft transmission of RNA silencing (cosuppression) from stocks to their scions, in tobacco. These investigators caused co-suppression of nitrate reductase and nitrite reductase in transgenic tobacco plants and used these plants in grafting experiments. The co-suppression moved from the silenced stocks into the scions. They concluded that the PTGS can move in the plant and across graft-junctions. As observed by R. Frankel, the transfer of silencing reported by Palauqui *et al.* (1997) was also unidirectional: from the stock to the scion.

In Chaps. 1 and 2, I have mentioned early observations of co-suppression in plants and discussed the defense of plants against viruses that led to the degradation of viral RNA. The respective studies provided useful information and led to means of antagonizing viral pathogenicity in crop plants. But the molecular details of the mechanisms of viral inhibition were not clarified by these early studies on PTGS and on TGS in plants. In this chapter, the focus will be on the intentional silencing of specific genes in plants and on procedures of "reverse genetics" that were developed in order to study the roles of known coding sequences in the metabolism and differentiation of plants. Because the role of genes in plant patterning became an intensively-studied subject and miRNA forming genes were revealed in plants (as they were revealed in other organisms as indicated in previous chapters), Chap. 12 will specially focus on miRNAs and patterning in plants. During rather early studies on co-suppression and PTGS in plants it was found that certain plant viruses code for proteins that negate the gene silencing in plants. Once discovered, it became clear that these proteins should be considered in studies on PTGS, especially when viruses are recruited as vehicles for the integration of transgenes or RNA species into plants for the induction of gene silencing. This subject will therefore be mentioned in this chapter and a special section will be devoted to viral suppression of gene silencing.

The title of this chapter truly represents the content but for two exceptions. The team of Ralph Quatrano of the Washington University in Saint Louis, MO, showed that RNA silencing is also applicable in the moss *Physcomitrella patens* (Bezanilla *et al.*, 2003). When a transgene coding for GFP was introduced into cells of this moss, the GFP gene could be silenced by co-integration of plasmids that caused RNA silencing either transiently (for 8 days) or constitutively.

A team of Heriberto Cerutti (Wu–Scharf *et al.*, 2000) from the *Beadle Center* of the University of Nebraska in Lincoln, Nebraska, dealt with transposon silencing in the green unicellular alga *Chlamydomonas reinhardtii*. Here, I would like to remind the readers that George Beadle was a pioneer of biochemical genetics who received the Nobel Prize for his studies in *Neurospora* genetics. He was born on a farm in Nebraska and did his undergraduate studies at the

University of Nebraska. He then moved to Cornell University, Stanford University, California Institute of Technology and finally went to the University of Chicago where he became President of the University (see the well-written biography of George Beadle by Berg and Singer, 2003).

Silencing of Plant Genes

Early studies: 1990 to 1998

One of the early publications on PTGS was that by the team of D. Grierson of the Nottingham School of Agriculture, UK. I have mentioned this study in Chap. 1 and discussed the reduction of the expression of the enzyme polygalacturonase (PG) in tomato, with the intention of delaying fruit softening by co-suppression. In a short publication (Smith *et al.*, 1990) these investigators provided some details of the silencing of the endogenous PG gene. They transformed tomato plants with a chimeric PG gene that was designed to produce a truncated transcript of PG constitutively. Consequently, the endogenous PG gene was inhibited during ripening: both the mRNA and the PG protein were reduced. The authors assumed that the truncated transgene caused degradation of the endogenous mRNA but also added: "Although it is possible to envisage that RNA duplex formation by sense-antisense interactions renders the mRNA unstable, extensive duplex formation would not be expected to occur with sense transgene transcripts." The mechanism of this co-suppression remained enigmatic at the time.

Van Blokland *et al.* (1994) of the Biocentrum in Amsterdam continued the study by Van der Krol *et al.* (1990) on co-suppression of the gene chalcone synthase (*chs*) in Petunia that led to white or variegated flowers. The former investigators introduced constructs that contained the full length of the *chs* cDNA or only parts of this cDNA into Petunia plants. In part of the resulting transgenic plants the transgene and the "resident" *chs* were silenced. This was manifested by lower respective mRNA while the rate of *chs* transcription was not reduced. The investigators were surprised to find that even *chs* coding sequences devoid of a promoter caused, in a few cases, co-suppression of the *chs* gene. A possible explanation I can provide is that in such cases, the *chs* sequence was inserted downstream of a strong promoter in the Petunia genome.

R.B. Flavell (1994), then at the John Innes Centre, near Norwich, UK, reviewed the information up to early 1994, on the induced inactivation of gene expression in plants. Some of the reviewed cases were "intentional" while others can be defined as "unintentional". Among the latter was a case in which, after *Arabidopsis* transformation, there were in the host genome, multiple, closely-linked copies of the hygromycin phosphotransferase gene (for resistance to hygromycin). The result was that some transgenic plants lost the hygromycin resistance (Schied *et al.*, 1991). Flavell came up with various plausible explanations for this phenomenon but none of these included the formation of short dsRNAs as triggers of specific gene-silencing.

When we were dealing with gene silencing in mammals by dsRNA in Chap. 9 the problem of using long dsRNA for the silencing of specific genes was elaborated. Such long dsRNA commonly caused, in differentiated mammalian cells, a cascade of reactions that elicits interferon (IFN)-induced phosphorylation of the PKR protein and consequently caused general silencing in these cells. This non-specific silencing, by long dsRNA, could be overcome in mammals by various means (one obvious way is to use dsRNA of less than 30 nt). It was considered a problem that is confined to vertebrates. But ... possibly a similar cascade also exists in plants. Investigators from the University of Wyoming, USA (Langland *et al.*, 1995) reported that plants contain an analogue of the vertebrate protein PKR. The phosphorylation of this PKR is stimulated in mammals by IFN formation. In plants the pPKR phosphorylation is stimulated directly by dsRNA during viral infection. The phosphorylated pPKR could cause apoptosis in plants. Contrary to certain reports there is probably no analogue of mammalian IFN in plants.

The possible problems that are involved in inducing silencing of genes in plants were exemplified by a study of the H. Vaucheret Laboratory in Versailles, France (Elmayan and Vaucheret, 1996). In this study the bacterial gene *UidA* that encodes the GUS reporter protein was engineered behind strong promoters and transported into the genome of tobacco by genetic transformation. The investigators obtained 11 transformants each of which had a single insertion of the transgene. While in some plants there was a strong expression of GUS, this expression was variable, spatially and temporarily in others. In some transgenic plants the GUS activity declined very early in plant development. The decline in GUS activity was found in homozygous progenies of the *UidA* insertion as well as in their derived haploids. The decline of GUS activity was not a result of a lower rate of transcription in the nuclei. The conclusion of Elmayan and Vaucheret (1996) was that the decline in GUS activity resulted from PTGS but the mechanism of this PTGS was not yet clear. An additional study of the Versailles Laboratory (Palauqui and Vaucheret, 1998) dealt with the movement of PTGS from stocks to scions, a system that I have mentioned in the first paragraph of this chapter. These authors previously found that a 35S (promoter)-*Nia2* (for nitrate reductase, NR) can impose, after due transformation into transgenic tobacco, the silencing of both the transgene and the endogenous NR. When such 35S-*Nia2* silenced plants were used as stocks and transgenic plants, in which the 35S-*Nia2* did not cause silencing, were grafted as scions, the scions became silenced. In their further experiments the investigators found that the silencing of *Nia2* could be maintained in plant organs, even after they did not contain the 35S-*Nia2* sequence. The authors thus concluded that the presence of a 35S-*Nia2* transgene is dispensable for the mRNA-degradation step of PTGS. Silencing in the absence of the original "silencer" elicits the memory of an anecdote about an Einstein lecture to the public. At the end of the lecture one member of the audience had a question. She understood how cable-telegraph worked but asked how wireless telegraph worked? Einstein's answer was that the cable telegraph is like a very long cat ... you pinch its tail in California and the cat sounds a "Miau" in New York. Wireless telegraph, Einstein explained, is the same as cable telegraph but without the cat. Back to the article of Palauqui and Vaucheret (1998). In this article the mechanism by which the transgene initiates the silencing was not specified although the article was submitted to the *Proceedings of the National Academy of Science, USA* (PNAS) 1 month after the *Nature* publication of Fire *et al.* (1998) appeared in print. In the latter publication it was shown that dsRNA is the trigger for PTGS.

It is expected of a tender to praise his merchandise ... David Baulcombe and associates of the John Innes Centre, near Norwich, UK, had a long experience with plant viruses. Hence, Angel and Baulcombe (1997, 1999) developed a virus-based procedure to cause specific gene silencing in plants. In their model they showed how a transgene, expressed in transgenic tobacco plants can be silenced by constructs that coded for a viral genome (Potato Virus X-PVX). The viral construct could be engineered in a way that the code for the coat protein (CP) of the PVX was replaced by the code of the gene that was subject for silencing. The system worked properly with transgenic tobacco plants that expressed the gene for GUS. In these plants the infection with the PVX/GUS construct caused viral replication and a strong silencing of the GUS transgene. The authors termed their procedure *Amplicon-mediated* gene silencing and recommended the procedure as an important new strategy for the consistent activation of gene silencing. The *Amplicon* system for high-level expression of alien genes in transgenic plants that was first developed by Angel and Baulcombe (1997, 1999) and mentioned above, was further developed by Mallory *et al.* (2002a). The latter study was done by investigators of the University of South Carolina, USA and the John Innes Centre, UK. In this study the investigators handled a problem that involved the use of *Amplicons*. The introduction and multiplication of the viral RNA in the infected plants initiates a PTGS, but the viral replication and its transgene "passenger" are gradually reduced. To overcome this situation the investigators also added to the *Amplicons* the TEV (tobacco etch virus) helper component-proteinase HC-Pro that negates the PTGS, and by that it allowed a high level expression of the *Amplicon*-contained transgenes.

The laboratory of Dominique Robertson of the North Carolina State University, Raleigh, NC, suggested Geminiviruses for the delivery of silencing constructs into target plants (Kjemtrup *et al.*, 1998; Peele *et al.*, 2001). Geminiviruses are plant viruses with a single-stranded DNA genome. Their DNA commonly exists in two partially identical rings. The geminiviruses replicate in the plant nuclei by using the plant replication machinery. In the genome of some of these viruses there is "space" for about 1 kb of alien DNA. Since they have a high rate of replication, geminiviruses are useful for the delivery of short DNA sequences into host cells. These investigators used modified tomato golden mosaic virus (TGMV) and infected *Nicotiana benthamiana* plants by the biolistic transformation procedure. They silenced either a transgene or an endogenous gene. The alien gene was the firefly luciferase gene and the endogenous gene was *SU* which is a coding for an enzyme that is required for chlorophyll biosynthesis. Homozygous *su* mutants are yellow ("Sulfur") and heterozygotes (*SU/su*) are yellow/green. For the silencing of *SU*, fragments of this gene were employed. Silencing was already achieved in their earlier study (Kjemtrup *et al.*, 1998) but the system was improved in a later study (Peele *et al.*, 2001) in which an additional endogenous gene was silenced with the geminivirus-derived vectors. It is noteworthy that while the geminivirus experiments were performed in the Robertson Laboratory in North Carolina, Vicki Vance of South Carolina participated in another plant virus-mediated silencing (by "*Amplicons*" of PVX), research. There is probably a gap between the two Carolinas as no mention of Vances research appears in Dominique Robertson's publications.

During the later years of the 1990s various laboratories continued to present cases of PTGS and TGS in plants and even suggested possible mechanisms for these silencings. I shall not detail all these studies but rather mention a few of them. The group of Richard Jorgensen (of UC–Davis, CA) proceeded to study the silencing genes involved in Petunia-flower pigmentation (Jorgensen *et al.*, 1996; Que and Jorgensen, 1998; Jorgensen *et al.*, 1998). Herve Vaucheret and associates (of Versailles, France) elaborated their gene silencing studies with the silencing of the GUS gene (*uidA*) but shifted to *Arabidopsis* transgenic plants and analyzed genes that are implicated with silencing, such as *sgs1* and *sgs2* (Elmayan *et al.*, 1998). The laboratory of Michael Wassenegger of the MPI in Martinsried, Germany (Schiebel *et al.*, 1998) isolated the cDNA for an RNA-directed RNA polymerase (RdRP or RdRp) in tomato and suggested that this RdRP renders a silencing RNA into dsRNA and that the later triggers the silencing. In another publication of this laboratory (Wassenegger and Pelissier, 1998) the authors proposed a model for RNA-mediated gene silencing in plants (Fig. 31). This model was published at about the same time as the results of Fire *et al.* (1998) were published in which it

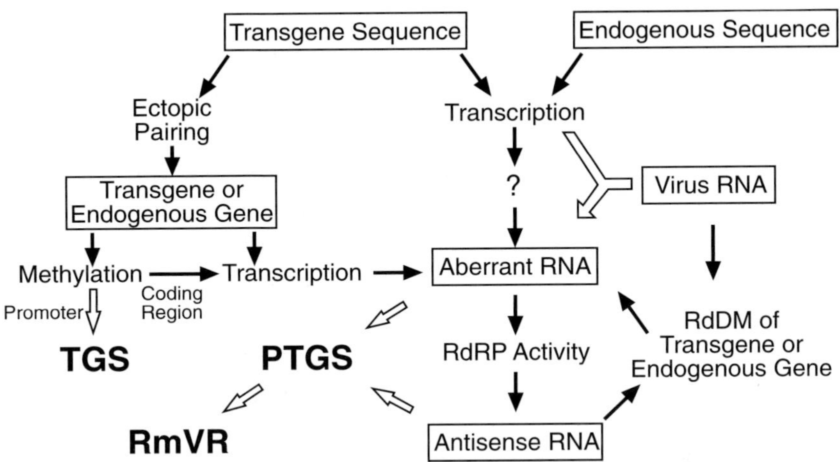

Fig. 31. Models used by Wassenegger and Pelissier (1998) for transgene mediated, homology dependent, gene silencing. Abbreviations: TGS, transcriptional gene silencing; PTGS, post transcriptional gene silencing; RmVR, RNA-mediated virus resistance; RdDM, RNA-directed DNA methylation; and RdPP, RNA-directed RNA polymerase. (From Wassenegger and Pelissier, 1998.)

became evident that dsRNA causes the degradation of homologous mRNA. I presented the Wessenegger model not as a representation of our present concept of the RNA silencing-mechanism in plants but rather as a "historical" suggestion that was since then modified extensively. Two additional laboratories came up with models of RNA silencing in plants before the specific role of externally applied dsRNA became evident (Fire *et al.*, 1998). One of these models was suggested by Metzlaff *et al.* (1997) of the R. Flavell Laboratory (John Innes Centre) and the other by Depicker and Van Montagu (1997) of the Department of Genetics of the University of Gent, Belgium. In both cases the authors admitted that the RNA silencing of plants was not sufficiently understood for providing a clear explanation for this phenomenon.

Following the awareness of the role of dsRNA in gene silencing

The laboratory of David Baulcombe had major contributions in the field of antiviral defence mechanisms in plants (see: Chap. 2), as well as in PTGS, during the early years of the study of these subjects in plants (see: previous section). This laboratory also continued its investigations after the publication of the Fire *et al.*'s (1998) article.

Hamilton and Baulcombe (1999) then examined the hypothesis that antisense RNA will hybridize with the sense RNA and will consequently cause the degradation of the respective mRNA. If such long dsRNA composed of a sense and an antisense (full length) RNAs did exist, they should have been detected as such long dsRNA in silenced plants. But no such long dsRNA was actually detected. The investigators thus devised procedures to detect short dsRNA. They first looked for short dsRNA in tomato plants in which the gene for the endogenous enzyme 1-aminocyclopropane-1-carboxylate oxidase (ACO) was previously silenced by PTGS, and had a very low level of ACO mRNA. In such plants the investigation revealed 25 nt RNAs that were homologous to the antisense of ACO mRNA. Such short RNAs were absent from transgenic tomato plants that were not silenced with respect to ACO. A similar search for small RNA was performed with tobacco plants that contained a transgene for β-glucuronidase (GUS) and in which this *GUS* gene was silenced by PTGS. There were two tobacco lines with this silencing. There was also a tobacco line in which the silencing of *GUS* was suppressed. Again, 25 nt long antisense RNA was found in the two silenced tobacco lines but not in the line in which the silencing was suppressed. Moreover, there was a correlation between the level of gene silencings and the amount of small 25 nt RNAs. In addition, the authors investigated the systemic spread of silencing and the systemic appearance of the respective small antisense RNA. For that, *Nicotiana benthamiana* plants were transformed with a construct that coded for the gene for green fluorescent protein (GFP). In some of these plants the fluorescence "faded" in the newly developing leaves, probably as a result of PTGS in the systemic leaves. The 25 nt antisense RNA for the *GFP* gene was detected and it was restricted to leaves in which the silencing took place. Similar detection of 25 nt RNAs, complementary to potato virus (PVX) RNA, were detected after inoculation of *N. benthamiana* with PVX. Interestingly, the investigators did not specify whether the 25 nt RNAs were single stranded or double stranded. Because the identification of the short RNAs was by northern blot hybridization the radio-labeled probe would detect both ssRNA and dsRNA on the blot.

A diversion from the regular PTGS procedure to silence specific genes was suggested by a team from the Boyce Thompson Institute

(Cornell University). This team (Beetham *et al.*, 1999) explored a specific type of gene silencing. The idea of these investigators was to cause a substitution of the codon for an amino acid in the wild-type sequence of a gene by a codon that will cause the replacement of this amino acid by another one. By that, a very specific mutation will be established. Take, for example, if the CCA codon is exchanged by a CAA codon the change will cause the replacement of proline by glutamine. The investigators designed for that self-complementary chimeric oligonucleotides composed of DNA and modified RNA residues and termed these constructs COs. They contained a 25 nt homology domain that will target the gene-region where the exchange of bases should occur. These 25 nt contained 5 nt for the actual replacement and were flanked by 10 nt on each side to cause a kind of site-specific homologous recombination. This approach was previously applied in mammals and resulted in specific phenotypical changes in mouse melanocytes. The Cornell investigators chose the gene that encodes the enzyme acetolactate synthase (ALS). Plants with the wild-type ALS are sensitive to sulfonylurea herbicides. But certain mutations in the ALS gene render the plants resistant to such herbicides. Consequently, the investigators used a specific CO to change the Proline-196 of ALS by another amino acid. This was done and tobacco cells with this change could be selected on a herbicide-containing culture medium. Similarly, a defective gene for GFP could be restored to wild type by the appropriate CO, causing reestablishment of fluorescence in transgenic plants that initially harbored the defective *GFP* gene. In the latter case it was not base substitution but rather a base addition that reestablished the wild-type coding sequence.

An interesting turn in the approach of silencing specific genes was brought forward by the laboratory of Peter Waterhouse and associates (Smith *et al.*, 2000). We mentioned this laboratory in Chap. 2 where anti-viral studies in plants were reported. Soon after the discovery that dsRNA is the trigger for PTGS in nematodes, this laboratory reported (Waterhouse *et al.*, 1998) that duplexes (i.e. dsRNAs) are much more efficient in gene silencing in plants than either sense (for cosuppression) or antisense RNA. In this publication Waterhouse and collaborators already indicated that the dsRNA can be synthesized *in vivo* when a transcript contains a self-complementary sequence that is annealed into a stem-loop structure. In their later publication (i.e. Smith *et al.*, 2000) this laboratory elaborated the dsRNA approach. They compared the rate of gene silencing by various constructs such as single-stranded sense and antisense RNAs, several stem-loop (or hairpin) structures and a specific construct in which two head-to-head RNA sequences are spaced by an intron that can be spliced. When this type of head-to-head construct with an intron as spacer between the two sequences was utilized to silence the *Arabidopsis* gene Δ12 desaturase (that catalyses the conversion of oleic to linoleic acid and see below the section on RNA silencing for crop improvement) a 100 per cent silencing was achieved. The same head-to-head construct but without the intron resulted in only 70 per cent silencing. Such constructs were also effective in silencing a pathogenic plant virus (potato virus Y, PVY).

The regeneration of functional transgenic wheat, maize and rice plants after genetic transformation is difficult; at least it was so in 2000. In a study intended to show that dsRNA induced gene silencing is possible in these important cereal crops, Schweizer *et al.* (2000) of the Institute of Plant Biology, Zurich, Switzerland, used leaf cells in their experiments. Leaves or leaf segments were transformed by bombardment with tungsten particles (of 1.1 μm in diameter) that were coated with supercoiled reporter plasmids and dsRNA (see: Galun and Breiman, 1997, for details on this transformation procedure). The investigators found that the reporter genes for GUS and GFP could be silenced in the cereal-leaf cells by co-bombardment with the appropriate dsRNA. Also, the endogenous genes for dihydroflavonol-4-reductases in maize and barley could be silenced (the silencing was observed by reduced red-anthocyanin pigmentation in the affected cells). The dsRNA for the dihydroflavonol-4-reductase of barley, *Ant18*, could be silenced by a plasmid that coded for an inverse-repeat sequence that, after transcription in the cells, formed an RNA hairpin. In addition, appropriate dsRNA also reduced the level of a gene of the fungus *Blumeria graminis (Mlo)* that is implicated in pathogenicity. This happened when the leaves of barley were infected by the fungus 40 hours after dsRNA delivery into these cells. The authors therefore suggested that direct delivery of dsRNA into cereal leaves should lead to a rapid and sequence-specific interference with gene function at the single-cell level.

It should be clear that studies on RNA silencing in plants paralleled such studies in animals. Moreover, while plant biologists pioneered in the studies of co-suppression and defence against viral pathogens that involved RNA silencing, there were by far more investigations in animal systems in this field than in plant systems. This "shower" of investigations in animal RNA silencing was accelerated after the publication of Fire *et al.* (1998). This book is divided in "drawers" of chapters that deal respectively with the various animals, fungal and plant systems, because it is easier to present the vast information in this manner. But actually plant, fungal and animal systems were investigated simultaneously. Features of RNA silencing that were revealed in certain animals served investigators of RNA silencing in plants and *vice versa*. It quickly became evident that RNA silencing in eukaryotes evolved early in the evolution and that many of the features of this silencing are shared by plants, fungi and animals. The similarity goes as far as between angiosperms, protists and mammals. In an early mini-review Bass (2000) already indicated these similarities and stressed the accumulation of information from various organisms that led to the gradual clarification of the RNA silencing mechanisms. Take, for example, the first reports on co-suppression were from plants. Then the role of dsRNA was reported in nematodes. The enzymatic activity that cuts dsRNA to fragments of about 21 nt was revealed in lysates of *Drosophila* cells that were transfected with dsRNA. Further valuable information on the RNA silencing (RNAi) was obtained when an extract from the syncytial blastoderm of *Drosophila* embryos served as an experimental system (Zamore *et al.*, 2000) as was detailed in chap. 9. All this tempted Vaucheret and associates (Fagard *et al.*, 2000) to look in more detail at some genes that encode proteins that are essential for RNA silencing in plants, fungi and animals. The three proteins SGS2, QDE-1 and EGO-1 were previously found to be required for PTGS in plants, quelling in fungi and RNAi in animals, respectively. Fagard *et al.* (2000) isolated *Arabidopsis* mutants that were defective in PTGS. These mutants were affected at the *Argonaute1 (AGO1)* locus. The protein AGO1 is similar to the QDE-2 protein of fungi and the RDE-1 protein of animals that are also required for RNA silencing. When the amino acid sequences of the AGO1 wild type and of the mutants were analyzed, the investigators found a change of one amino acid (leucine) that resides in a highly conserved motif. This motif also exists in QDE-2 and in RDE-1, and also in these proteins a change in this amino acid impairs RNA silencing. But there is a difference: fungi and animals with mutated QDE-1 and RDE-1, respectively, are impaired in RNA silencing but they are morphologically normal and viable. On the other hand, plants with an *ago1* mutation had several developmental abnormalities, including sterility. The authors thus suggested that RNA silencing in plants is required for normal differentiation. This was a prophetic suggestion, as we shall see in Chap. 12 where miRNAs were subsequently found to be required for plant patterning.

It should be noted that genes controlling steps of PTGS and RNAi are common members of gene families and the number of these family members varies in different organisms. Take, for example, the EGO-1/SGS2 gene family has at least seven family members in plants (e.g. *Arabidopsis*) and four family members in nematodes (*C. elegans*).

While previous studies of the Baulcombe Laboratory in defence mechanisms of angiosperms against pathogens involved mainly regular plant viruses, the publication of Papaefthimiou *et al.* (2001) that is the result of a collaboration between the John Innes Centre and the University of Crete, Greece, focused on the defence of plants against a viroid. Viroids are mainly pathogens of plants but there are animal viroids like the hepatitis delta virus. Viroids have a circular ssRNA genome of about 250–400 nt that replicates in the nucleus of the cell. While being a continuous circle, there are stretches of complementing nucleotide sequences that cause the circle to fold into dsRNA with bulges, in which the bases are not complementing. Papaefthimiou *et al.* (2001) chose the potato spindle tuber viroid (PSTVp) for their study. There are several isolates of PSTVp that differ in their virulence (mild, severe and lethal). Such isolates were used to infect tomato plants. After infection, short RNAs of 22 nt and 23 nt representing different domains of the viroid genome were produced in the plants. The short RNA species peaked at about 30 days after inoculation. Such a fragmentation was observed with all the different isolates of the PSTVp. There is an increase of ~22 nt fragments of RNA that are homologous to the viroids and later a reduction of such fragments led the investigators to suggest that a PTGS, against the viroid is initiated after a certain level of viroids is accumulated in the host cells. Based on these results and additional studies with plant viruses that inhibit PTGS the Vaucheret team, in collaboration with Peter Waterhouse (of the CSIRO in Canberra, Australia), suggested a model of a branched pathway for transgene-induced RNA silencing in plants. Because further studies in plants and animals provided additional information and consequently more novel models, I shall not detail the Beclin *et al.*'s (2002) model. In several studies, independent of the Versailles Laboratory, Peter Waterhouse and collaborators continued their studies with RNA silencing in plants (e.g. Wesley *et al.*, 2001; Stoutjesdijk *et al.*, 2002; Finnegan *et al.*, 2003; Helliwell and Waterhouse, 2003). The Wesley *et al.*'s (2001) study is actually a continuation of a previous research (Smith *et al.*, 2000) that was noted above to increase the efficiency of PTGS in plants. The emphasis was on constructs that encode hairpin RNA (hpRNA) in which the sequence homologous to a target gene will be arranged in inverted repeats so that these sequences will be repeated in a "tail-to-tail" (or "head-to-head") manner. For those who recall the Mother Goose Rhymes, the rhyme on the two-headed horse comes to mind:

> See, See! What shall I see?
>
> A horse's head
>
> Where his tail should be.

The study by Smith *et al.* (2000) showed that transcripts that will fold into hairpin structure can cause PTGS in plants and that the efficiency for this silencing is further increased when the space between the inverted repeats is an intron that will be spliced in the cells. In the Wesley *et al.*'s (2001) study the investigators designed a standard vector that they termed pHANNIBAL. In it the sense-intron-antisense middle region is flanked by restriction sites (*XhoI, EcoRI, kmpI, ClaI, HindIII, BamHI* and *XbaI*) so that this middle region, based on PCR syntheses from the target gene, can be inserted in a desired plasmid and serve as vector for genetic transformation. Technically, mass-silencing required some amendments of pHANNIBAL so that for large-scale silencing a vector termed pHELLSGATE was designed.

With the development of the *in planta* genetic transformation of *Arabidopsis* pHELLSGATE could be used for hundreds of silencings of specific genomic targets.

In a further study by the CSIRO in Canberra, Australia, Stoutjesdijk *et al.* (2002) used the *Arabidopsis* Δ12-desaturase (*FAD2*) gene (that was mentioned above) to compare the silencing efficiencies of three methods:

- co-suppression, CS
- hairpin RNA (hpRNA) or HP
- intron-spliced hairpin RNA (ihpRNA) or iHP

They also employed specific promoters in order to express the silencing in certain components of the plant (e.g. a promoter of *Brassica* that induces expression in the embryo and endosperm of developing seeds) and focused on different regions of the transcript of *FAD2* such as the 3′ UTR. The most efficient silencing was that by the iHP construct with the 3′ UTR of the *FAD2* transcript. One of the silenced transgenic *Arabidopsis* was propagated sexually for five generations and the silencing was maintained. Of the plants transformed with the HP construct 75 per cent had dramatically reduced Δ12-desaturase whereas only 50 per cent of the plants transformed by the CS (co-suppression) showed silencing of the *FAD2* gene. A detailed description on how the iHP vectors should be constructed was provided in a later publication (Helliwell and Waterhouse, 2003). The latter publication is recommended to those who plan to use RNA silencing in plants. I shall only present the schemes for the construction of pHANNIBAL, pKANNIBAL and pHELLSGATE (Fig. 32). For those who wonder about the pHANNIBAL/pKANNIBAL terms, the pKANNIBAL is not misspelled as pCANNIBAL but as pHANNIBAL in which the selectable gene encodes *Kanamycin* resistance rather than a gene encoding ampicillin resistance ...

Plant genes, especially those involved in differentiation, may receive long and unusual names. This happens particularly when a mutation was termed before it was well characterized. A typical case is the *Arabidopsis* gene termed *CARPEL FACTORY* or also *SHORT INTEGUMENTS*. It was only later found that the carpel and integument abnormalities were the result of a mutation in a gene that

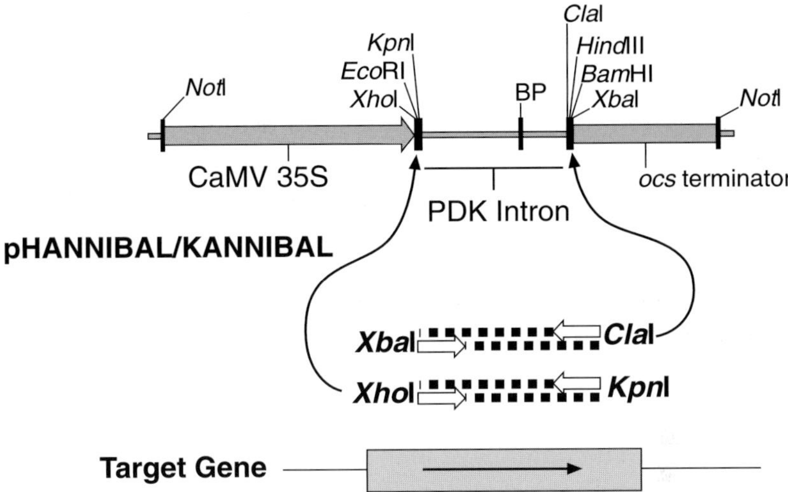

Fig. 32. (A) Outline of the use of the pHANNIBAL or pKANNIBAL vectors for gene silencing. A PCR product is amplified from the target gene with restriction sites incorporated into PCR primers. The products are digested with appropriate restriction enzymes and ligated into the appropriately cut vector. The two PCR products are usually ligated into the vector sequentially rather than simultaneously. (From Helliwel and Waterhouse, 2003.)

encodes a *Dicer*-like enzyme complex, and thus the *CARPEL FAC-TORY* gene of *Arabidopsis* is actually a homologue of the *Dicer-1* gene of *Drosophila*. Consequently, its name was changed to *DICER-LIKE1* (*DCL-1*). *Drosophila* mutants, defective in *Dicer-1* do not fragment dsRNAs into ∼21 nt. Finnegan *et al.* (2003) of the Waterhouse Laboratory and the Universidade Federal, Rio de Janeiro, Brazil, analyzed the role of *DCL-1* in RNA silencing of *Arabidopsis*. These investigators constructed a mutant of *DCL-1* (lacking the second dsRNA-binding poypeptide) and found that the intact (not-mutated) *DCL-1* is essential for silencing endogenous miRNAs of *Arabidopsis* but not for PTGS that is taking place after an introduction of a hairpin structure with homology to a transgene. Since there are four family members of DCL-1 in *Arabidopsis* that differ in their domain-compositions (e.g. *DCL-1* has a nuclear-localization domain that is lacking in *DCL-2*), the authors speculated that while *DCL-1* is active in the miRNA process, another *DCL* of *Arabidopsis* received the role of defence against viruses and against the transposition of TEs. I shall return to this question in Chap. 12 that deals specifically with miRNAs in angiosperms.

A distinction between two processes of the RNA silencing was suggested by the team of David Baulcombe of the John Innes Centre in the UK (Hamilton *et al.*, 2002). These investigators observed two distinct groups of small RNA species during RNA silencing processes: one of a size of 21–22 nt and the other of the size of 24–26 nt. They consequently related these two size-groups to "local" and "systemic" silencing, respectively. The summary of their experimental work was as follows. For "local" silencing they used *N. benthamiana* plants and applied pressure-injected *Agrobacterium* plasmids that included the gene for GFP that was activated by a strong promoter (35S) ligated upstream of the GFP gene. The patches of leaf-transformation resulted in strong green fluorescence that peaked 2–3 days after transformation. Analyses of the resulting small RNA indicated that there were two groups: of 21–22 nt and of 24–26 nt. The abundance of the two groups was about the same. When at a later stage the systemic spread of silencing took place samples were removed from the upper leaves that were not directly treated with the 35S-GFP construct. In the latter samples of leaves there were less small RNAs but the 21–22 nt group was much more abundant than the 24–26 nt group. The investigators also employed several virus suppressors of RNA silencing. Some of these (e.g. P1 of the rice yellow mottle virus) reduced the small RNAs differentially (e.g. the 24–26 nt size-group was much more suppressed than the 21–22 nt size-group). The levels of GFP mRNA in the transformed patches (local silencing) were inversely-related to the levels of the abundance of the 21–22 nt group. There was no correlation between the level of mRNA suppression and the abundance of the 24–26 nt group. In "systemic" leaves it was different. When the 35S-GFP was applied together with the viral suppressor of RNA silencing (e.g. P1) that at local silencing suppressed the appearance of the 24–26 nt group, the level of GFP in the upper (systemic) leaves was strongly suppressed. While if co-infected with a viral suppressor (e.g. AC2 of the African cassava mosaic virus, ACMV) that does not reduce the 24–26 nt group at the location of infiltration, there was only a slight systemic silencing of GFP. The investigators also analyzed the small RNA species that are homologous to retroelements that reside in the plants. These retroelements are considered to be contained by the RNA silencing mechanism. Only the 24–25 nt RNAs had homology to the retroelements. From these and additional experimental results the authors concluded that the long (24–26 nt) RNA species are required for systemic RNA silencing and possibly also for specific methylations that reduce transcription, but not for sequence-specific degradation of mRNA. The short (21–22 nt) RNA species correlate with mRNA degradation but not with systemic signaling, neither with methylation. It is beyond the scope of this book to provide all the experimental results of Hamilton *et al.* (2002). Though rather elaborate, this study left some questions unanswered. One of these is an obvious question: are the two size-groups of small RNAs resulting from one *Dicer* or are there different *Dicers* that generate each size-group? Thus, the two size-groups of siRNAs require further investigations, possibly after more details on the chemistry of silencing become available. The existence of two size-groups and their interpretations bring up a famous Talmudic story (Chagiga, page 15/1) about mysticism. It tells about four respected scholars who explored the celestial seat of God (the Divine Chariot). Two of them were so agitated by their vision that one died and another went crazy. A third (Elisha Ben Avuya) saw an impressive image sitting on a celestial throne and concluded that it was another image of God. This convinced him of the duality of God, according to the Persian faith of that time. Consequently, he left the Jewish faith. Only the fourth scholar (Rabbi Akiva) had the correct interpretation of what he saw: the sitting image was Metathron (the Biblical Enoch, son of Jared who was taken to heaven and for 1 hour each day served as God's secretary). The sitting image could not be an angel, as according to Jewish mysticism angels have stiff knees and cannot sit. The moral of the Talmudic story is that even high-level scholars who do not have all the relevant information may arrive at wrong interpretations.

The Talmudic moral for the suggestions of Hamilton *et al.* (2002) was brought to light by a publication from the laboratory of Vicki Vance (Mallory *et al.*, 2003) who participated earlier in the development of the *Amplicon* system with the laboratory of David Baulcombe. Mallory *et al.* (2003) referred to the results (Hamilton *et al.*, 2002), that indicated that the capacity for systemic silencing correlates with the accumulation of a particular class (24–26 nt long) of small RNAs.

The Vance team studied this proposed correlation further by a series of grafting experiments. In these experiments stocks of transgenic tobacco that had a stably integrated transgenes (for GUS) were used. These transgenes caused specific silencing of GUS. The silencing, and especially the systemic silencing that was able to move into the leaves of a non-transformed scions (systemic silencing) was analyzed. The transgenes were either of several kinds. Some caused a simple transcript of GUS while one construct had a "tail-to-tail" (inverted repeat) of the transgene (GUS) so that it will cause the formation of dsRNA in the plant. To increase the amount of information from these grafting experiments the investigators used two types of transgenic stocks: with the HC-Pro (silencing suppressor) or without HC-Pro. Previous knowledge indicated that the grafting by itself did not induce any silencing and that RNA silencing in the scion was always specific to the gene (or transgene) silenced in the stock. The scions used in these grafting experiments (T19) were also transgenic and expressed GUS but without silencing. The rare silencing that happened in the scions were monitored and those scions which showed such silencing were excluded. Stocks with the T4 (tail-to-tail, inverted repeat) transgene were able to form the respective dsRNA of the GUS coding sequence, in spite of the presence of HC-Pro. When the T19 scions were analyzed 4 weeks after grafting on T4 stocks, there was no indication of any silencing that moved into them from the stock but at 7 weeks there was already systemic silencing and it was expressed by a strong reduction of GUS and the appearance of 21–24 nt short RNAs that were homologous to the GUS coding sequence. This result clearly showed that the dsRNA-producing stock generated a silencing signal that can cross the graft junction. Root stocks that produce the (simple) sense transcript for GUS are also capable of transmitting the silencing to T19 scions and do so even much quicker: at 4 weeks after grafting the silencing reached the scions. The grafting was repeated with two kinds of stocks that contained *Amplicons* with the GUS gene and were themselves silencing GUS but the *Amplicons* that were used were defective in their movement so that they themselves could not move from stock to scion. In these grafts the silencing did not move into the scions, neither was there an accumulation of the GUS-homologous dsRNA.

The HC-Pro suppressor of silencing was then added to the picture. The stocks with either the T4 or one of the *Amplicons* also contained HC-Pro. This did not prevent the formation of GUS mRNA, neither the formation of full-length dsRNA with GUS sequence but the short dsRNAs (21–24 nt) that are indicative of silencing did not appear in these stocks. When T19 scions were grafted on T4 x HC-Pro stocks it took about 7 weeks for the silencing to reach the scions, and the results were similar to the T19/T4 grafting experiments. When this grafting was done there was a surprise: *Amplicons* (that did not move themselves) in the presence of HC-Pro caused the silencing to move and reach the scions quickly (within 4 weeks). The silencing was very effective (reducing GUS mRNA at least 10-fold). When a transgene as GUS is introduced into tobacco plants and this transgene causes silencing (co-suppression) of the expression of this gene, there is no specific methylation of the DNA that encodes this gene. But in the transgenic tobacco line T4 (with an inverted repeat of the GUS-coding sequence) DNA-methylations do take place. It was also previously found by the Vance Laboratory that HC-Pro that suppresses the RNA silencing does not interfere with the methylation of the respective coding DNA. When the methylation of two sites of the GUS coding sequence was analyzed in T4 x HC-Pro plants, no elimination of the methylation was revealed. The same was observed when *Amplicons* (with GUS gene) were introduced into plants together with HC-Pro: no change in methylation. Whenever the silencing signal traveled from the stock to the scion — methylation also took place in the scion. Methylation of DNA can cause reduction of transcription in the methylated gene. This reduction of transcription was termed TGS. But there is another kind of methylation in the chromatin that could have an even stronger effect on the transcription from a respective DNA sequence. This is the methylation of lysine 9 of histon 3 in the nucleosome (see: Appendix II in Galun, 2003). The methylation of lysine 9 in the H3 histone of nucleosomes can now be analyzed by immunological techniques. Mallory *et al.* (2003) did not follow the latter methylation.

The chemistry of TGS silencing is not yet clear and especially, as indicated in the previous paragraph, the relation between DNA methylation and the methylation of histone amino acids in the plant's chromatin requires further investigation. There are two aspects of TGS which can be studied with respect to "taming" TEs (see: Galun, 2003) and it can also be studied from the viewpoint of silencing either transgenes (after genetic transformation) or silencing (temporarily and/or spatially) endogenous genes and by that affecting the metabolism and/or the patterning of plants. Marjorie and Antonius Matzke who first worked at the Austrian Academy of Science in Salzburg and later moved to Vienna, are pioneers in the study of TGS in plants. The early work of the Matzke team was summarized in reviews (Kooter *et al.*, 1999; Matzke *et al.*, 2001a, 2001b). Notably, in these reviews the Matzke team already made a prediction that the silencing may have a role beyond protection against pathogens and TEs, namely, to regulate normal plant development. In a later study (Mette *et al.*, 2001), the investigators of the Matzke team looked at the suppression by the viral protein, HC-Pro, of transgene-induced silencing. The transgene that was chosen was nopaline synthase (encoded by an *Agrobacterium* gene that is inserted into the plant genome during agroinfection). The target for silencing was the promoter of the gene (NOSpro). When HC-pro was introduced into plants that contained NOSpro, the expression of NOSpro was not changed. The silencing of NOSpro by the methylation of the DNA that encodes this promoter could be induced (by RdDM) by introducing an RNA hairpin sequence that is homologous to this promoter. When HC-pro was added this RdDM system of silencing (that is TGS) was not reduced. This was in contrast to the situation in PTGS, where the silencing as well as the accumulation of the homologous siRNA (of $\sim$22 nt) were strongly reduced by HC-pro. The authors had suggestions to explain how the same HC-pro could affect PTGS and TGS differentially but no definite answer was provided. In another study (Mette *et al.*, 2002) the Metzke team turned to TEs in plants. They elaborated a procedure to "fish" "unknown" TEs by using siRNAs. It was found that when short siRNAs were isolated from *Arabidopsis* and then cloned and sequenced, these sequences could serve a computer search of the *Arabidopsis* genome (that is now sequenced) in order to identify "novel" (or yet unknown) TEs. The investigators provided several examples in which such TEs were revealed. The Metzke team also came up with additional information on RNA-directed methylation of DNA in plants (Aufsatz *et al.*, 2002a, 2000b).

I have mentioned above that Hamilton *et al.* (2002) detected two classes of siRNA in silenced plants: a 21–22 nt class and a 24–26 nt class. These two classes bring us to an additional finding of the Metzke team (Papp *et al.*, 2003) of nine members and which included investigators from the USA. The results of the latter study suggested that there are different nucleus-located *Dicer*-like (DCL) enzymes in *Arabidopsis* (and probably also in other plant's nuclei). One of the DCLs that is apparently nuclear-located (i.e. DCL1) was suggested to be implicated in the formation of miRNAs from transcripts that have inverse repeats and fold in the nucleus into stem-loops. The investigators asked whether there are indeed several nuclear DCLs, each generating a different size group of dsRNAs. The viral silencing suppressor P19 of which there are two versions, a nuclear and a cytoplasmic version, was used in this study. The P19 protein is assumed to bind specifically to DCL-generated 21–25 nt dsRNAs that have a 2 nt 3′ overhang and suppresses the accumulation of siRNAs. Indeed, the nuclear P19 caused a significant reduction of 21–22 nt siRNA as well as 21 nt miRNA. But this P19 had a much smaller effect on 24 nt siRNA. The cytoplasmic version of P19 did not decrease the quantity of siRNA but caused a 2 nt truncation of siRNAs and miRNAs. The results of this study also indicated that the nucleus-located DCL1 is required for the production of miRNAs (from their precursors) but not for the production of siRNAs. The results also hinted the existence of another nuclear DCL that generates siRNA from longer dsRNA. The proposed roles of DCL1 and other (nuclear) DCLs and the impact of the nuclear and the cytoplasmic versions of P19 is summarized in a model of Papp *et al.* (2003) that is presented in Fig. 33. Obviously, this model is shown here to present the view of these authors and not as a presently generally-accepted process.

Is there a connection between human premature aging and the PTGS and TGS of plants? While the question is apparently strange the answer is in the affirmative. The answer came as expected from a team of investigates affiliated with companies of drug manufacture: Novartis in Switzerland and Syngenta of North Carolina, USA. Glazov *et al.* (2003) were interested in the fate of mRNA after it is cleaved (by RISC) by the PTGS of *Arabidopsis*. PTGS does not lead to the accumulation in the cells of long mRNAs fragments that

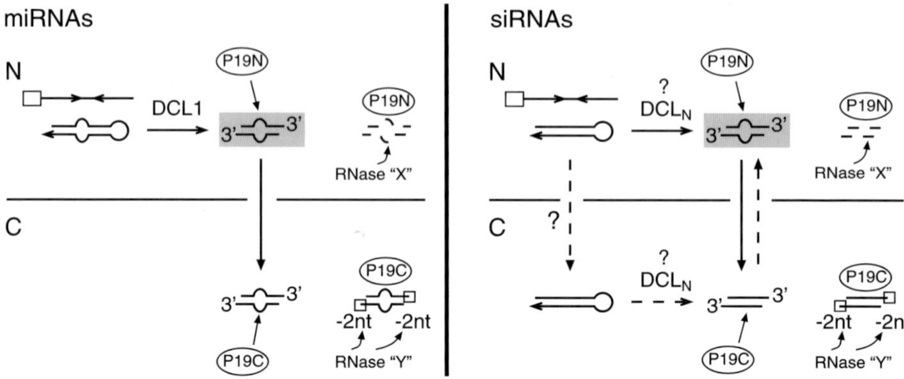

Fig. 33. Model for different effects of P19N and P19C on DCL-generated short RNAs (21–25 nt with 2-nt 3′ overhangs) in plants. Left, miRNA: the precursor, an imperfect duplex RNA that is transcribed in the nucleus (N) from an imperfect inverted DNA repeat (arrows), is processed by DCL1, a nuclear activity in *Arabidopsis*. Right, siRNAs: NOSpro dsRNA, a perfect RNA duplex, is transcribed from an inverted DNA repeat downstream of the 35Spro (box). Processing of at least some precursors of 21–22 nt siRNAs possibly requires a nuclear *Dicer* (DCL$_{N}$) that is not DCL1. If dsRNA precursors are transported to the cytoplasm ©, a cytoplasmic *Dicer* activity (DCL$_{C}$) could processes them to siRNAs, which could relocate to the nucleus (dotted arrows). Nuclear products of DCL cleavage could also be transported to the cytoplasm (solid vertical arrow). In their respective compartments, P19N and P19C bind to the immediate products of DCL-catalyzed cleavage. It is hypothesized here that this leads to complete degradation by an unknown RNase (X) in the nucleus, and clipping of the 2 nt 3′ overhands by a different RNase (Y) in the cytoplasm. Nuclear pathways for processing precursors of miRNA and 21–22 nt siRNAs are inferred from the effects of P19N, indicating the presence of DCL cleavage products in the nucleus (shaded gray). Whether all miRNA precursors are processed in the nucleus remains to be determined. Alternative modes of P19 action include inhibiting or deregulating DCL activity. (From Papp *et al.*, 2003.)

were cleaved in only one site. Therefore, it seemed reasonable to these investigators that after the PTGS cleavage there should be further scavenging of the mRNA, probably by exonucleases. But no such exonucleases were previously identified in angiosperms. The *C. elegans* MUT-7 protein is a member of the RNase D family and is implicated with PTGS of this nematode. Thus, the investigators looked in the genome of *Arabidopsis* for sequences that have similarity to the coding sequence for MUT-7. They found several! One *Arabidopsis* sequence was closely related to MUT-7 and had an RNase D-like domain that is most similar to a sequence in the gene for the human Werner syndrome (WRN). WRN patients are inflicted by very early aging. The *Arabidopsis* gene was termed *WEX*. Sequences similar to *WEX* were also revealed by search of the rice genome. There is an *Arabidopsis* mutant line that has a T-DNA insertion in the *WEX* gene. This was termed *wex-1*. This mutant produces very little or no mRNA of *WEX*. In spite of the lack of *WEX* mRNA the *wex-1* plants appeared morphological normal and even flowered *earlier* than wild-type *Arabidopsis* plants. The investigators used a "silent indicator line" in which there is a gradual increase of silencing during development (visualized by the fading of GFP). The silencing in this indicator-line is caused by PTGS. But this PTGS-derived silencing did not happen when the tested plants were also homozygous for *wex-1*, clearly showing that WEX protein is required for PTGS. The WEX was not required for TGS.

Movement of the Silencing Signals in Plants

An impressive demonstration that informative nucleic acids can be trafficked a long way is the appearance of *Mouse ears* in scions of tomato plants! The *mouse ears* mutation (*Me*) is imposed by a transcriptional fusion of a Hox gene to another transcript and the fused transcript was claimed to be able to move in the phloem from the *Me* mutant stock to the wild-type scion, imposing a *Me* phenotype (of "mouse ears") in the scion (Kim *et al.*, 2001). The problem with this impressive movement is that it was not verified by other investigators. But several other studies did indicate that the RNA silencing signals can move long distances in plants (see: short review of Jorgensen, 2002). Are these signals "naked" RNA sequences or are they "passengers" in specific vehicles during their trafficking? There are still no clear answers to these questions. If vehicles are involved, what are they and how is the signal amplified during this voyage? The Frederick Meins Laboratory of the Friedrich Miescher Institute in Basel, Switzerland (Klahre *et al.*, 2002), found that introducing dsRNA into plants can cause not only local silencing of genes that have transcripts with homology to the dsRNA. The silencing can move great distances. Moreover, there was most probably a RdRP stage that amplified the silencing signal because after traveling dsRNA had sequences that were still homologous to the silenced target transcript but they were 3′ or 5′ away from the homology to the originally applied dsRNA. Some examples for the movement of the silencing signed were provided in the previous section.

The laboratory of Oliver Vionnet at the CNRS in Strasburg, France, and associates (Himber *et al.*, 2003) studied the movement of PTGS in *Nicotiana benthamiana*. Their goal was to compare the "local" spread of silencing, from cell to cell of up to about 10 to 15 cells, to the "systemic" silencing that can travel a long distance as from lower leaves to the leaves at the top of the plant. This study was based on previous knowledge obtained in the Vionnet Laboratory and in other laboratories. This knowledge included data that indicated that silencing signals contain a nucleic acid component. It was also previously known, as indicated above in this book, that plants as well as nematodes (but not *Drosophila* neither mammals) have a mechanism to replicate silencing nucleic acids in which RNA-dependent RNA polymerases (RdRPs) are involved and there is a "transitivity", meaning that the replicated RNA can have sequences that are upstream or downstream of the original silencing sequence. Actually, in this respect plants do "better" than nematodes. While in nematodes transitivity proceeds mainly (or only) form 3′ to 5′, in plants it is bidirectional as detailed in Vanitharani *et al.* (2003). Also, as noted above, there was evidence that in plants only the ~21 nt siRNA is active in PTGS while the ~25 nt siRNA is involved in TGS. As for long-distance movement there was indirect evidence for the movement of this silencing through the phloems and then upon unloading in the leaves, it spreads again from cell to cell *via* plasmodesmata. On the level of genetics it was known that long-range transmission of silencing requires two genes. One is *SDE1* (encoding a putative RdRP) and the other is *SDE3* (encoding a putative helicase). The experiments of Himber *et al.* (2003) were performed with transgenic *N. benthamiana* plants that expressed GFP. These were 16c-line plants which can be silenced by co-suppression: when a leaf blade was infiltrated with agrobacteria that contain a T-DNA with a GFP sequence, the green fluorescence is first increased at the location of infiltration, but after about 10 days it is reduced drastically. The local silencing has a defined border that is visualized by its lack of green fluorescence and could be also defined by the co-infiltration of another reporter gene for GUS. Typically, the border between the infiltrated area and the spread of local silencing had a width of about 13 cells, meaning that the silencing moved beyond the infiltration area. The P1 suppressor of silencing that was known to suppress the silencing by ∼25 nt siRNA but not by ∼21 nt siRNA did not affect the ∼13 cell border while other suppressors of silencing (that suppress the 21 nt siRNA silencing) eliminated this border. The investigators also obtained experimental evidence that there is no need for replicating the siRNA for the short-range (border) movement of silencing. Adding more components to their experimental procedure, as the use of PVX-GFP that is deleted with respect to a sequence that encodes a 24–26 kDa protein (that is essential for a cell-to-cell movement of the virus) provided more information on the movement of the silencing signal. It was also apparent that for some mRNAs, as the endogenous mRNA encoding the Rubisco small subunit (RbcS) there was no transitive RNA silencing. The authors arrived at several conclusions that were based on their elaborated study. They found it likely that silencing can spread from a single-silenced cell to cells that are 10–15 cells away from the former cell. The short-range spread of silencing signals does not involve DNA, hence it should involve RNA. The candidates for this short-range spread are siRNAs and the cell-to-cell movement is probably through plasmodesmata. *Extensive* cell-to-cell movement of the silencing was linked to transitivity and required the SDE1 and SDE3 proteins. Secondary accumulated siRNAs were apparently required for this long-range silencing. The authors suggested a possibility that 21 nt secondary siRNA are involved in long-range silencing but that the 25 nt siRNA represents the signal that moves in the phloem. The authors presented a model for the cell-to-cell movements of the silencing signal (both the regular movement and the *extensive* movement). This model is provided in Fig. 34. There are additional unanswered questions with respect to the movement of the silencing signal in plants. Take, for example, Feinberg and Hunter (2003) identified a gene, *SID-1*, that is essential for systemic RNAi in *Ceanorhabditis elegans*. The SID-1 was found to be a multispan transmembrane protein that enables passive cellular uptake of dsRNA. Plant cells have plasmodesmata for cell-to-cell movement of the silencing signal. But do plants also produce an SID-1 like protein?

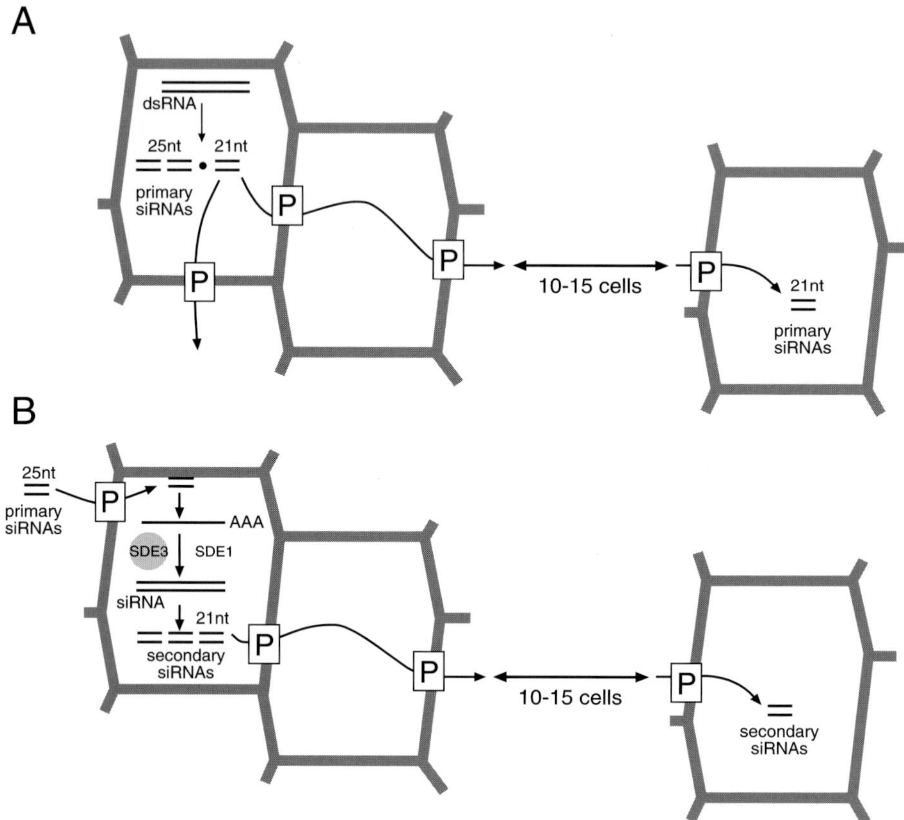

Fig. 34. Model for cell-to-cell movement of RNA silencing in plants. (A) Silencing can spread over 10–15 cells in the absence of amplification, through movement of 21 nt primary siRNAs. (B) Extensive cell-to-cell movement requires 21 nt siRNA-induced *de novo* synthesis of dsRNA by the action of SDE1 and SDE3 using transgene mRNAs as template. This leads to the production of secondary 21 nt siRNAs that spread over a further 10-15 cells. P: plasmodesmata. (From Himber *et al.*, 2003.)

Important additional information on RNA silencing and the movement of the silencing signal came from the publication of Guilliang Tang and collaborators from the Massachusetts Medical School in Worcester (Phillip Zamore) and the Whitehead Institute of MIT in Cambridge, MA (David Bartel and Brenda Reinhart). The main experimental system that was used in this study (Tang *et al.*, 2003) were extracts from wheat germ. This extract is used since about three decades ago, for *in vitro* translation of proteins from plant transcripts. The investigators therefore asked whether wheat germ extracts are appropriate for the *in vitro* study of the RNA silencing mechanism in angiosperms. Is there only one *Dicer*-like complex in plants or are there different *Dicer*-like complexes that generate two size classes of siRNA (of ~21 nt and of 24–25 nt)? Furthermore, the investigators searched for an RdRP activity in plants that can synthesize dsRNA from a single-stranded RNA template without an exogenous primer. If such an activity does exist, will it generate the (24–25 nt) siRNAs? The investigators also intended to look for a RISC complex in the wheat germ extract and its involvement in the miRNA-guided cleavage of specific mRNA.

The question about two-size classes (generated by the *Dicer*-like cleavage) was addressed by the addition of hundreds of small RNAs (including miRNAs) to the wheat germ extract. Indeed, the distribution of sizes of the fragmented RNA was bimodal: with two peaks of 21 nt and 24 nt. This result was compatible with the results of other investigators who studied intact plant tissue, as indicated above. This also means that the wheat germ is a reasonable research tool. Notably, there was also a bias with respect to the base at the 5′ end of the RNA. Those with a 5′-uridine predominated in the smaller class (of ~21 nt) while those with a 5′-adenosine predominated in the longer class (of ~24 nt). Wheat germ caused the "direct" cleavage of the RNA into the two-class sizes without intermediates of longer sequences. There were about 2.5-fold more of the longer class than of the shorter class. When the miRNA precursors were excluded the ratio increased to over 4-fold (because the miRNAs will be in the shorter class of ~21 nt). The two-class sizes were also generated by extracts of cauliflower (a dicot and in the same plant family, Cruciferae, as *Arabidopsis*). The cleavage into these classes requires ATP which was known for the *Dicer*-cleavage in *Drosophila* and the nematode *C. elegans*. By using ^{32}P-radiolabeled dsRNA the investigators found that the fragments from the cleavage in the wheat germ were indeed double stranded with 2-nt, 3′ overhangs and 3′-hydroxyl termini. Such cleavage products were found previously in animal extracts (e.g. *Drosophila* embryo lysate). The wheat germ-generated siRNA also had the 5′ monophosphate termini as in animal systems. It appears that the *Dicer*-like complex level is a bottleneck of dsRNA cleavage so competition experiments are feasible; in competition experiments either 21 nt or 24 nt long dsRNAs can be added to reveal whether these dsRNAs cause differential suppression of cleavage by the *Dicers*. The results of such experiments supported the suggestion that there is more than one *Dicer*-like enzyme in wheat germ and that different *Dicer*-like enzymes generate cleavage-derived products of different size classes. In the *Arabidopsis* genome there are coding sequences for four *Dicer*-like enzymes; how many *Dicer*-like genes there are in wheat is yet not known.

The investigators also found RdRP activity in the wheat germ. The RdRP could synthesize the RNA copy without or with an appropriate primer RNA. Hence, single-stranded RNA could trigger the *de novo* synthesis of 25 nt small RNAs in the wheat germ extracts and the investigators surmised that the production of this longer dsRNA (of 24 nt) by the RdRP is related to the orthologue of *Dicer*-like enzyme that "specializes" in the generating of this longer small RNA. Thus, it was reasonable to the investigators that the following takes place in the wheat germ extract. Single-stranded RNA provides the template for the synthesis of cRNA by RdRP and the resulting template-RNA: cRNA hybrid (dsRNA) is then cleaved into $\sim$24 nt siRNA by a specific *Dicer*-like enzyme. We should recall that it is this $\sim$24 nt siRNA that is assumed to be implicated in the long-range movement of the silencing signal. Hence, the experimental results of Tang *et al.* (2003) are also relevant to this transmission of RNA silencing in plants.

Development of Efficient RNA Silencing Techniques

In this section I shall deal primarily with the improvement of RNA silencing techniques for their use in biochemical genetic studies such as those in reverse genetics in which the genomic coding sequence is known but the function of the respective gene is unknown. Means to improve the RNA silencing in plants for the use in crop improvement will be discussed in the next section. The improvements in experimentally applied RNA silencing in plants were reviewed by Horiguchi (2004) who provided ample references on this subject.

For the experimental construction of an efficient siRNA one has to recall that dsRNA can be derived by several ways. One possibility is that an RdRP can convert a single-stranded RNA into a dsRNA. It is also possible that a sense (ssRNA) and an antisense (ssRNA) are transcribed in the same nucleus (or are synthesized *in vitro*) and then hybridized to form dsRNA that is then processed *in vivo* into siRNA. A precursor for a hairpin RNA (hpRNA) can be formed in the nucleus that contains a transcript in which a given RNA sequence is arranged as inverted repeats which may then fold into the hpRNA. If there is a spacer between the inverted repeats the folding product will be a stem-loop RNA. Vectors based on these considerations (e.g. pHANNIBAL, p-HELLSGATE) were already mentioned in a previous section of this chapter. The hairpin RNA and the stem-loop RNA can be cleaved *in vivo* into dsRNA. The various actual procedures to design constructs for RNA silencing in plants were based on the above mentioned possibilities. Take, for example, if a strong promoter is put upstream of a given inverted-repeat sequence it will result in an hpRNA. But the efficiency of this hpRNA can be increased if the spacer between the inverse repeats is an intron that is spliced after transcription. This will lead to a more efficient silencing sequence termed ihpRNA. It is possible to use a promoter (in front of the inverse repeats) that is activated by a specific compound (e.g. 17β-estradiol) so that the silencing RNA will be transcribed only upon the addition of this compound. The use of a promoter that is specifically induced by 17β-estradiol for regulated RNA silencing in plants was detailed by Nam-Hai Chua and associates (Guo *et al.*, 2003) of the Rockefeller University, NY and The National University of Singapore. Inducible promoters are especially useful when the gene that is to be silenced is a vital one and should be silenced for only a short time and/or in a specific location of the plant. The use of a promoter that can be activated by ethanol was reported by Chen *et al.* (2003). These authors suggested using such a promoter from the *ale* gene. When induced an intron-containing inverted-repeat sequence will be transcribed and the intron will be spliced *in vivo* in tobacco tissue.

The vectors for RNA silencing can be swiftly integrated into the plant genome by *Agrobacterium*-mediated genetic transformation (see: Galun and Breiman, 1997 and Galun and Galun, 2001). Obviously, the combination of a binary *Agrobacterium* plasmid with a construct that will transcribe a ihpRNA, the transcription of which is driven by a promoter that can be activated experimentally, could lead to the desired RNA silencing that is not only sequence specific but also efficient and defined in time and location.

By now, the construction of vectors for PTGS and for RNAi research has reached commercial companies as well as multinational collaborative efforts. Take, for example, the Promega Corporation (www.promega.com) in its Promega Notes Magazine No. 87 (2004) featured two articles on RNAi; one of which is devoted to the construction of hairpin-forming vectors that are based on *siStrike*™ (www.promega.com/sirnadesigner/). The Chang Bioscience, Inc. offers a Support Vector machine (SVM), an RNAi learning program in an electonic protocol book. Vectors intended for RNAi knockouts (PTGS) in *Arabidopsis* were developed by the Agrikola project that is a European Union supported activity that provides detailed protocols (www.agrikola.org). It is feasible that in the future several commercial manufacturers of products for biochemical-genetic research will come up with specific PTGS and RNAi tools. Dealing with detailed protocols is outside the scope of this book. Notably, such detailed protocols were provided in some of the chapters of the book "RNAi -A Guide to Gene Silencing" that was edited by Hannon (2003).

Beyond the detailed protocols there remains a general question regarding the silencing elicitated by transgenes. It was frequently reported that silencing was induced when a strong promoter (as the 35S promoter of Cauliflower Mosaic Virus) was used or when two or more copies of a gene (as for GUS or GFP) were integrated in a given genome. Lechtenberg *et al.* (2003) analyzed this silencing by transgenes in *Arabidopsis*. They found that the mere presence of multiple promoters and/or transgene sequences did not result in gene silencing. Also, tandem repeats of transgenes and/or inverted repeat T-DNA structures (of *Agrobacterium* plasmids) were not by themselves sufficient to trigger silencing of reporter genes. Instead, PTGS was correlated with the copy number of highly expressed transgenes. Take, for example, 12 copies of the gene encoding streptomycin phosphotransferase (SPT) and four copies of the gene encoding GUS did trigger silencing. This indicated to these investigators that the silencing may be attributed to the high transgene *dose* rather than the *repeat arrangement* themselves. In practice it is problematic to calibrate the number of integrations of a transgene by genetic transformation but, in general, this number is much lower after *Agrobacterium*-mediated transformation than after particle bombardment. The latter method of genetic transformation is the method of choice in some plants (e.g. soybeans).

While the PTGS is still not understood in all its details there is now enough information and proper methods are available to use PTGS as an efficient and standard tool for biological research in plants and for crop improvement. The latter subject will be detailed in the next section. As for biological research I shall give two examples. The first is a relatively recent study (Hoffmann *et al.*, 2004) by the Louis Pasteur University, Strasbourg and the INRA in Thiveral-Grignon, France. This study dealt with the pathway of phenylpropanoid biosynthesis. This pathway starts with phenylalanine and leads to numerous compounds as well as to the production of lignin. Many enzymes are involved in the different metabolic steps. To test the roles of these enzymes one can silence specific genes that encode these enzymes. Thus, Hoffmann *et al.* (2004) silenced, by PTGS, the gene encoding acetyltransferase in *Arabidopsis* and in *N. benthamiana*. In *Arabidopsis* this silencing caused dwarfing and a change in lignin composition. Also, in *N. bethamiana* the silencing of the gene that encodes acetyltransferase caused changes in lignin content and composition, revealing also a marked decrease in syringyl units and an increase in p-hydroxyphenyl units. The other example of the utilizing of RNA silencing as a tool in molecular biology in plants is the study by Ron and Avni (2004) of the Tel Aviv University. Adi Avni and associates were studying the plant-genes encoding proteins (leEIX proteins) that interact with fungal elicitors (EIXs) through specific binding to the plant cells and consequently trigger a hypersensitive response in the plant. This response confines the fungus and prevents the further spread of the fungus. This is a defence mechanism of plants against pathogenic fungi. Ron and Avni (2004) caused the silencing the *LeEix* genes by PTGS (e.g. the investigators amplified a 684 nt segment of the *LeEix1* gene and inserted the sense and the antisense orientations of this segment into the pKANNIBAL silencing vector). Silencing of the

LeEix genes prevented the binding of EIX to cells of an EIX-responsible plant and thus inhibited the hypersensitive response.

These are merely two examples for the use of RNA silencing in molecular-biology studies in plants. The tool of gene silencing by PTGS will most probably become routine so that, as in the article of Ron and Avni (2004), the term RNA silencing will not appear in the titles, just as methodologies as Southern-, Northern- and Western-blot hybridization are not mentioned anymore in the titles of articles.

RNA Silencing for Crop Improvement

Before providing examples on how RNA silencing can be utilized for crop improvement, we should note that there is a problem involved. Any imposed silencing of a plant gene involves the introduction of nucleic acids into the respective plant. This means performing what is popularly called "genetic engineering" and obtaining genetically modified (GM) plants. There is a tendency to avoid GM plants, especially in some European countries and some governments even banned the cultivation of GM crops and restricted their import. I shall not discuss this issue here but note that much of the fear from GM is due to ignorance (in one of the European countries about half of the people who participated in a poll stated that they never consumed DNA). On the other hand, the percentage of GM cultivars of crops as maize, soybeans and cotton is steadily increasing in North American countries. It is my estimate that plant breeders will come up with GM crops that are safe for the environment and the consumers. This will in the future eliminate the fear of GM crops and put this subject on a rational tract. I elaborated this subject in my books on transgenic plants (Galun and Breiman, 1997; Galun and Galun, 2001) and a recent review on part of this subject was provided by Ellstrand (2004). RNA silencing, by definition, deals with the reduction (or total elimination) of gene-expression rather than with the expression of alien transgenes. Therefore, this silencing was not used to render crops resistant to pathogens or to herbicides for which the expression of transgenes is required (crops resistant to a given herbicide can be sprayed with this herbicide to reduce the competition by weeds). However, the use of RNA silencing for interfering with plant-pathogen interactions should not be completely neglected. The example of the EIX-LeEIX interaction, in the study of Ron and Avni (2004) that was summarized in the previous section suggests that RNA silencing could render crop plants resistance to specific pathogenic fungi. Existing reports on crop improvement by RNA silencing discussed mainly improvement of crops with respect to the quality of their products.

Decaffeinated coffee beans

Coffee beans devoid of caffeine sounds paradoxal. *Caffea arabica* was probably first cultivated in Yemen, at the south-eastern corner of the Arabian peninsula. A legend tells that Christian mission-aries who brought Christianity to Ethiopia, as early as the 2nd and 3rd century, provided coffee to their audience to keep them alert to the preachings. Coffee as a stimulating beverage spread thereafter to all over the world. But now there is an increasing demand for decaf-feinated coffee because of the concern that caffeine can adversely affect sensitive individuals by triggering palpitations, increasing blood pres-sure and insomnia. Chemically reducing the caffeine of coffee beans may leave undesired components in the decaffeinated beans. There-fore, coffee plants that produce caffeine-free beans is a reasonable aim. However, it should be noted that the level of caffeine varies consider-ably among cultivars so that traditional breeding can also reduce the caffeine levels (e.g. Silvarolla *et al.*, 2004). But it takes about 25 years to breed a new coffee variety. Ogita *et al.* (2003) of the Nara Institute of Science and Technology, in Japan, used RNA silencing for the reduc-tion of caffeine in *Caffea* trees. Three N-methyltransferase enzymes are involved in caffeine biosynthesis in *Caffea* trees: CaXMT1, CaMXMT1 (theobromine synthase) and CaDXMT1 (caffeine synthase). These enzymes successively add methyl groups to xanthosine, converting it to caffeine. Ogita *et al.* (2003) aimed their efforts to reduce the expres-sion level of the gene that encodes theobromine synthase, CaMXMT1. They chose a sequence of the 3′ untranslated region (3′ UTR) of *CaMXMT1* mRNA and constructed "short" and "long" RNAi homologous to this 3′ UTR. The "short" RNAi sequence contained a 139 nt and a 161 nt, from the same 3′ UTR with a spacer of 517 nt (encod-ing GUS). The "long" RNAi contained two identical 332 nt, separated also by the same spacer. The investigators did not specify whether the sequences were engineered in tandem or in a tail-to-tail configuration. The constructs were inserted into an *Agrobacterium* vector and the later was used in *Agrobacterium*-mediated genetic transformation of *Caffea canephora*. The choice of *C. canephora* rather than *C. arabica* was not explained. *C. canephora* is probably more amenable for genetic transformation than *C. arabica*. The transformation was not easy but transformed seedlings were finally obtained. Leaves of 1-year old plants were analyzed for purine alkaloids and it was found that the leaves of plants with the *CaMXMT1* RNAi had a 30–80 per cent reduction in theobromine and a 50–70 per cent reduction in caffeine content, as compared to leaves of control plants. The investigators assumed that a similar reduction in these alkaloids will be found in the beans of these plants and that RNA-silencing mediated reduction of caffeine will also be achieved in the future in *C. arabica* from which most of the world coffee is derived.

Improvement of grain proteins

While reducing caffeine in coffee beans is an approach that is intended to avoid health problems certain individuals wish to go on with the consumption of this beverage. Other efforts were aimed to improve the nutritional value of major crop plants. An example of the latter approach is a study by Joachim Messing and associates (Segal *et al.*, 2003) of the Waksman Institute of the Rutgers University in New Jersey. The latter study discusses the nutritional value of maize grains. While being a successful and widely cultivated crop the endosperm proteins of maize are deficient in essential amino acids, particularly in lysine, tryptophan and methionine. Therefore, maize is nutritionally problematic when used as the main food or feed for non-ruminant animals (ruminants are much less dependent on the outside supply of essential amino acids). Corn breeders in the USA came up with a recessive maize mutant, *opaque-2* (*o2*), that had more nutritious proteins in the grains endosperms; but cultivars with this mutation had undesirable agronomic features and corn growers did not like these maize cultivars. Segal *et al.* (2003) had in mind to use the *Dicer* (or RNA silencing) in order to cleave the mRNA of a specific gene that encodes a 22-kD protein as this protein has an undesirable amino acid ratio. The investigators intended to get a line of transgenic maize that will have the *opaque* phenotype but without the deleterious effects of the homozygous *o2/o2*. They focused on α-zein. The zeins of maize endosperms are classified into four sub-families: α-, β-, γ- and δ-zeins. Of these α-zein accounts for about 70 per cent of the endosperm proteins. It is encoded by multiple active genes in several chromosomal locations, and this protein was sub-divided into 19-kD and 22-kD sub-families. From the previous information regarding the impact of *o2* it was known that reducing the 22-kD protein will retain a more balanced amino acid composition in the endosperm protein. Only *o2* had additional undesired effects. Segal *et al.* (2003) therefore intended to silence specifically genes encoding the 22-kD α-zein. The study was rather elaborate and so only a brief summary of it will be described here. The investigators used sequences of the mRNA for the 22-kD and built RNA-silencing constructs. Among these were constructs that after transcription in the maize cells will fold into hairpin configurations. Using appropriate marker and reporter genes and particle-bombardment of callus cultures that have the capability to regenerate functional plants, the investigators obtained transgenic maize plants that behaved as dominant mutants with respect to silencing the genes encoding the 22-kD α-zein. Such transgenic maize plants indeed produced grains in which the amino acid composition of zeins was balanced. Specifically, the level of lysine in the endosperm-protein was very much increased over that of wild-type maize. The effect of hairpin RNAi was much greater than silencing by an antisense construct. Thus, the investigators recommended to use RNA silencing biotechnology to improve the nutritional value of crop plants.

Messing and associates were not the only team that recommended the use of RNA silencing to improve the nutritional value of crops. Gad Galili and associates of the Weizmann Institute of Science, Rehovot, Israel, conducted a long-range study on the metabolic pathways in plants that lead to the syntheses of three essential amino acids: lysine, threonine and methionine (see: details and references in Galili and Höfgen, 2002). Tang and Galili (2004) reviewed the subject of RNA silencing for improving the nutritional value in crops and suggested strategies for the application of RNAi technologies for this purpose. We should not be surprised that what investigators intended to do, the rice plant did already very long ago! Glutelins account for about

60 per cent of rice grains. Here, I stop for a moment .... In the articles dealing with cereals such as maize and rice the authors frequently wrote about *seed* proteins. But in botanical terms these cereals (as well as wheat) produce *caryopses* rather than seeds. There is a difference because the former also contain maternal tissue derived from the fruit (pericarp). Back to the kernels or grains of rice: a specific rice mutation *LGC-1* (for Low Glutelin Content) is reduced in glutelin content but has an elevated prolamin content. Glutelins accumulate in protein body II while prolamins accumulate in protein body I. Humans barely digest protein body I. Thus, a shift to protein body I (more prolamins, less glutelins) will convert *LGC-1* mutants into a practically low-protein rice. This is good for those people who, because of kidney malfunction, should have a low-protein diet. Kusaba *et al.* (2003) who are from various research institutions in Japan investigated the mode of action of the *LGC1* mutation. They found that in the homozygous *Lgc1* mutant there is a 3.5 kb deletion between two highly-similar *glutelin* genes that causes a tail-to-tail repeat. This may lead to the formation of a stem-loop structure and the dsRNA stem may then serve as a potent siRNA to silence the *glutelin* genes.

Changing the composition of oils in cottonseed

Liu *et al.* (2002) of the laboratory of Allan Green at the CSIRO in Canberra, Australia, undertook a research that was aimed to change the oil composition of cottonseed. Cotton is considered primarily as a fiber crop. The oils in its seeds are regarded as a "by product". This is unjustified because cottonseed oil is the world's sixth largest source of vegetable oil. The typical fatty acid composition of cottonseed oil is about 26 per cent palmitic acid (C16:0), 15 per cent oleic acid (C18:1) and 58 per cent linoleic acid (C18:2). There is very little variation in the composition of oils among cotton cultivars. Therefore and because of the tetraploidy of standard fiber cotton (*Gossypium hirsutum*), changing of oil composition by standard breeding is not practical. On the other hand, because oil composition can have an impact on the health of human consumers the changing of this composition by other biotechnologies could be beneficial to a considerable fraction of the human consumers since this fraction of the population has a tendency to accumulate health threatening levels of the undesirable low-density lipoprotein cholesterol in their blood. Obviously, the composition of consumed oil has an impact on the ratio of high-density lipoprotein cholesterol over low-density lipoprotein–cholesterol in the consumer's blood. Changing the oils for specific purposes as hydrogeneration for the production of margarine results in the production of trans-fatty acids that have cholesterol-raising properties, equivalent to those of saturated fatty acids. A scheme for the biosynthetic pathway of fatty acids is shown in Fig. 35. There

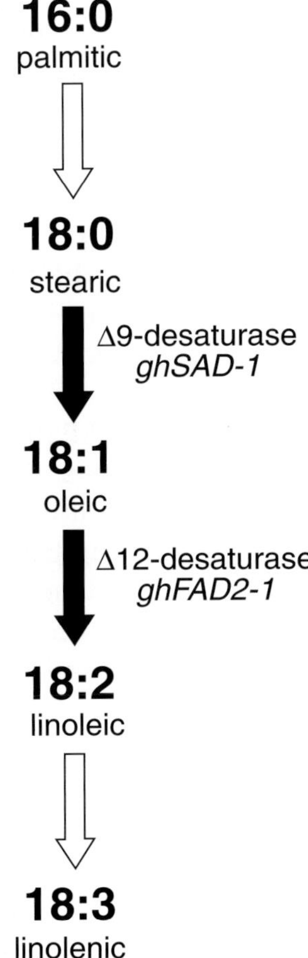

Fig. 35.   Diagramatic representation of fatty acid pathway in cottonseed. (From Liu *et al.*, 2002.)

are two important enzymes in this pathway. One is Δ9-desaturase, encoded by the gene *ghSAD-1* that converts stearic acid to oleic acid; the other is Δ12 desaturase, encoded by the gene *ghFAD2-1* that converts oleic acid to linoleic acid (as mentioned already in a previous section of this chapter). In the past Knutzon *et al.* (1992) succeeded to drastically increase the level of stearic acid in *Brassica* seeds (rape seeds) by antisense biotechnology. Kinney (1996) changed the fatty acid composition in seeds of soybeans and rapeseeds, by the co-suppression biotechnology. But both antisense and co-suppression technologies were of low effectivity and also provided unpredictable results. Because regeneration in cotton is difficult, Liu *et al.* (2002) decided to turn to a more effective technology to change the fatty acid composition of cotton seeds, they thus turned to RNA silencing biotechnology. For that they constructed hairpin forming sequences with homology to either *ghSAD-1* or to *ghFAD2-1*. Such constructs were inserted into the cotton genome. This introduction into the genome of the Coker 315 cultivar was performed by *Agrobacterium*-mediated genetic transformation using the appropriate constructions of the transformation vectors. The hairpin forming sequence with homology to *ghSAD-1* caused the respective transgenic plants to produce seeds in which the stearic acid increased from 2–3 per cent to about 40 per cent. Silencing the *ghFAD2-1* gene increased the level of oleic acid from about 15 per cent in Coker 315 to up to 77 per cent in the silenced transgenic plants. The palmitic acid levels were reduced after both types of silencing. By crossing transgenic plants that were silenced with respect to either of the two desaturase genes, plants could be selected in which both genes were silenced. Some silencing could be achieved by using the antisense technology but the PTGS biotechnology was much more efficient.

The manipulation of fatty acid levels in cotton seed provides an interesting lesson. There are several possible approaches to affect the oil composition of these seeds. In cases in which there is sufficient genetic variability among the cultivars of a given crop or in its wild relatives that can be crossed with the cultivars, the choice of regular plant-breeding is a reasonable option to achieve a change in fatty acid composition. One may be able to introduce genes encoding desaturating enzymes from very different organisms as cyanobacteria and rats into plant cultivars to cause this change (see: Kinney, 1996). It is also possible to change the fatty acid composition by co-suppression (Kinney, 1996) or by antisense sequences (Knutzon *et al.*, 1992) and finally, as was detailed above (Liu *et al.*, 2002), such changes can be imposed by PTGS. It seems that at least in specific cases, the last approach is the most promising one. It is plausible that in the future we shall witness the replacement of regular breeding by the addition of transgenes and co-suppression efforts by the RNA silencing biotechnology. This is because the methodologies such as the use of better vectors for the RNA silencing are improving quickly. Nevertheless, the bottleneck of public acceptance of crops derived from GM should not be neglected.

Viral Suppressors of RNA Silencing

A person strolling in a peaceful meadow is rarely aware of the fact that he is walking among an endless number of battle fields. The battles are constantly going on within himself as well as inside the plants in the meadow. Only if the stroller is a biologist would he know that constant wars are taking place in the apparently tranquil environments. First, there is not a single organism (the stroller included) in the meadow that is free of TEs. Active TEs may be inserted into vital genes and can be detrimental to the organism. True enough, the great majority of the TEs lost the capability to transpose or were even mutated during the ages to such an extent that their original image as active TEs can hardly be recognized. The organisms developed means to tame the TEs and one of the means is by RNA silencing: the transposases of TEs can be silenced by either or both PTGS and TGS. A TE that lost its transposition capability is a "sitting duck". Mutations will accumulate in it rendering the TE to a "skeleton" of the original element (I have discussed this issue in detail, previously in Galun, 2003). Organisms are also hosts to pathogens. Among these are plant viruses. Again, plants developed defences against these pathogens and the most prevalent defence is by RNA silencing, as detailed in Chap. 2. But some of the pathogenic viruses are fighting back. They developed *suppressors* of the RNA silencing. These suppressors were already mentioned above but in this section I shall focus on these suppressors and summarize the present information on this subject.

Before entering the subject of viral suppressors of RNA silencing in plants, I have a philosophical remark. In an invited review in the Annals of Botany, Anthony Trewavas of the University of Edinburgh, Scotland, wrote about "Aspects of Plant Intelligence" (Trewavas, 2003). Using the term "plant intelligence" is obviously permissible as long as the term is defined. It was defined. The review was intentionally written in an apparently controversial manner and it brings to light several aspects of plants responses to their environment (such as pathogens). It is therefore a recommended reading material. But it should be read critically. I disagree with bestowing *intentions* to plants. Such intentions were attributed by Trewavas in several places in his review. Take, for example, there is the statement that "animals and plants … evolved to optimize fitness". I would phrase this as: "evolution optimized the fitness of animals and plants". To attribute "intentions" and "purposes" to plants, in order to enhance their chance of survival on the basis of empirical results can be misleading. To attribute intentions to plants is almost the same as attributing a rubber ball the intention to bounce when thrown on the floor. It should be easy for Trewavas to walk along St. David street in Edinburgh and to contemplate on the teachings of this St. David (the philosopher/historian, David Hume). Trewavas may then wish to amend his statements about purposes and intentions of plants.

The awareness of viral suppressors

It is impossible for an "outsider" like me to determine who the first investigator who discovered these suppressors was. The readers who are eager to identify the original discovers are referred to the authoritative review of Vicki Vance and associates (Roth *et al.*, 2004). The awareness of the existence of such viral suppression of RNA silencing in plants came from several directions. When Vance (1991) reported on the synergism between PVX and PVY mixed infection of tobacco plants, she had already noted observations that were made about 35 years earlier; that in mixed infection by two taxonomically unrelated viruses the two viruses may "help" each other and cause more severe disease symptoms than a single-virus infection. Then Vance *et al.* (1995) found that this synergism is not limited to the PVX/PVY mixed infection. Such synergism was also found when PVX was coinfected with either of three other members of the potyvirus group: tobacco vein mottling virus (TVMV), tobacco etch virus (TEV) and pepper mottle virus (PepMoV). What exactly triggered the higher replication of PVX-RNA was not revealed by Vance *et al.* (1995) but they already put their finger on the proximal region of the genomes of TVMV and TEV that encodes the P-1 and HC-Pro polypeptides. Additional information (Pruss *et al.*, 1997; Shi *et al.*, 1997) from the collaborative research by Vicki Vance and James Carrington, of the Texas A & M University, focused on the HC-Pro (rather than on the P1 and P3) polypeptide as the helper for accelerated replication of PVX-RNA. Moreover, it was suggested that the central domain of HC-Pro rather than another (zinc-finger) domain is causing the synergism. The synergism effect was also detected when plants were co-infected by potyviruses with other viruses as CMV and TMV. At about the same time, the concept of the resistance of plants to virus infections was already well developed (e.g. Mueller *et al.*, 1995; Ratcliff *et al.*, 1997 and see: Chaps. 1 and 2 in this book), and even earlier, Roger Beachy and associates (e.g. Beachy *et al.*, 1990) were able to render plants resistant to viruses by introducing into them the viral coding sequence for the coat protein. Hence, viruses could help one another or cause plants to be resistant to viruses. Interestingly, both of these two opposing effects turned out to involve RNA silencing.

Silencing stages suppressed by HC-Pro and by other viral suppressors

The results of three studies that were submitted for publication at about the same time (July 20, August 11 and August 21, 1998) by three respective research teams (Anandalakshmi *et al.*, 1998; Kasschau and Carrington, 1998; Brigneti *et al.*, 1998) already coined the HC-Pro with the same term: suppressor of gene silencing in plants. Each of the three teams used different experimental procedures but their conclusions were similar: HC-Pro caused the suppression of PTGS. The team of Shou-Wei Ding (Singapore) and David Baulcombe (Norwich, UK) also examined another polypeptide, *2b*, encoded by the CMV (Ding *et al.*, 1995) and found that 2b, like the HC-Pro of PVY, is a suppressor of PTGS. However, they suggested a different mode of suppression of RNA silencing. While HC-Pro acts by blocking the maintenance of

PTGS in tissues where silencing had already been set, the 2b protein prevents the initiation of RNA silencing. From these two suppressors the investigations went on to reveal numerous additional suppressors of RNA silencing encoded by the genomes of various groups of plant viruses as listed in Table 3.

A more recent suppressor of RNA silencing was revealed in the turnip yellow mosaic virus (TYMV), a plus-strand RNA virus prevalent in the dicot family Brassicaceae (Chen *et al.*, 2004b). The TYMV encoded p69 is a virulence factor that suppresses the siRNA (derived from dsRNA) but it promotes (in *Arabidopsis*) the miRNA pathway. Conspicuously, p69 does not suppress silencing by IR-RNA but does suppress silencing by sense-RNA transgenes. The p69-expressing plants contain elevated levels of *Dicer* mRNA and of miRNAs consequently, and the miRNA-guided cleavage of two plant genes is enhanced. Thus, even in the absence of TYMV transgenic plants that express p69 have symptoms that are characteristic of TYMV infected plants. The p69 thus maybe mechanistically similar to the NS of tomato spotted wilt virus (TSWV) as noted in Table 3.

Approaches to identify suppressors of RNA silencing

In previous sections of this chapter, several cases were described in which the suppression of RNA silencing was involved. Now, I shall discuss viral suppressors in a more orderly manner. There are three main approaches to identify, in plants, virus-encoded suppressors:

- Transient expression assays — *Agrobacterium* co-infiltration
- Reversal of silencing assays
- Stable expression assays

I shall summarize these three approaches that were reviewed by Roth *et al.* (2004).

In the *transient expression assay* approach the plant is co-infected with two *Agrobacterium* strains, each of these harboring a different T-DNA (vector). In one strain there is a vector that induces the silencing of a reporter gene (e.g. a gene for GFP). The second strain of *Agrobacterium* harbors a T-DNA that encodes a candidate for suppression.

Table 3.   Plant viral suppressors of RNA silencing.

Genus	Virus	Suppressor	Evidence	Reference
Carmovirus	Turnip crinkle virus (TCV)	CP	TCV infection does not reverse silencing. In agro-coinfiltration assay, CP blocks sense and antisense induced local silencing and prevents systemic silencing.	Qu *et al.* (2003) and Thomas *et al.* (2003)
Closterovirus	Beet yellows virus (BYW) Beet yellow stunt virus (BYSV)	p21 p22	Suppresses inverted repeat (IR) induced local silencing in agro-coinfiltration assay. BYV p21 corresponds to BYSV p22.	Reed *et al.* (2003)
CucumovirusCucumovirusa	Cucumber mosaic virus (CMV) Tomato aspermy virus (TAV)	2b	Infection with CMV or with PVX-2b vector blocks silencing. Interferes with systemic signal.	Li *et al.* (2002)
Furovirus	Beet necrotic yellow vein virus (BNYVV)	P14	Agro-coinfiltation assay with sense induced silencing. BNYVV P14 corresponds to PCV P15.	Dunoyer *et al.* (2002)
Geminivirus	African cassava mosaic virus (ACMV) Tomaro yellow leaf curl virus-China (TYLCV-C)	AC2 C2	Infection with ACMV, PVX-AC2, or PVX-C2 reverses silencing. Blocks sense induced silencing in agro-coinfiltration assay. AC2 and C2 are homologs.	Dong *et al.* (2003), Voinnet *et al.* (1999) and van Wezel *et al.* (2002)
Hordeivirus	Barley stripe mosaic virus (BSMV) Poa semilatent virus (PSLV)	γb	RNA mediated cross protection between PVX-GFP and TMV-GFP vectors is eliminated when γb is expressed from the PVX vector.	Yelina *et al.* (2002)
Pecluvirus	Peanut clump virus (PCV)	P15	PCV infection blocks silencing. P15 blocks local and delays systemic sense-induced silencing in agro-coinfiltration assay.	Dunoyer *et al.* (2002)

Table 3. (*Continued*)

Genus	Virus	Suppressor	Evidence	Reference
Polerovirus	Beet western yellows virus (BWYV) Cucurbit aphid-borne yellows virus (CABYV)	PO	BWYV PO suppresses local but not systemic sense-induced silencing in agro-coinfiltration assay. CABYV PO tested only on local silencing.	Pfeffer *et al.* (2002)
Potexvirus	Potato virus X (PVX)	p25	PVX infection does not suppress silencing. In agro-coinfiltration, p25 blocks systemic but not always local silencing.	Roth *et al.* (2004)
Potyvirus	Potato virus Y (PVY) Tobacco etch virus (TEV)	HC-Pro	Evidence from multiple types of assay. Does not block systemic silencing in stable expression grafting assay, but does in agro-coinfiltration assay.	Roth *et al.* (2004)
Sobemovirus	Rice yellow mottle virus (RYMV)	P1	Infection with PVX-P1 viral vector reverses silencing.	Voinnet *et al.* (1999)
Tenuivirus[a]	Rice hoja blanca virus (RHBV)	NS3	Agro-coinfiltration assay of sense induced local silencing.	Bucher *et al.* (2003)
Tombusvirus	Tomato bushy stunt virus (TBSV) Cymbidium ringspot virus (CymRSV)	P19	Limited activity in reversal of silencing; strong activity in agro-coinfiltraton. AMCV (artichoke mottled crinkle virus) P19 also works as a suppressor.	Voinnet *et al.* (2003), Qu and Morris (2002) and Takeda *et al.* (2002), see: Roth *et al.* (2004)
Tospovirus[a]	Tomato spotted wilt virus (TSWV)	NS_s	TSWV infection reverses silencing. In agro-coinfiltration, NS_s suppressed sense, but not IR, induced local and systemic silencing.	Bucher *et al.* (2003) and Takeda *et al.* (2002)

[a]Tospoviruses and tenuiviruses replicate in their insect vectors and in plants. (From Roth *et al.*, 2004.)

The two strains are co-infiltrated at a patch of a leaf in a plant (commonly *Nicotiana benthamiana*). The level of the reporter (GFP) at the infiltrated patch is monitored for a certain number of days. When the second strain does not encode an active suppressor, the GFP-induced fluorescence will gradually increase but then it fades due to the silencing by the plants RNA-silencing mechanism. But when the second strain does encode a viral suppressor, the patch where co-infiltration was performed will retain its bright green (GFP-induced) color under UV light (examples of this approach were provided by the studies of Llave *et al.*, 2000 and Voinnet, 2002). This approach can have several modifications. One of these is the use of a transgenic *N. benthamiana* plant that has already expressed the reporter gene (e.g. Ruiz *et al.*, 1998b). This approach was also used to detect the ability of viral suppressors to interfere with the systemic spread of RNA silencing.

The *reversal of silencing assay* approach is using an already silenced plant, meaning that a plant in which a given (reporter) gene is already silenced. Then this plant is infected with a virus that may contain a suppressor of the silencing. If indeed the silencing is reversed this serves as an indication that the virus indeed encodes a suppressor (e.g. Voinnet *et al.*, 1999).

The *stable-expression assay* uses a stable transgenic plant that expresses a candidate suppressor of silencing. This plant is then crossed to a plant of another line that was silenced for a reporter gene. The reversal of silencing in the progeny of such crosses will indicate the existence of a suppressor of silencing in one of the parental plants (e.g. Anandalakshmi *et al.*, 1998; Kasschau and Carrington, 1998). This approach was also employed to investigate the role of suppressors in systemic silencing; frequently grafting experiments were included in such studies (e.g. Mallory *et al.*, 2001, 2003).

The mechanisms of suppression

We can look at the mechanisms of suppression from two viewpoints. We may ask which stage of the RNA silencing process is suppressed by a given suppressor. It could be the initial fragmentation of long dsRNA into short sequences (of ∼21–25 nt) or the suppression could occur at a later stage, preventing the destruction of the mRNA target or even only preventing the systemic spread of silencing. Another way to look at mechanisms of silencing is to reveal the molecular interactions between the suppressor and specific components of the RNA silencing process. Up to now there was more progress made with the first viewpoint than with the elucidation of the intimate interactions between a suppressor and components of the silencing process. But the "story" is complicated. First, there are apparently conflicting results with respect to the same suppressor that was used by different research teams. Moreover, the interpretation of even similar results by different research teams may differ. Using different host plants in studies on viral infectivity and viral suppressors of RNA silencing can also lead to apparently conflicting results. Many of these studies were performed with *N. benthamiana* rather than with tobacco (*N. tabacum*) or any other of the 64 species of *Nicotiana*. Probably, investigators preferred *N. benthamiana* that is more sensitive to some viruses than the other *Nicotiana* species. The team of Richard Nelson, of the Samuel Roberts Noble foundation in Oklahoma, USA (Yang *et al.*, 2004), found that *N. benthamiana* is defective in a gene that encodes the RdRP that is essential for viral defence by RNA silencing. This defect probably rendered *N. benthamiana* hypersusceptible to viruses. We should note the vast difference between *hyper susceptibility* and *hyper sensitivity* (HS). The latter (HS) is of advantage to the host .... It causes quick death of cells around the cell that was infected by the pathogen. These dead cells prevent the pathogen from spreading further into the healthy hosts tissue. The HS response is reminiscent of the tactics of Field Marshall Mikhail Ilarionovich Kutuzov, the Russian General of Czar Alexander I who defended Russia against the invasion of Napoleon. Kutuzov used the barren-land tactics, thus preventing any supplies form the Grand Armee and finally burned down all of Moscow. This inflicted hunger and freezing on this Armee and defeated Napoleon. Only ... the "idea" of defence by *barren-land* was adopted by plants at least 100 million years before it was utilized by Kutuzov.

Roth *et al.* (2004) provided several examples for contradictory results. Here is one example. The silencing suppressor HC-Pro reduced the accumulation of siRNA but it also reduced the production of mature miRNAs in plants (see Kasschau *et al.*, 2003). But it seemed that it can also enhance the accumulation of miRNAs (Mallory *et al.*, 2002b). These and other apparently conflicting observations could result from different experimental components but they may be resolved after the exact interaction of a suppressor with the silencing components is known. In spite of this problem, for some suppressors there is a consensus; for example, for the 2b suppressor from CMV. The suppression of RNA silencing by 2b was one of the first suppressions that was revealed (Ding *et al.*, 1995). The 2b and other suppressors were recently reviewed by Ding *et al.* (2004). It was found that while 2b prevented the initiation of RNA silencing it did not reverse silencing that was already initiated. It was also confirmed that 2b blocks the systemic movement of the silencing signal. One typical grafting experiment provided the evidence: when a "spacer" stem that contained the 2b suppressor was grafted between a root stock and a scion the silencing did not pass this spacer (Guo and Ding, 2002). Nevertheless, Hamilton *et al.* (2002) suggested that 2b only delays but does not block the systemic movement of RNA silencing. If the information provided above was not confusing, here is another "twist". As reported in Ding *et al.* (2004) there is an animal-virus encoded suppressor of RNA silencing. It was revealed in the flock house virus (FHV) and termed B2 which has no sequence similarity to 2b but the former can serve as a suppressor of RNA silencing in plants. B2 can substitute 2b in the suppression of silencing and its suppression is stronger than 2b.

The tombusvirus suppressor P19

The P19 suppressor of RNA silencing in plants deserves special attention because in this suppressor (also termed p14 by some researchers), the investigators revealed the interaction between the suppressing protein and components of the RNA silencing process. The P19 was detected rather early in the study of RNA silencing (e.g. Voinnet *et al.*, 1999). It first appeared to be a weak suppressor: it reversed RNA silencing only in the regions of the plant's veins. But then, in certain transient-expression experiments it had an impressive suppressive effect (e.g. Hamilton *et al.*, 2002; Silhavy *et al.*, 2002; Takeda *et al.*, 2002; Voinnet *et al.*, 2003).

A team of Jozef Burgyan from Gödöllo, Hungary and from ENEA, Casaccia, Italy (Silhavy *et al.*, 2002), studied the mode of suppression of RNA silencing by P19 in great detail. They found that this tombusvirus protein is a potent silencing suppressor that prevents the spread of mobile silencing signals. *In vitro* the P19 binds 21–25 nt dsRNAs that are generated from the PTGS process; these dsRNAs all have 2 nt 3′ overhanging ends. The P19 was also found to bind synthetic dsRNAs that are 21 nt long with the same ends as PTGS-derived dsRNA. Much less or no binding was recorded when the dsRNA was longer or when it was blunt-ended. On the phenotypical level Silhavy *et al.* (2002) introduced the coding sequence for P19 into plants. Their attempts to introduce the *Cymbidium* P19 into *N. benthamiana* failed but the P19 gene from the artichoke mottle crinkle virus (AMCV) could be expressed in the respective transgenic plants. In the latter plants and their sexual progeny there were clear morphological abnormalities. This caused the investigators to suggest that functional PTGS is essential for normal differentiation in plants. We shall discuss this question of RNA silencing and plant patterning in the next chapter.

A further elucidation of the mechanism of suppression by P19 came from an unexpected location: the laboratory of Dinshaw Patel in the Memorial Sloan-Kettering Cancer Center in New York (Ye *et al.*, 2003). These investigators first verified the main results of Silhavy *et al.* (2002) and were satisfied that P19 indeed binds specifically to RNA fragments that are double stranded along 19 nucleotide pairs and have, on each end, a two-nucleotide overhang. They then determined the structure of homodimeric P19 in complex with a 21 nt (having 19 base-pairs) dsRNA (siRNA) by using X-ray crystallography (at 1.85 Å resolution). They found that the 19 bp RNA duplex is cradled within the concave face of a continuous 8-stranded β-sheet, formed across the P19 homodimer interface. Direct and water-mediated intermolecular contacts are restricted to the backbone phosphates and sugar 2′-OH groups, consistent with sequence-independent P19-siRNA recognition. One could say that the P19 is "blind" with respect to the base-sequence in the 19 nt double-stranded oligonucleotide — P19 will bind any such sequence. There are two α-helical "reading heads" that project from opposite ends of the P19 homodimer and position pairs of tryptophans for stacking over the terminal base-pairs, therefore the P19 is capable of "measuring" and bracketing both ends of the siRNA duplex. One may thus say that once the P19 finds its "victim" that has the proper dimensions, the P19 will embrace it "mortally". One month after the submission of Ye *et al.*'s article (2003) to *Nature*, the team of Burgyan in collaboration with J.M. Vargason and T.M. Tanaka Hall of the National Institute of Environmental Health Sciences, North Carolina, USA, submitted their manuscript (Vargason *et al.*, 2003) to *Cell*. The two papers were published at about the same time. The Vargason *et al.*'s (2003) publication was more detailed than that of Ye *et al.*'s (2003) but both publications provided almost identical information. The former (Vargason *et al.*, 2003) studied the 2.5 Å crystal structure of P19 from the carnation Italian ringspot virus (CIRV) that was bound to a 21 nt siRNA. Biochemical and *in vivo* assays demonstrated that this P19 protein acts as a molecular califer to specifically select siRNAs, based on the size and the duplex region of this RNA. These investigators also found that tryptophans of the P19 had a specific role in the binding between P19 and siRNA. On the other hand, the P19 is indifferent to the base sequence in the siRNA, the specificity of the binding is only based on the number of base-pairs (19 nt are optimal) and the characteristics of the 3' overhangs. In other words, any *Dicer* product is attractive to the P19; the *Dicer* has a "measuring-rode" between its two endonucleases and the P19 has a "reading-head" to recognize the diced products (Fig. 36). Even before the two above mentioned publications (Ye *et al.*, 2003, and Vargson *et al.*, 2003) were submitted the Hungarian team (Lakatos *et al.*, 2004) submitted a detailed article on the mechanism of silencing-suppression mediated by P19. Only the latter article was published after the reports on the crystal structure of P19 bound to siRNA. Lakatos *et al.* (2004) reported that the P19 of the *Cymbidium* ringspot virus (CymRSV) is not only suppressing the RNA-silencing in *N. benthamiana*; it can also suppress the RNA silencing in the heterologous *Drosophila* in an *in vitro* system where P19 of CymRSV prevented the siRNA from transferring an RNA strand to the RISC. In the presence of this P19 the amount of free siRNA was markedly diminished and thus rendered it inaccessible for the next stage of the RNA-silencing process. These investigators also found that when a CymRSV that had a mutation in the ORF that encodes P19 was used to infect plants, there was a recovery from the viral infection. Such a recovery does not occur with non-mutated P19. The results of *in vivo* experiments with *N. benthamiana* plants and the *Drosophila*

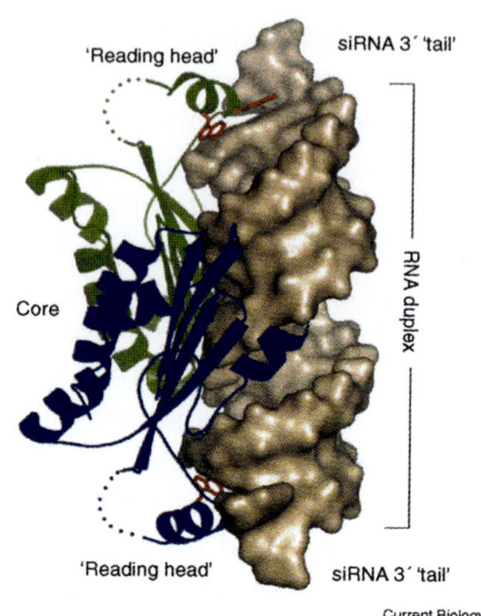

Fig. 36. The structure of the p19 silencing suppressor bound to siRNA. The p19 dimer binds one face of an siRNA duplex (brown). Contacts between the 'core' and the RNA phosphate groups contribute to the proteins high affinity for dsRNA, while a pair of tryptophan residues (red, Trp42, Trp39) in the 'reading head' measure siRNA length. Because each p19 monomer (blue & green) contributes a 'reading head' the protein has been described as a 'molecular califer' that sizes up double-stranded RNA so as to bind best to canonical siRNAs. (From Zamore, 2004, based on J. Vargason and T.M. Tanaka-Hall.)

in vitro experiments were fully compatible. The *in vitro* studies also enable a quantitation and indicated that there was a positive correlation between the level of P19 and the level of siRNA/P19 complexes.

In his "Dispatch" Zamore (2004) integrated the information on the binding of P19 to diRNA, furnished by several publications (e.g. Silhavy *et al.*, 2002; Ye *et al.*, 2003; Vargson *et al.*, 2003; Lakatos *et al.*, 2004) into one coherent picture. This review of Zamore (2004) is therefore recommended to readers who wish to obtain updated information of the P19 protein and how it suppresses RNA silencing.

The biochemistry and molecular structure of other viral suppressors of RNA silencing

Up to now, the P19 is the only viral protein for which biochemical and structural information regarding its interaction with components of the RNA silencing is available. Because such information should shed light on the whole subject of RNA silencing it is anticipated that similar information to that now available for P19, will also become available in the future for other viral proteins that suppress RNA silencing such as HC-Pro, 2b and other suppressors that were listed in Table 3.

It should be noted that while the basic mechanism of RNA silencing is shared by almost all eukaryotic organisms, the mode of the *suppression* of this silencing differs considerably in different plant-virus genera. These genera of viruses "invented" their unique tools to overcome the plants' defence mechanism.

A viral suppressor of RNA silencing may confer enhanced resistance to pathogens

The seemingly paradoxal possibility that a suppressor of RNA silencing can impose resistance in plants, to viruses and other pathogens, was analyzed by the team of Vicki Vance and investigators from the University of Kentucky, USA (Pruss *et al.*, 2004). This study was on the HC-Pro suppressor. It showed that when tobacco plants with the *N* resistance gene and an ability to express HC-Pro were infected with tobacco mosaic virus (TMV), there were fewer and smaller lesions on these plants than on plants that did not express HC-Pro. Plants that expressed HC-Pro were also more resistant to tomato black ring neprovirus (TBRV) and to the oomycete *Perenospora tabacina* than control tobacco plants. The system that involved the enhanced resistance is rather elaborate and will not be detailed here. For our deliberation it is noteworthy to indicate that the same suppressor of RNA silencing can act in two opposing directions. It can suppress the plant's defence against pathogens but can also enhance the plant's defence against other pathogens.

RNA Silencing in Angiosperm Plants II: MicroRNA and Control of Differentiation

The miRNAs were already defined and described in previous chapters that dealt with metazoan organisms. Here, I shall focus on miRNAs in plants but first I shall remind the reader about the nomenclature of these short RNAs. The genomic sequences that code for the miRNAs are now termed miRs. Since miRs are genes, the specific mammalian genomic sequence that encodes the miR130 should be written as *miR130*. Likewise, a plant gene that encodes the miRNA that targets to the trancript of the gene *DCL1* should be written as *miR162*. The transcript of a *miR* gene (commonly in the form of an inverse-repeat structure, with a spacer between the repeats) is processed (in the nucleus) into a *pri-miRNA* and the latter is processed further into a *pre-miRNA*. Finally, a single-stranded *mature miRNA* (of ~21–25 nt) will bind to a protein complex and guide this protein complex to its target. Instead of using the term microRNA, the short term miRNA is also used.

When RNAi and miRNAs are involved, investigators "move" very fast in their research. As indicated above, it took only a short time from the findings in nematodes that *lin-4* and *let-7* encode miRNAs till miRNA genes were revealed in other metazoa, including man. Not much later, four laboratories submitted their respective findings about plant miRNAs for publication. In fact, the teams of James Carrington and associates of the Oregon State University (Llave *et al.*, 2002a), the team of Marjory and Antonius Matzke, of the Austrian Academy of Science, Salzburg (Mette *et al.*, 2002), the team of David and

Bonnie Bartel of the MIT and the Rice University in Texas (Reinhart *et al.*, 2002) and the team of Joachim Messing and Xuemei Chen of the Rutgers University in New Jersey (Park *et al.*, 2002) submitted their papers in March, April, May and June 2002, respectively. These four teams of investigators looked at plant miRNAs from different angles. Together, these four publications "opened" the era of miRNAs in plants. Another remark is due before I proceed. Although this chapter is entitled "RNA Silencing in Angiosperm Plants II", miRNAs were also revealed in other plants. In a brief communication Floyd and Bowman (2004) of UC-Davis, California, reported on sequences encoding miRNAs in two gymosperms (*Pseudotsuga menziesii* and *Taxus globosa*) in a fern (*Ceratopteris richardii*), in a lycopod (*Selaginella kraussiana*), in a moss (*Physcomitrella paterns*), in a liverwort (*Marchantia polymorpha*) and in a hornwort (*Phaeoceros carolinianus*). The authors concluded that conserved miRNAs negatively regulate a given type of genes (class III HD-Zip genes that are required for meristem-derived leaf symmetry in angiosperms) in all tested land plants. Hence, *miR165/166* is conserved since about 400 million years ago. Still, detailed information on miRNA of plants is available primarily in angiosperms; I shall therefore retain the title of this chapter.

This chapter will deal with several aspects of RNA silencing by plant miRNAs. The mechanism of silencing by three avenues will be discussed: silencing by degrading mRNAs, silencing by the inhibition of translation and silencing by chloromatin remodeling. These avenues of gene-silencing by plant miRNAs will be compared to those known in metazoa.

This chapter will also discuss the means by which plant miRNA were identified, provide partial lists and guide the readers to data bases that are continuously updated.

One of the prime roles of plant miRNAs is in the regulation of plant patterning. This subject deserves a short introduction. Undifferentiated cells in the meristem of a plant's apex ("stem cells") that are destined to participate in patterning an organ (e.g. leaf, floral members) have to receive cues to direct them "where" to go. For that, the cells should first have a perception of where they are. This elicits the association with a passage from "Alice in Wonderland" by Lewis Carroll, in which there is an intriguing dialogue between Alice and the Cheshire Cat, that is sitting on a tree at a road junction:

> *Tell me, please, (said Alice) which way I ought to walk from here.*
> *That depends a good deal on where you want to get to (said the Cat).*
> *I don't care much where ... (said Alice)*
> *Then it doesn't matter which way you walk (said the Cat).*

The Cat then informed Alice that they are located in the Land of Madness.

In some ways the undifferentiated cells in the apex are in a similar situation to that of Alice. But like the Cat these cells "know" their location. If the cells do not have a clear destination, they will keep their undifferentiated division and reach "no-where". But if the cells have a correct perception where they are and receive the proper cues to destine them toward the participation in a given pattern, they can "walk" on the right track and reach their goal. In one of the early publications of the team of Elliot Meyerowitz on flower development in *Arabidopsis* (Bowman *et al.*, 1989), it was phrased: "... Each cell must somehow determine its position relative to others and must differentiate accordingly." Only the differentiation of a plant organ, starting from undifferentiated "stem-cells", is immensely complicated and scientists are only beginning to understand its mechanism. Take, for example, the "cues" can be of very different types as gradients of proteins or hormones, activity of transcription factors or small RNAs (e.g. miRNA). The perception of space and where in this space the cells are located is another issue. The latter issue may have various "solutions". One such solution is being revealed in bacteria. Bacteria of the genus *Vibrio* evolved a way to sense whether they are "alone" or there is already a "crowd" of the same bacteria in their vicinity. This is the "Quorum Sensing" whereby the bacterium sends out small molecules into the medium and then measures the level of these molecules that are also contributed to the medium by its neighbors (see: Gottesman, 2004a and 2004b, for review and literature). A somewhat similar sensing of the surrounding was found many years ago in cells of the cellular slimemold. In his book Enrico Coen (1999) elaborated the sensing of location by embryogenic cells but how exactly this sensing of space is operating in plant cells is still under investigation and most probably there are many mechanisms for this sensing. I shall add a note about "differentiated" and "undifferentiated" cells. These terms are problematic and wherever the term *undifferentiated* appears in this book it should be read as flanked by quotation marks. This is because even cells in the plant's apex are already "differentiated" — they are on the way to contribute to plant organs. Truly undifferentiated plant cells are very rare. Such a state may be attributed to the descendants of the pollen-mother-cells, after meiotic division and before the male-gametophyte is established (see: Frankel and Galun, 1977). In such cells there is a stage in which the pre-existing mRNA and the polysomes are "cleared". These cells "forgot" their sporophytic past but are not yet gametophytes. Conceptually and amazingly each meristematic plant cell should have a complete and detailed "road map" for patterning (not only of a specific organ but also of the whole plant because these cells are totipotent). The goal to fully understanding the complicated mechanism of patterning may be so far away that we shall not be able to achieve it in the near future. As I have already pointed out above, this should not discourage us because we can start the journey on a "road" toward a full understanding. Overcoming the hurdles of this road may be more rewarding than reaching the final goal.

A final introductory note concerns not only plant miRNAs. Until recent years gene silencing by small RNAs was considered a mechanism that is exclusively operating in eukaryotes. In recent years it became evident that small RNA sequences in bacteria (mostly studies in *Escherichia coli*) can base-pair with transcripts and either repress or activate the translation from these transcripts. More than 50 such small regulatory RNAs (denoted non-coding RNAs or *ncRNAs*) were revealed in *E. coli*. This book will discuss eukaryotic RNA silencing only. Readers interested in ncRNAs (of bacteria) are referred to the review of Storz *et al.* (2004).

Generation of MicroRNAs in Plants and Mechanisms of RNA Silencing by Plant MicroRNAs

Some features of the generation of miRNAs and the mechanism of silencing by plant miRNA were revealed in the four pioneering publications on these miRNAs. In the first submitted publication,

Llave *et al.* (2002a) had already provided some clues. The genomic sequences in *Arabidopsis* from which the miRNAs were derived, were found in several locations in the genome. Some were clustered in intergenic regions, others were inside introns of protein-coding genes, in transposon-like sequences and even in or around the repeats that code for the structural 5S rRNA genes. The sequences that encoded miRNAs could be computer-folded into hairpins (or stem loops) of various sizes. Because the mature miRNAs that were revealed in plant cells were much shorter (predominantly 21–24 nt) than the genomic sequences that encode them, it was clear that (probably after *in vivo* folding) the transcripts of the miRNA genes have to be processed. Mette *et al.* (2002) who focused on a group (the "40" family) of genomic sequences in *Arabidopsis* also found that these sequences that were revealed in intergenic regions were transcribed to RNAs of about 200 nt. This means that the sizes of the transcripts that are precursors of plant miRNAs can be about 3-fold longer than the precursors of animal miRNAs. The precursor-transcripts found by Mette *et al.* (2002) had terminal inverted repeats and could be computer-folded into structures that were mostly double-stranded "stems" but have "loops" or "bumps" too. The folded structures had at least about 21 nt of base-pairs (with a bump) that after processing should yield miRNAs.

Reinhart *et al.* (2002) cloned endogenous (small) RNAs from *Arabidopsis* and described 16 plant RNAs that had features of miRNA. They were encoded by genomic sequences. The transcripts of these sequences could be computer-folded into fold-back secondary structures (hairpins and stem-loops) predicted to be precursors of miRNAs. The total length of these transcripts varied considerably and some reached a total length of several hundred nt. A sample of these transcripts in their fold-back configuration is shown in Fig. 37. For the processing of the transcripts into miRNA the wild-type gene for the *Dicer* homolog of *Arabidopsis, CARPEL FACTORY* was required. This gene (*CAF*) was also termed *SHORT INTEGUMENT* (*SIN1*) and later given the name *DCL1* for *Dicer*-like 1. It is one of the four homologs of the *Drosophila's Dicer* that exist in the genome of *Arabidopsis*. Mutation in *CAF/SIN1* was known to cause morphological abnormalities (e.g. in

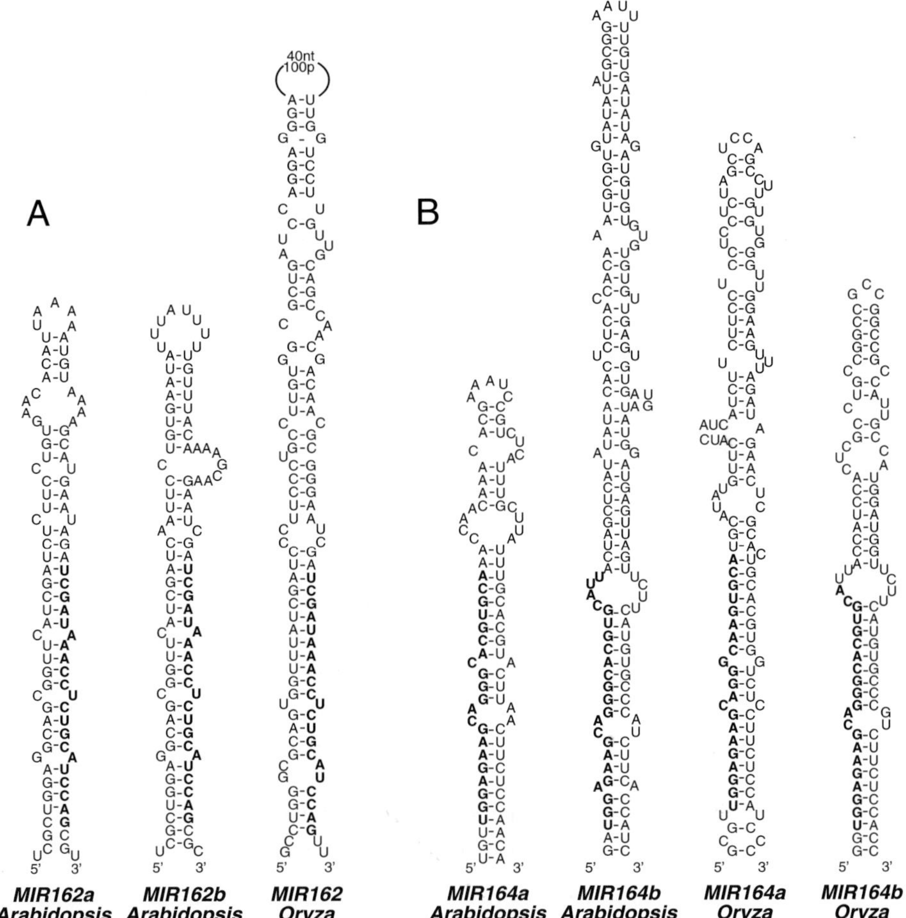

Fig. 37. Conservation between the *Arabidopsis* and *Oryza* predicted stem-loop precursors of miRs. (A) miR 162 homologs. (B) miR164 homologs. Sequence homology is seen within the miRNA (bold) as well as its paired sequences and a few base-pairs adjacent to the miRNA. The remainder of the sequence has drifted considerably with the main constraint being the formation of a stem-loop structure. (From Reinhart *et al.*, 2002.)

the carpel and in the integument). The requirement for wild-type *CAF/SIN1* for processing of miRNA clearly indicated that miRNAs are essential for normal differentiation. The mature single-stranded miRNAs were of the sizes of 20–22 nt. They were derived from the "stem" region of the precursors and frequently started (at their 5' end) with U (uracil) or even with U-G (uracil-guanine). Reinhart *et al.* (2002) found that the mature miRNAs of *Arabidopsis* that they had analyzed could interact with the transcripts of their respective targets (e.g. transcripts encoding transcription factors) by near-perfect base-pairing. This was different from what was revealed in animals, where the interactions between the miRNAs and their targets are almost never in the coding region but rather in the 3′ UTR and the pairing is not perfect. Another finding of Reinhart *et al.* (2002) was that the various plant miRNAs are rather conserved: very similar miRNAs were found in *Arabidopsis* and rice (e.g. *miR162* and *miR164*). This indicated that these miRNAs have been evolutionarily conserved for more than 250 million years. Similar findings to those reported by Reinhart *et al.* (2002) were reported by Park *et al.* (2002). The latter article was submitted about 1 month after the former article, as noted above. Park *et al.* (2002) started with cloning 230 unique putative small RNA sequences from *Arabidopsis*. Then they subtracted most of these clones. Take, for example, 176 sequences corresponded to known non-coding RNAs (e.g. rRNA, tRNA) were subtracted. They also subtracted those that had more than 2 nt mismatches with sequences of the *Arabidopsis* genome. They kept 39 clones that were homologous to exons, introns or intergenic regions. Of these 29 were capable (by a computer program) to fold into stem-loop structures. Further subtractions left the investigators with 11 "confirmed" miRNAs. Of these three were also detected by Reinhart *et al.* (2002). Most of the miRNAs found by Park *et al.* (2002) were transcribed from multiple, identical or related sequences scattered in several locations in the *Arabidopsis* genome. The latter authors found that the production of miRNAs was strongly reduced in plants that were mutated in *HEN1* but as in the case of mutants in *CAF* this did not cause the accumulation of the precursors of the mature miRNAs. It was known that both CAF and HEN1 have putative nuclear localization signals. This suggested that the respective processing of the precursor-transcripts of miRNAs takes place in the nucleus. The *Dicer* that cuts animal dsRNA into siRNA is operating in the cytosol. Hence, CAF and HEN1 probably differ from *Dicer*.

Carrington and associates of the Oregon State University in Corvallis, OR (Llave *et al.*, 2002b), made a significant contribution to the gene regulation in plants by miRNA. They found that miRNA39

is homologous to several regions of the family of *Scarecrow-like* genes
that are putatively encoding transcription factors and act in floral
buds. This homology results in specific cleavage of the mRNAs of
these genes. Thus, miRNA39 has probably an essential role in floral
patterning.

As indicated in Chap. 11, Tang *et al.* (2003) of the Phillip Zamore
Laboratory, at the Massachusettes Medical School, in Worchester, MA,
in collaboration with David Bartel of the MIT in Cambridge, MA,
developed the wheat-germ extract system to analyze the biochemistry
of RNA silencing in an *in vitro* plant system. If indeed, as revealed later
(see: Fig. 27) that more than one RNase stage is required in plants to
generate mature miRNA from the respective transcripts, Tang *et al.*'s
(2003) analyses would not be able to reveal the first stage. This would
be because the first stage of the generation of pri-miRNA is assumed
to take place in the nucleus by a specific *Dicer*-family member (DCL1)
that has a nuclear-localization signal. The wheat-germ extract represents primarily plant cytoplasm rather than nuclear activity.

What is the role of Argonature-type proteins in the generation
of mature plant miRNAs? In *Drosophila* it was found that one (non-
mutated) gene encoding AGO2 is defective in the generation of siRNA
but AGO2 is apparently not required for the activity of miRNAs.
On the other hand, AGO1 is dispensable for siRNA-directed dsRNA
cleavage but is required for the production of mature miRNAs. Are
there also Argonaute proteins in plants that are specifically required
for the generation of mature miRNAs? It was proposed that in *Arabidopsis AG01* is indeed required for miRNA-directed organ development (see: Kinder and Martienssen, 2003, 2004) but intact *AGO1*
also seems to be essential for plant PTGS. Genetic studies are very
useful in clarifying some aspects of miRNA production but the exact
role of the Argonaute proteins in this generation awaits detailed biochemical studies. As was mentioned above, the primary transcript for
miRNA in animals is cleaved by Drosha in the nucleus, resulting in a
pre-miRNA that has a stem-loop structure with a 3′ overhang of
2 nt. The pre-miRNA is then exported to the cytoplasm where a
Dicer cleaves this pre-miRNA further into a dsRNA and one of the
strands (the antisense strand) of this latter dsRNA is (after helicase-
induced separation of the strands) complexed with a RISC-like

complex. Is DCL1 (and/or DCL4) of *Arabidopsis* replacing the Drosha of animals? This possibility was suggested as seen in the review by He and Hannon (2004). The domains of DCL1 and DCL4 (Fig. 38) are rather similar and both proteins include domains for nuclear localization signals, helicase and dsRNA binding.... But genetic data indicate that DCL1 in plants may also fulfill the task of the animal *Dicer*. Thus, the conversion of pre-miRNA in plants into the 20–22 nt dsRNA may happen (also?) in the nucleus. Is there a mechanism that determines

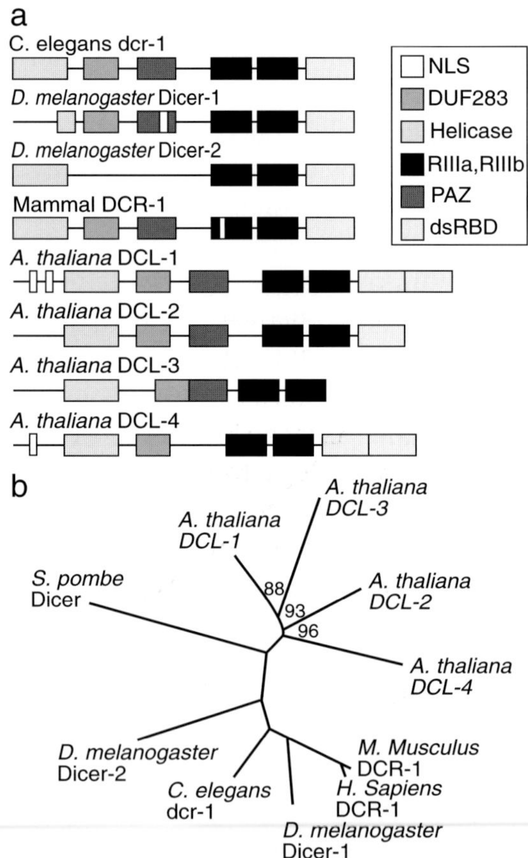

Fig. 38. The structure and function of the *Dicer* family. The domain structure of *Dicer* homologues in worms, flies, mammals and plants is schematically presented. The *Dicer* homologues that function in the microRNA (miRNA) maturation pathway have the PAZ (Piwi-Argonaute-Zwille) domain. (From He and Hannon, 2004).

which of the two stands, derived from the pre-miRNA, will serve as the guide for the RISC-like complex to the target mRNA? Apparently, there is such a mechanism. The strand of choice is most probably the one with a less stable 5′ end where there are relatively more A-U base-pairs than G-C base-pairs. This choice probably reflects the relative ease of unwinding (by helicase) from one end of the miRNA:miRNA* (antisense:sense*) duplex. It also seems that in cases where there are no differences in the stability of the two ends, either of the two strands can join the RISC-like particle. This consideration is also valid for siRNAs that serve in regular PTGS.

The vast majority of plant miRNAs that were characterized with respect to their interaction with their respective mRNA targets guide the RISC-like particle to cleave their mRNA-target rather than to cause translational repression. A notable exception was studied by Chen (2004c). The latter study dealt with miR172. miR172 has high complementarily with a coding region in APETALA2 (not with the 3′ UTR of the transcript of this gene), but rather than cleaving the respective mRNA, miR172 causes repression of translation.

There is an apparent clear distinction between plant and animal miRNAs. The latter bind to the 3′ UTR of their target (at several sites), have a low level of complementarily to the target and cause translational repression. The plant miRNAs bind to the coding region of their target, have near perfect homology to this region and cause cleavage their target. But this distinction may not hold true after further investigations. It is possible that there are many more plant miRNAs and several of these have low complementarily with their targets. But when the sequences of small RNAs of plant cells serve for computer-search of the plant genome and the matching sequences serve to identify miRNAs — plant miRNAs with low homologies to genomic sequences will not be identified.

If indeed, the processing of transcripts to pri-mRNAs as well as the processing of pri-miRNAs to pre-miRNAs in plants takes place in the nuclei — the miRNA precursors in plants should be exported from the nuclei to the cytoplast. A gene, *HASTY*, was identified in *Arabidopsis* (Bollman *et al.*, 2003) that encodes a protein that is orthologous to the mammalian exportin. The mammalian exportin is a nucleocytoplasmic transport receptor that facilitates the transport of macromolecules across the nuclear pore. *Arabidopsis* plants in which the *HASTY (HST)* gene is mutated and where no HASTY protein is formed have morphogenesis defects that suggest that the miRNA control of differentiation is defective in the *hst* mutants.

Once the miRNA: miRNA* duplex is transported (through the nuclear pores?) from the nucleus to the cytoplasm, the two ~21 nt strands are separated and one of these strands (mature miRNA) binds to the protein complex. The other single-stranded miRNA* is probably degraded. In PTGS this protein complex was termed RISC (RNA-induced silencing complex). In the miRNA systems of animals this complex was termed miRNP and as indicated in Chap. 10, contains several specific proteins (e.g. elF2C2, Gemin3, Gemin4 and RNase). The composition of the protein complex that is led by the plant miRNA to degrade its target is yet not fully known. Bartel (2004) suggested terming this complex RISC although it could differ in its composition from the RISC that is active in PTGS. "RISC" could therefore fulfill very different tasks: (1) In most miRNA silencing in plants as well as in siRNA silencing in plants and in animals, "RISC" cleaves the mRNA at the location of homology (or near-homology) with the respective mature miRNA or siRNA; (2) In most animal miRNA silencing "RISC" causes suppression of translation from the transcript in which the 3′ UTR has short complementary segments with the mature miRNA; (3) In some cases "RISC" may mediate methylation of histone, rendering active chromatin to silent chromatin; the latter process is "transcriptional gene silencing" termed TGS. The mechanism of TGS is not fully understood and will not be detailed in this book. Is the very same "RISC" active in all three tasks? There is no reliable answer to this question yet. The notion that the same "RISC" of plants will either degrade its target or cause suppression of translation and the "decision" of which silencing will happen, depends only on the homology between the miRNA and its target — is problematic. As was mentioned above, the plant miR172 has near-perfect homology to its target (in the transcript of *APETALA2*) and causes suppression of translation rather than degrading of its target. Notable when the miRNA-bound to "RISC" is causing cleavage of the targets' transcript (as in almost all investigated cases of plant miRNAs) the cleavage is at a specific location. It happens (in the transcript) between the nucleotides pairing to residues 10 and 11 of the miRNA. After cutting its mRNA target, the same "RISC" (with its miRNA guide) remains functional and can cause the degradation of additional mRNAs that have the same sequence.

Interestingly, a feedback inhibition was revealed in *Arabidopsis* (Xie *et al.*, 2003; Bartel and Bartel, 2003). The target of miR162 is the transcript of *DCL1* that is required for the process of miRNA-induced mRNA degradation. This suggests that when mature miR162 is produced it may lower the activity of all(?) other miRNAs. Thus, in tissues in which miR162 is relatively abundant there may be a reduction in the maturation of other miRNAs. This could happen in plant tissues where miRNA regulation of differentiation and/or regulation of metabolic pathways are not required. Conversely, if all (or most) existing mRNA are erased in order to start a completely new fate for the cell (as in the very early phase of the initiation of the gametophyte mentioned above) the cell may degrade non-selectively mRNAs and polysomes with bound mRNA. The generation of specific mature miRNA may provide a selective degradation of certain mRNAs. In the following sections we shall look into more details at the known targets of plant miRNAs. But it should already be noted here that while relatively more miRNA targets were revealed in plants than in animals, the number of plants' targets may even be far greater than presently identified.

While several steps in the generation of plant miRNAs and the basic modes of gene silencing imposed by these miRNAs were recently revealed our understanding of these processes is far from complete. Talented investigators are focusing on these processes and surely valuable information is expected. An example of such efforts is the study of the Nina Fedoroff Laboratory (Han *et al.*, 2004) at the Pennsylvania State University. These investigators focused on an *Arabidopsis* gene termed *HYPONASTIC LEAVES 1 (HYL1)*. As befitting Fedoroff, the mutant (*hyl1*) was obtained by the insertion of the *Ds* transposon into *HYL1*. The double recessive *hyl1* mutant shows a pleotrophic affect on several morphological fetures of *Arabidopsis*, as well as on its reaction to plant regulators. The HYL1 protein has two dsRNA-binding domains. In the *hyl1* mutants the level of three analysed miRNAs (miR159, miR167 and miR171) is strongly reduced. Also, the levels of target mRNAs of these miRNAs were elevated. Over-expression of HYL1 could be achieved by introducing into plants, the coding sequence for HYL1 behind a strong promoter. The HYL1 is predominantly a nuclear protein and is found in *Arabidopsis* nuclei as nuclear bodies and ring-like structures. Half or more of the HYL1 protein in extracts are in the form of 300 kD complexes. But ... some HYL1 were also occasionally found in the cytoplasm, and in *hyl1* the generation of functional miRNAs was not completely eliminated. So what is the role of HYL1 in the generation and functionality of miRNAs? We expect the Fedoroff team to provide additional information.

Screening for Plant MicroRNAs and the Identification of Their Targets

I have mentioned the study by the Bonnie and David Bartel team (Reinhart *et al.*, 2002) in the previous section as the latter study was one of the early four studies on plant miRNA. Reinhart *et al.* (2002) used a similar approach to that that was used previously to identify animal miRNAs. Briefly, small RNAs of the size-range of 18–26 nt from extracts of *Arabidopsis* organs were cloned and run on a gel and hybridized to end-labeled antisense DNA. The sequences of the hybridized RNA clones were then used for comparison with genomic sequences of *Arabidopsis*. When the hybridized genomic sequences indicated that they had an inverse repeat, this indicated that they contain a code for a miRNA. By this criterion and the similarity to animal miRNA the authors identified 16 apparent plant miRNAs. Already in this early study it was found that among the 16 miRs five were transcribed from single copies while 11 miRs corresponded to several loci in the *Arabidopsis* genome. It was not clear if all the loci that encoded the same miR were actually transcribed. Some could be pseudogenes. Eight out of the 16 *Arabidopsis miR* genes had matches in the rice (*Oryza sativa* ssp. *indica*) genome (this number was reduced from eight to seven in a later publication). In an "update" of the above mentioned study, Bartel and Bartel (2003) revealed that the 16 miRNAs of *Arabidopsis* identified by them differ from animal miRNA with respect to their homology to the respective targets. The plant miRNAs matched the coding sequence rather than having several short complementary segments in the 3′ UTR of their targets (transcript). The matching between these miRNAs and their respective targets was near perfection and the plant miRNAs commonly manifested their silencing by the degradation of their target-mRNA rather than by suppression of the translation. These findings of the Bartel and Bartel team were summarized in Table 4. In another study the Bartel and Bartel team (Rhodes *et al.*, 2002) looked closer at the targets of the miRNAs that were identified by them. They detected 49 targets of 14 identified miRNAs in *Arabidopsis*. Of these 34 were members of transcription-factor gene-families that are involved in developmental patterning or in cell differentation. The authors suggested that the targeting of developmental transcription factors indicates that many plant miRNAs function during cellular differentation and by that eliminate key regulatory transcripts from daughter cells. Such an elimination should be required especially for transcripts that have a long "half life", unless specifically degraded.

The conservation of miRNAs and their targets among angiosperms is amazing. As indicated above, eight (later seven) of the *Arabidopsis* miRNA/target systems were also found in rice. When the authors used a three-mismatch cutoff, then six of the *Arabidopsis* miRNAs had at least one target in the rice genome. This means that miRNA/target systems were conserved since the early evolution of angiosperms.

Other research teams added plant miRNAs and targets for these miRNAs, thus considerably extending the number of identified plant miRNAs. It is expected that this number will gradually grow further because in the past the investigators looked primarily "under the street lamp". The identification was biased with respect to miRNAs that were more abundant and/or had near homology to genomic sequences. In order to reveal additional plant miRNAs Jones-Rhoades and Bartel (2004) developed a computational identification process that included several steps. In some of the analyses the apparent *Arabidopsis* miRs were compared to rice homologs (or homeologs). The start was a computer analysis of 133 864 *Arabidopsis* and 410 167 rice inverted repeats detected in the respective genomic sequences. They then looked at the (computer) folding of these inverted repeats

Table 4.   *Arabidopsis* microRNA targets.

MicroRNA	Target family	Predicted target genes	No. of mismatches[a]
miR156	SQUAMOSA-PROMOTER BINDING PROTEIN (SBP)-like proteins	10 *SPL* genes[b]	1–2
miR157	SBP-like proteins	9 *SPL* genes[b]	1–3
	Putative DEAD-box RNA helicase	At5g08620 (=AtRH25)[b]	3
	Unknown proteins	At1g22000, At3g47170[b]	3
miR158	Unknown protein	At1g64100[b]	3
miR159a	MYB transcription factors	5 *MYB* genes[b,c]	2–3
	Unknown protein	At1g29010[b]	3
miR159b	MYB transcription factors	3 *MYB* genes	3
miR160	Auxin Response Factors	*ARF10, ARF16, RF17*[b]	1–3
miR161	Pentatricopeptide repeat proteins	9 genes[b]	3
miR162	DICER	*DCL1 (=CAF=SIN1=SUS1)*	1–nt bulge
miR163	SAM-dependent methyltransferases	5 genes[b]	0–2 with 1–nt bulge
miR164	NAC domain proteins	*CUC1, CUC2, NAC1, 2 others*[b]	2–3
miR165	HD-Zip transcription factors	*PHV, PHB, REV, ATHB-8*[b]	3
miR166	HD-Zip transcription factor	*ATHB-15*[b]	3
miR167	Auxin response factors	*ARF6*[b,c], *ARF8*[c]	3–4
miR168	ARGONAUTE	*AGO1*[b]	3
miR169	CCAAT-binding factor (CBF)-HAP2-like proteins	At1g17590, At1g54160[b]	3
miR170	GRAS domain transcription factors (SCARECROW-like)	*SCL6-II, SCL6-III, SCL6-IV*[b]	2
miR171	GRAS domain transcription factors (SCARECROW-like)	*SCL6-II, SCL6-III, SCL6-IV SCL6-IV*[b,d,e]	0
miR172	APETELA2-like transcription factors	*AP2,3 AP2-like genes*[c]	1–3

[a]G:U wobbles are included as mismatches in this analysis; [b]Rhoades *et al.* (2002); [c]Park *et al.* (2002); [d]Reinhart *et al.* (2002); [e]Llave *et al.* (2002a). (From Bartel and Bartel, 2003.)

and focused on 20 mers within the predicted hairpins of the folded structures. The various steps of the elimination of sequences that were not predicted to be miRs retained 87 *Arabidopsis* miRs and 122 rice miRs. These authors also refined the detection of targets. Targets that had less homology to the respective miRNAs were also taken into account. Take, for example, the miRNA JAW has four to five mispairs to the mRNAs of several TCP transcription factors (TCP is an abbreviated term for a family of proteins, that include T̲BI, C̲YC and P̲CFs that control plant patterning; this term was given by Cubas *et al.*, 1999). Such miRNAs were retained as apparent miRNAs provided that the miRNA-complementarity is preserved in homologous *Arabidopsis* and rice miRNAs. By this rather elaborated and computer-assisted procedure that was followed by experimental validation, Jones-Rhoades and Bartel (2004) added 23 miRNAs and brought the number of *miR* genes in plants to 92 (representing 22 families). This study also indicated that plant miRNAs have conserved functions beyond the regulation of plant development. Some plant miRNA targeted superoxide dismutases, laccases and ATP-sulfurylases. Moreover, the expression of *miR395* was increased by environmental stress (sulfate starvation). It seems that this procedure was also affective in the identification of rare miRNAs. On the other hand, this procedure will not detect some specific miRNA. Take, for example, those *Arabidopsis miRs* that are *not* conserved in rice will be "lost" in this search. Other miRNAs may be "lost" due to the stringencies of the cutoffs of homologies. Also, when the matching sequences between the miRNA and the coding region of the mRNA are replaced by several short recognition sites at the 3′ UTR of the respective target-transcripts, as was commonly found with animal miRNAs and their targets, some plant miRNAs may escape easy detection. It is therefore expected that more miRNAs of plants will be identified in the future and their respective targets shall be revealed.

Investigators who study miRNAs reached an agreement for a uniform system for miRNA annotation (Ambros *et al.*, 2003). To avoid confusion, with respect to the identification of new animal and plant miRs, a miRNA repository/registry center was established at the Wellcome Trust Sanger Institute in Hinxton, Cambridge, UK (Griffiths-Jones, 2004). This registry can be approached through the Rfam (UK) website at http://www.sanger.ac.uk/Software/Rfam/mirna. Queries and feedback concerning *Caenorhabditis*, Drosophila, man, mouse and *Arabidopsis* can be sent to the email address: microrna@sanger.ac.uk

Differentiation and Plant miRNAs

A comprehensive treatment of plant patterning is much beyond the scope of this book. Such a treatment requires an extensive text that will combine traditional concepts with updated approaches. But in this book the central theme is RNA silencing and it became evident in recent years that this silencing has an important role in plant patterning. The following section will therefore present basic information and concepts of plant patterning that are required as an introduction to a discussion on the impact of miRNAs in the patterning of plants. This section will then deal with a selected number of cases in which the role of miRNAs in the patterning of specific plant organs and tissues as flowers, leaves and vascular bundles, was analyzed.

Plant patterning: General considerations

The common statement that most of the plant's body-cells are formed postembryonically by projection from apical meristems (i.e. Ori *et al.*, 2000) is basically correct. But this does not mean that the shoot apical meristem (SAM) and the root apical meristem (RAM) are the only locations from where differentiated plant-cells are projected. Readers who are not familiar with plant biology may perceive the wrong notion that in animals cell-differentiation occurs at the embryo stage while in plants differentiated cells are only derived from SAM and RAM. There are ample other plant organs where controlled cell differentiation takes place. Also, postembryonic cell differentiation takes place in animal organs. The pluripotent bone-marrow cells in mammals is only one of many examples. True enough, the SAM of plants can serve as a useful system to investigate plant patterning because of its relatively simple organization. The SAM was therefore used in studies on SAM-derived cell differentiation in the model plant *Arabidopsis* (see: Bossinger and Smyth, 1996). The SAM organization was elegantly described by Weigle and Jurgens (2002). The primary shoot meristem is formed early in embryonic development. The SAM is thought to contain two major functional-domains: the central zone and the peripheral zone. The former zone consists of a few cells that divide very slowly. The latter zone consists of cells from which leaf-initials are derived in a regular pattern (typical for each plant species). After the transition from the vegetative to the reproductive phase, the peripheral zone initiates whorls of floral members, as will be detailed below.

After the leaf-initial is established, several conceptual kinds of differentiation takes place. There is a change from a radial symmetry to a bilateral symmetry in which the cell-layer, adjacent to the shoot (adaxial) differs from the layer that is away from the shoot (abaxial). There is a further differentiation of the shape of the lamina of the leaf and its margin and there is additional differentiation such as the formation of trichomes and stomata. Each of these patternings has a large range of plasticity imposed by genetic and environmental effectors. An example is the distribution of stomata in the epidermis of the leaf lamina; these can be close to one another or far away from one another. Bergmann *et al.* (2004) found that a null-mutation in the *Arabidopsis* gene, *YODA*, causes an excess in stomata whereas the constitutive expression of *YODA* eliminates stomata. Environmental effectors and additional plant genes can also influence the density of stomata. The plasticity of floral members is well known since many years ago, and the impact of homeotic genes and miRNAs on the differentiation of floral members will be discussed below in some detail. I shall note two cases of such plasticities in the stamens of *Nicotiana* that emerged from my own studies here. When protoplasts are isolated from *N. tabacum* (tobacco) leaves and X-irradiated before *in vitro* culture, functional plants can be obtained. These can be self-pollinated to obtain a sexual progeny. Frequently plants of this progeny have stigmatoid anthers (Fig. 39). Stigmatoid anthers can also be obtained by interspecific transfer of mitochondrial components in *Nicotiana*. Regeneration of plants from cybrid protoplasts which have *N. sylvestris* nuclei and *N. alata* mitochondria will result in flowers that have stigmatoid (and sterile) anthers (Aviv *et al.*, 1984). An additional case of flower plasticity, from my own work (see: Frankel and Galun, 1977), is the differentiation of stamens and pistils in flowers of cucumber. Specific genes, environmental factors (day-length and temperature) as well as growth regulators (gibberellins, anti-gibberellins, auxins, ethylene, suppressors of ethylene effects) can all affect the further development of initials of stamens and pistils, leading to a range of mature flowers from

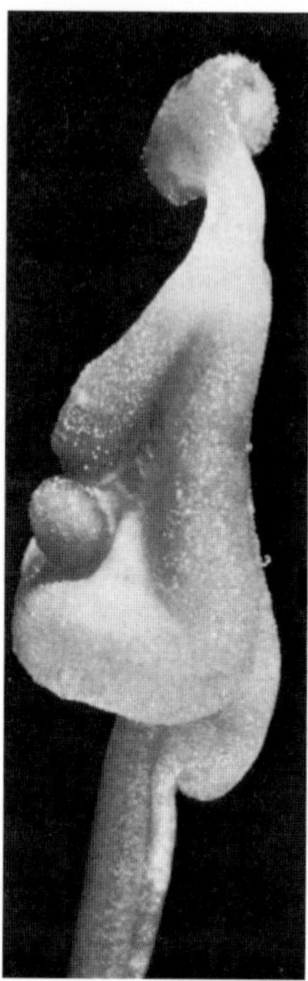

Fig. 39. Stigmatized anther of a *Nicotiana tabacum* mutant regenerated from an X-ray-radiated mesophyll protoplast cultured *in vitro*. (From Frankel and Galun, 1977.)

staminate flowers (devoid of pistils) through perfect flowers (with stamens and pistils) to pistillate flowers (devoid of stamens).

I would also not state, as Schwarz-Sommer *et al.* (1990) indicated: "Morphogenetic processes in plants, therefore, unlike in animals, cannot easily be related to *maternally determined* position information ...." True enough, the polar flow of maternally derived components in the fertilized egg of animals determine a (decisive) early gradient, while such a flow was not revealed in the fertilized egg-cell in plants. But such a flow cannot be excluded because the first division of the egg cell in plants already determines which cell will divide perpendicularly, forming a line of cells — the suspensor, and which cell will divide in several angles and form the pro-embryo from which the young embryo will emerge. Indeed, in the egg cell of the brown alga *Fucus*, a clear polarity is established soon after fertilization (in this case the polarity is light-induced) and this polarity is "remembered" by the young embryo till after several cell divisions (Galun and Torrey, 1969; Torrey and Galun, 1970). Hence, a flow-gradient in the early differentiation of embryos in plants should not be dismissed.

As indicated above I shall detail only a few examples where miRNAs are involved in patterning; such as flower development, leaf symmetry and vascular differentiation. In these examples homeotic genes are involved. They cause the differentiation to occur "out of place". *Homeosis* is the term used by developmental biologists to describe the development of the "wrong" organ at the "wrong" place. When a mutated gene is causing homeosis the wild-type gene can be identified as a *homeotic gene*, meaning that the wild-type gene leads the cells to normal differentiation. It was revealed that homeotic genes involved in flower development encode proteins that have a motif of amino acids at the amino-terminal end that is shared by these proteins as well as by DNA-binding transcriptional regulation in yeast and in vertebrates. The yeast gene encoding such a DNA-binding protein is *MCM1*, the ortologous gene from vertebrates is *SRF*, the *Arabidopsis* gene is *AGAMOUS* and the snapdragon (*Antirrhinum*) gene is *DEFICIENS*. The functional motif shared by these four genes was thus termed by Schwarz-Sommer *et al.* (1900) as *MADS-box* (<u>M</u>CM1-<u>A</u>GAMOUS-<u>D</u>EFICIENTS-<u>S</u>RF). These, as well as other regulatory proteins, seem to have a very "ancient" source, possibly even before multicellular structures were established in eukaryotes. This leads to a general assumption that the use of "ancient" proteins for different (and novel) regulatory processes, such as the control of patterning in plants and in animals, is a common choice of nature. It has a kind of parallel even in human activity .... When one visits the main mosque in Acre and looks closely at the many marble pillars that support this mosque, it becomes evident that the pillars are not uniform. Archeologists traced the origin of these pillars and found that they came from the ruins of the Roman cities, along the Mediterranean shores, south of Acre (Ashkelon, Apollonia, Caesarea) where they served in the construction of temples for Roman gods. Achmad (G'azar) Pecha, the (fierce) Turkish ruler of Acre, transferred these marble pillars to Acre when he built his impressive mosque (in 1781). Well, in a way the Roman pillars served again a similar purpose: to support a house of worship of god; only the gods were very different.

Back to MADS. Scholars of plant differentiation have a craving for this acronym. They used it in impressive titles of their publications. Here are some examples:

- *Plant development going MADS* (Jack, 2001)
- *A short history of MADS-box genes in plants* (Theissen *et al.*, 2000)
- *MADS-box genes reach maturity* (Causier *et al.*, 2002)

A thorough review on MADS-boxes (although it is entitled "short") was provided by Heinz Saedler and associates of the Max Planck Institute in Cologne, Germany (Theissen *et al.*, 2000). This excellent review (although due to ample recent progress in the molecular-genetics of plant patterning, it is slightly outdated) emphasizes the theme that changes in developmental-control genes are a major aspect of evolutionary changes in morphogenesis and that understanding the phylogeny of these genes (e.g. MADS-box genes) may help to understand the evolution of plant form. To render their theme memorable these authors provided a name: *Evodevotics*, for evolutionary developmental-genetics. Much of this review is devoted to flower development in various plant taxa and deals with "MADS-box genes and evodevotics of the flower". The MADS-box genes were also traced to non-flowering plants (ferns) and even to fungi and animals, suggesting that the "root" of the phylogeny of MADS have appeared about one billion years ago. At the "top" of the phylogenetic tree in flowering plants, the "old" MADS-box genes were utilized as homeotic selector-genes, determining floral organ identity and floral meristematic identity genes (see above on the re-use of the marble pillars for the construction of the mosque in Acre). The authors suggested that changes in MADS-box gene structure, expression and function have been a major cause for innovations in reproductive development during land-plant evolution. Much of the above mentioned summary of the review of Theissen *et al.* (2000) probably appears rather abstract to non-professionals of plant patterning but it will, hopefully, become clearer when we deal with specific cases of plant patterning. It may be noted that while the term *evodevotics*, invented by the Saedler team, did not stick in the literature, the term *MADS-box* that was also invented earlier by the Saedler team (Schwarz-Sommer *et al.*, 1990), was widely accepted.

The patterning of flowers

We shall begin with flowers and the roles of miRNAs in their development. Two major model flowers were utilized to study floral development: the snapdragon (*Antirrhinum majus*) flower and the flower of the Brassicaceae weed, *Arabidopsis thaliana*, that we shall abbreviate below as snapdragon and *Arabidopsis*, respectively. Snapdragons were studied in Germany since almost 100 years ago (e.g. by E. Bauer). In latter years this plant and its flowers were studied in parallel at the Max Planck Institute in Cologne (Saedler's department) and at the John Innes Institute, near Norwich, UK, by R. Carpenter and E. Coen. The two teams even collaborated on the study of TEs in snapdragon. While some studies on flower development in *Arabidopsis* were conducted in the 1960s, intensive research on flower development of this plant started in the laboratory of Elliot Meyerowitz, at the Division of Biology of the California Institute of Technology in Pasadena, only in the late 1980s (e.g. Bowman *et al.*, 1989). The breakthrough that initiated the under-standing of floral patterning was provided by two publications, one on snapdragon flowers (Schwarz-Sommer *et al.*, 1990) and another on both snapdragon and *Arabidopsis* flowers (Coen and Meyerowitz, 1991). We shall first deal with the publication of Schwarz-Sommer *et al.* (1990), albeit, in an abbreviated manner. These authors first iso-lated and characterized phenotypically a large number of mutants that affect floral initiation and alter the shape of floral members (e.g. sepals, petals, carpels and stamens). The mutants were frequently obtained by transposon tugging — this should not surprise us because for years the Saedler department was engaged with TEs. The process of flower formation was then divided into two basic stages: floral evocation and floral development. By floral evocation these investigators mean the transition of the vegetative apical meristem into a floral meristem. They then asked what the mechanisms by which the meristematic cells *sense* and *interpret* their position are, with respect to other cells, on the *way* to differentiate into the correct floral member (recall the Cheshire Cat and Alice story). But this question could not be answered. The morphogenetic mutants were then divided into three *classes*. Here, we stop for a moment to avoid confusion because the nomenclature may be perplexing. When a mutant termed, for example, *sterilis* was identified it means that the wild type of *sterilis* is producing a factor (protein) that is required to initiate a normal floral primordium. In the mutant the deficiency of this factor will prevent further normal development. Bacterial geneticist would term the active gene flp^+ (for floral primordium) and the mutated gene flp^-. Hence, the names of genes affecting floral development can be confusing: the wild-type gene *sterilis* is not causing sterility. On the contrary, it is essential for normal reproductive development. The same holds true for other genes involved in floral patterning: the meaning of the name of a gene involved in flower development does not necessarily indicate the role of the wild-type allele of the gene. Notably, these names seem to be perplexing only to the novice in plant patterning, professionals are not bothered.

The three *Classes* of mutants are as follows (see Fig. 40):

Class I — mutants involved with the initiation and with the formation of the floral primorida (e.g. *sterilis, squamata*).

Class II — mutation involved in the symmetry of flowers (i.e. whether they have a radial symmetry or are zygomorphic — this concerns snapdragon flowers that have no radial symmetry).

Class III — mutations of homeotic genes that specify floral-organ identity.

The Saedler Department (MPI) in Germany divided Class III again into three different categories ... of *types* (the human mind, especially in some cultures, adores "drawers"). Type 1 mutants affect the first and the second whorls (sepals and petals). Type 2 mutants affect the third and the fourth whorls (stamens and carpels). Type 3 mutants affect the second and the third whorls (petals and stamens). The names of the three types of mutants and the whorls affected by them are shown in the scheme of Fig. 40. The unexpected and very significant finding was that each of these mutants affected two adjacent

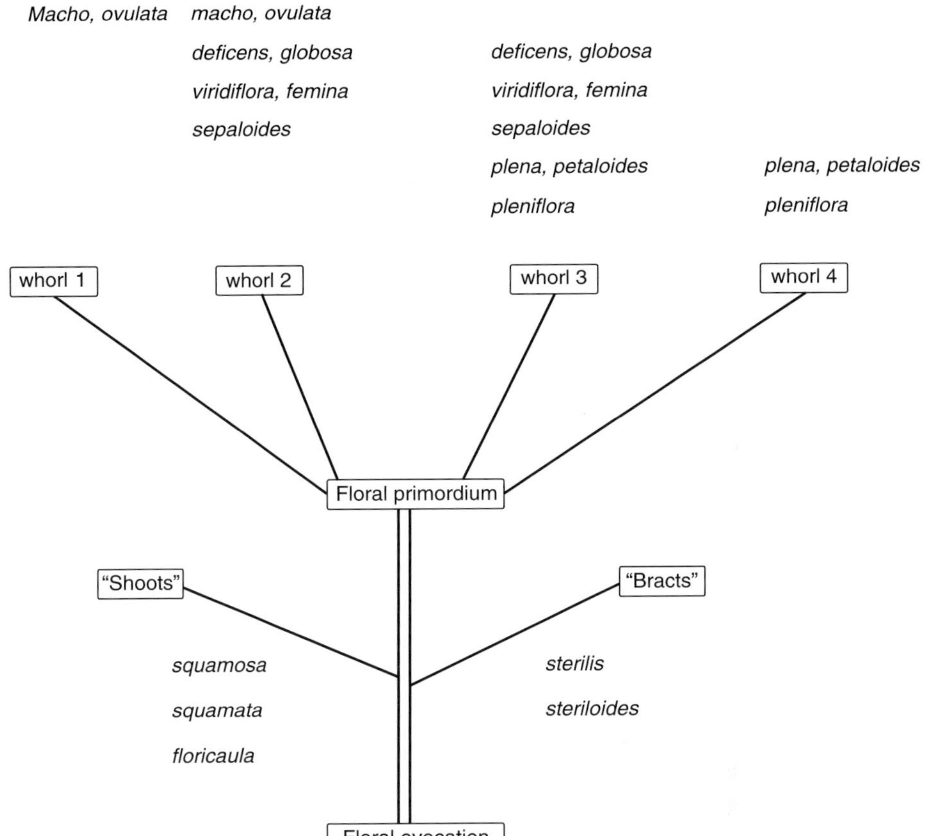

Fig. 40. Spatial and temporal activity of morphogenetic mutants in snapdragon; the quotation marks indicate the change of floral-primordium development in the corresponding mutants; whorls 1, 2, 3 and 4 are actually sepals, petals, stamens and carpels, respectively. (Modified from Schwarz-Sommer *et al.*, 1990.)

whorls. In no case did a mutation affect a single whorl and also the affected whorls were always adjacent to each other (meaning that no mutation affected the first and the fourth whorls, neither the second and the fourth whorls). This dual affect on two whorls was a riddle. How can the same gene (e.g. *deficiens*) specify two different whorls (petals and stamens)? The way to explain this phenomenon was to assume that there is an interaction between the genes. Thus, for example, if the gene *deficiens* is expressed together with the gene *ovulata*, petals will develop but if *deficiens* is expressed together with *sepaloides* then stamens will develop. But how can the interaction lead to the determination of organ identity? Schwarz-Sommer *et al.* (1990) solved parts of this puzzle and made very reasonable suggestions for a more far-reaching solution. First, some of the Class III genes (e.g. *deficiens*) were cloned and sequenced so that the amino acid sequence of the encoded protein was obtained. It became obvious that several of the Class III genes encode DNA binding proteins of the MADS-box type, hence they encode transcription factors. As such these MADS-box proteins could also interact with other MADS-box proteins and each "pair" of MADS-box proteins can have a combined and unique effect. The MADS-box proteins could also be affected by other effectors, hence changes in temperature can affect the final form of the floral members. The interaction with proteins that do not belong to the Class III coding genes could be by motifs that are outside of the MADS-box of these genes. The Class III genes were indeed expressed in the whorls that were controlled by these genes (as indicated by the scheme in Fig. 40). This means the Class III genes "know" where they are and are activated (i.e. generated their respective transcriptions) in the correct location. How does a cell in the floral meristem "know" where it is? Should we assume that these cells "invented" a GPS-like system hundreds of million years before high technology developed this device? There are two main theoretical means for the regulation of protein production by these genes. Their promoters may be activated in a given cell–environment, causing the respective transcription, followed by translation; or/and the production of proteins may be inhibited in locations where these proteins are not required. This inhibition can ensue by RNA silencing (e.g. miRNA mechanisms that will be discussed below), by specific protein-degradation or through remodeling of chromatin that will cause suppression of transcription (TGS). The specific spatially-limited expression of transcripts indicated that the activation of promoters is taking place and is probably a major kind of regulation but probably not the only one. Checks and balances may have evolved in patterning long before it was applied in the USA constitution. While Schwarz-Sommer *et al.* (1990) developed their scheme they were informed on the findings of Enrico Coen's team on snapdragon in Norwich. Moreover, information from the studies of the

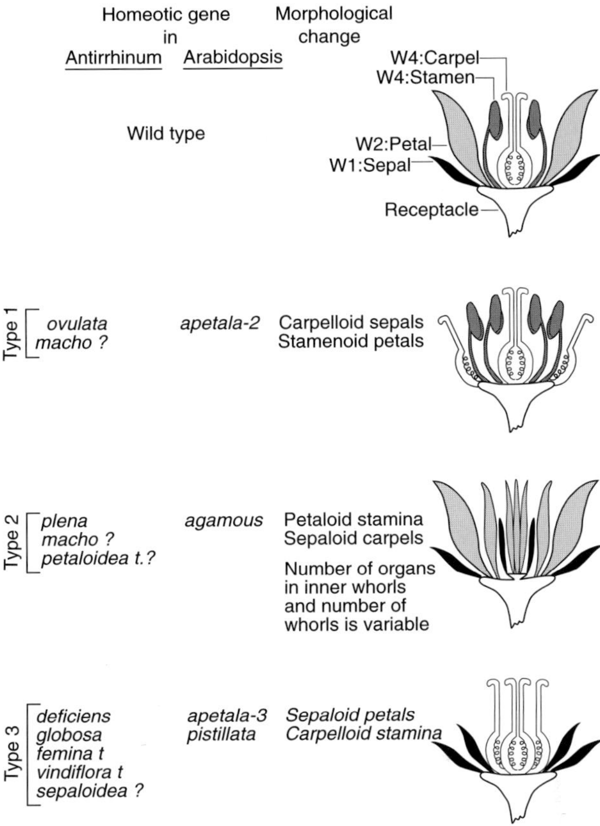

Fig. 41. Compilation of the three types of morphogenic genes (class III) that control floral organ identity in *Antirrhinum* and *Arabidopsis*. The idealized schemes show the direction of transformation of organs. However, in the mutant flowers not all organs in a whorl are equally transformed or transformed in the same direction. (From Schwarz-Sommer *et al.*, 1990.)

Meyerowitz Laboratory on the patterning of *Arabidopsis* flowers also reached the MPI investigators. The latter could therefore integrate the homeotic mutants of *Arabidopsis* and of snapdragon and come up with a scheme that described the roles of flower-morphogenesis genes in snapdragon and in *Arabidopsis* (Fig. 41). The study by Schwarz-Sommer *et al.* (1990) was a very important step in the elucidation of the mechanism of pattern formation. It also indicated that the overall "rules" of flower differentiation are amazingly similar in snapdragon and *Arabidopsis* in spite of the considerable phylogenetic distance between these two plants.

Less than a year after the publication of the article of Schwarz-Sommer *et al.* (1990), Coen and Meyerowitz (1991) published their review article entitled "The war of the whorls". It should be noted that both E. Coen and E. Meyerowitz were already, at that time, veterans in developmental genetics and well versed in *Drosophila* differentiation. Before the publication of their common review they also had ample experience with mutations affecting flower formation in snapdragon and in *Arabidopsis*, respectively (e.g. Coen *et al.*, 1990; Carpenter and Coen, 1990; Bowman *et al.*, 1989; Yanofsky *et al.*, 1990; Bowman *et al.*, 1991; Drews *et al.*, 1991). Moreover, they followed the work on snapdragon floral differentiation in Saedler's department and were aware of the comprehensive publication of Schwarz-Sommer *et al.* (1990). Nevertheless, Coen and Meyerowitz (1991) phrased their own model for the genetic control of whorls and floral-member identity for both snapdragon and *Arabidopsis*. This model became popular in subsequent years and was commonly termed the "ABC Model". It did not differ substantially from the model suggested by the Saedler Department. Schematically, the model of Coen and Meyerowitz can be drawn in the following manner:

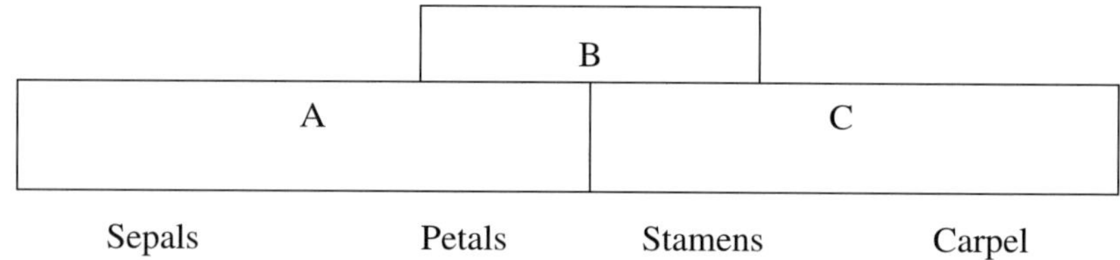

The three *functions* of **a, b** and **c** were assigned to A, B and C respectively.

The scheme is intended to indicate that there are three "regions": A comprises whorls 1 and 2; B comprises whorls 2 and 3 and C comprises whorls 3 and 4. This means that when only **a** is expressed, sepals are formed; when both **a** and **b** are expressed, petals are formed; when both **b** and **c** are expressed, stamens are formed and when only **c** is expressed, carpels will be formed. As Schwarz-Sommer *et al.* (1990),

Coen and Meyerowitz (1991) also based their model on phenotypes of flowers in homeotic mutants of *Arabidopsis* and snapdragon. While the final shape of the snapdragon flower differs substantially from the flower of *Arabidopsis* (the former is zygomorphic while the latter is radially symmetric), the flower initials are similar and both produce four whorls (of sepals, petals, stamens and carpels). The number of floral members in each whorl do differ (e.g. there are four sepals in *Arabidopsis* and five sepals in snapdragon). Coen and Meyerowitz (1991) were dealing with two kinds of homeotic genes: those that control the identity of meristems and those that control the identity of organs (i.e. floral members). We shall deal only with the latter genes.

The overall features of the models of Schwarz-Sommer *et al.* (1990) and of Coen and Meyerowitz (1991) are rather similar but there were differences in details. Take, for example, Coen and Meyerowitz (1991) provided experimental evidence that there is an interaction between the **a** and the **c** function in *Arabidopsis*. Moreover, the functions of **a** and **c** seem to be antagonistic and establish mutually exclusive domains of action. There are also differences between *Arabidopsis* and snapdragon with respect to functional mutations. In snapdragon a given mutation may inhibit the **a** function while in *Arabidopsis* the parallel mutated gene causes only loss of function. Coen and Meyerowitz (1991) also provided evidence that the spatial regulation of flower patterning is at the RNA level. Using probes from cloned homeogenes in *in situ* hybridizations indicated that the transcripts of these homeogenes are formed when and where they were expected according to the model. This means that they are activated in a regulated manner (and they "know" where they are).

There are several experimental evidences that show that the mechanism of patterning is rather elaborate. Take, for example, a homeogene for a function may affect in *Arabidopsis* the *number* of floral members in the third whorl (stamens). Some of the homeomutants are stronger or weaker than others. Take also, for example, that the *cycloidea* mutants in snapdragon have different effects toward changing the snapdragon flower from zygomorphic to radially symmetric. The *DEFICIENS* gene of snapdragon and the *AGAMOUS* gene of *Arabidopsis*, as well as other homeotic genes involved in flower patterning, were cloned and sequenced. In these genes, the same

MADS-box motif was revealed, suggesting that these genes serve as transcription factors. Some of the genes had another common motif or region that was the k-box but the function of this box was not known to Coen and Meyerowitz (1991).

It is noteworthy that homeogenes that have similar functions in the two distinct taxa (snapdragon and *Arabidopsis*) also have homologous DNA sequences. This clearly points to an evolutionary-early development of flower patterning. For more than 10 years since the models on the control of flower development were developed, proposed by Schwarz-Sommer *et al.* (1990) and by Coen and Meyerowitz (1991), this subject was thoroughly studied by Meyerowitz, Coen and other investigators. Reviews by Thomas Jack (2001a, 2001b) that had the intriguing titles "Plant development going MADS" and "Relearning our ABCs: new twists on an old model" summarized much of the 10 years of flower-development investigations (only it ignored the pioneer publication of Schwarz-Sommer *et al.*, 1990). The role of *SEPALLATA (SEP)* genes were added to the ABC model as follows:

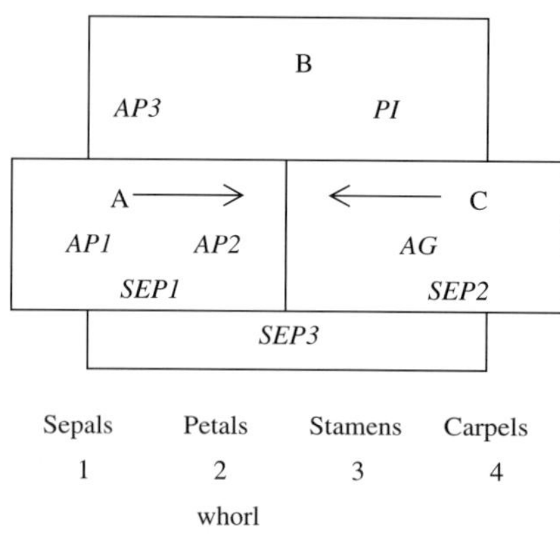

ABC and *SEP* genes

AG: *AGAMOUS* (*Arabidopsis* C class)

AP1: *APETALA*1 (*Arabidopsis* A class)

AP3: *APETALA3* (*Arabidopsis* B class)

DE: *DEFICIENS* (snapdragon *AP3 ortholog*)

GL: *GLOBOSA* (snbapdragon *PI* ortholog)

PI: *PISTILLATA* (*Arabidopsis* B class)

PLE: *PLENA* (snapdragon *AG* ortholog)

SEP1: *SEPALLATA 1*

SEP2: *SEPALLATA 2*

SEP3: *SEPALLATA 3*

SQUz: *SQUAMOSA* (snapdragon *AP1* ortholog)

The roles of the three SEP genes in *Arabidopsis* seems to be complicated and not fully understood. Take, for example, neither single *SEP* mutation (i.e. *SEP1, SEP2* nor *SEP2* alone) nor even two mutations out of the three *SEP* genes cause a dramatic phenotype of flower development. But when all three *SEP* genes are mutated, the change is dramatic, similar to a defect in BC double mutants (e.g. *pi & ag* or *ap3*

& *ag*), causing the development of only sepal-like floral members. But other indications imply that the three *sep* mutants are not identical in spite of the apparent redundancy in the roles of the three *SEP* genes.

During the years since the ABC model was first suggested several other important findings were made but because most of these are not relevant to the understanding of the role of miRNA in flower patterning, I shall not detail these findings. Readers who like to follow this subject in detail are referred to reviews of Goto *et al.* (2001) as well as those of Irish (2003) and of Jack (2004).

The review of Irish (2003) deals mainly with the evolution of MADS-box genes that serve in the developmental process culminating in their recruitment for flower patterning. The author suggests that the evolution of new morphologies (i.e. new patterns) relied on a large extent on co-option of a relatively small set of genetic pathways to new development roles. The genes underwent modifications that caused the formation of modified proteins that led to novel function (again ... the pillars of the mosque in Acre ...). The phylogeny of MADS-box genes was followed intensively with respect to changes in MADS-box sequences. Genetic diversifications within the angiosperms are quite considerable but their "roots" go much further back. By the time this review was written (2003) 82 MADS-boxes were located in *Arabidopsis* and 71 were estimated in the rice genome but the number of detected MADS-box genes are still growing. Compared with the wealth of MADS-box genes in angiosperms, *Drosophila* and the nematode *C. elegans* have only two MADS-box genes each (but they have other genes for transcriptional regulation). Interestingly, when the phylogeny of MADS-boxes is followed the accepted phylogeny of the angiosperms emerges and these two phylogenetics are compatible with the taxonomy of angiosperms that was suggested by old masters such as Carolus Linnaeus (1707–1778). It appears that these old masters had exceptionally good intuition.

The molecular and genetic mechanisms of floral patterning were reviewed by Thomas Jack of the Dartmouth College in new Hampshire (Jack, 2004). Jack was formerly in the laboratory of Elliot Meyerowitz (at Caltech). In this review the flower development was divided into four steps, consisting of the change from vegetative to reproductive growth, specification of floral identity, activation of floral organ identity and the "building" of the four floral members. I shall focus on floral-organ identity genes and the further evolution of the ABC models because this will lead us to the involvement of miRNA in patterning of plants.

The update of Jack (2004) added several important issues to the traditional ABC model. The appearance and disappearance of specific transcripts of homeotic genes was followed in greater detail. Take, for example, it became evident that the transcript of *AP1* is first present throughout the floral primordium. At later stages this transcript is limited to whorls 1 and 2. The transcript-appearance of *AP2* is noteworthy. While the function of AP2 is required only in whorls 1 and 2, the transcript is produced in all four whorls. We shall see below that in spite of the transcription being generated in all whorls, the protein is restricted to specific locations by the activity of specific miRNAs. The discovery of the SEP genes mentioned above (and designated as E-class genes) led to the change in the name of the model to the "ABCE model". In this "new" model it is postulated that sepals are specified by A activity, petals by A + B + E activity, stamens by B + C + E activity and carpels by C + E activity. In petunia another class of genes was defined: D class genes. These include several *FBP* genes (e.g. *FBP7*, *FBP11*; the loss-of-function alleles are respectively *fbp7* and *fbp11*). One of these genes, *FBP11*, was found to require ovule identity. The ortholog of *FBP11* in *Arabidopsis* is *STK*.

While most of the ABCE genes are members of the MADS-box family, there is an exception — the A class gene *AP2*. It also seems to be a transcription factor but of another kind. The MADs-box proteins are not only able to bind DNA (as dimers) but also to bind to each other by protein-protein interactions. Thus, complexes of four MADS-box proteins were suggested as the active transcription factors in some cases of floral member specification. There could also be an activation of floral-organ identity by floral-meristem identity genes. A candidate of the latter genes is LFY which could be an activator of *AP3* as *AP3* expression is reduced in the *lfy* mutant and *in vitro* the LFY protein can bind to AP3. There are additional genes that are involved in the activation of ABCE homeogenes. One of these has the long name *UNUSUAL FLORAL ORGANS*, shortened to *UFO*. There is also a *WUS* gene, the product of which seems to bind and activate AG. Furthermore, the product of *AG*, the AG protein, seems to be able to move in the floral primordium from the inner whorls to the outer whorls. How this movement is performed (through plasmodesmata?) is still not known. In short, there is not only an interaction between the proteins produced by the ABC (or ABCE) homeogenes but also proteins from other genes are taking part in the interactions. These interactions may cause activation as well as suppression and also possibly degradation. Finally, while quite a number of genes involved in floral-member development were mentioned above there are surely many more genes of this kind. This assumption is derived from a detailed study by the Meyerowitz Laboratory (Wellmer *et al.*, 2004) in which the spatial gene expression in *Arabidopsis* flowers was compared to expression profiles (cDNA microarray).

Examples of Plant Patterning Regulated by MicroRNA

The involvement of microRNA in the patterning of flowers

After the extensive "introduction" to the patterning of flowers we come back to RNA silencing; specifically to miRNA. Xuemei Chen (2004c) of the Waksman Institute, Rutgers University, New Jersey, focused on *APETALA2 (AP2)* that is a class A gene involved in the identity of sepals (whorl 1) and petals (whorl 2) in *Arabidopsis*. She found that miR172 is highly complementary to a region of *AP2* and therefore this miR172 could be involved in flower patterning. Interestingly, the homology to the mRNA of *AP2* was *within* the coding region but at the 3′ end of this region and outside of the consensus sequence of the *AP2* family of genes. She also found that in mutants that are defective in the generation of miRNAs (e.g. *hen1* and *dcl1*) there was a several-fold higher level of AP2 proteins than in w.t. *Arabidopsis*. Chen then overexpressed several versions of the miR172 family in transgenic plants. Enhanced expression of *miR172a-1* caused homeotic phenomena that were similar to those of *ap2* loss of function. This and other results of Chen's study suggested that miRNAs have a role in flower patterning. In the case of *AP2*, miR172 keeps the level of AP2 protein low at the locations and times when a level too high would disturb the normal flower patterning (note that the transcript is not suppressed but the protein level can be reduced in certain locations). As dealt with by Aukerman and Sakei (2003) and detailed below, the *AP2/miR172* interaction also affects flowering time. The complementarity between specific *AP2* genes (e.g. *AP2m1*) and miR172 family members could be studied in a clever way. The third base in some triplets could be changed without affecting the encoded amino acid sequence but such changes did prevent homologous pairing between the homeogene (*AP2*) and the miRNA, thus preventing the suppression of translation by the miRNA. Interestingly, Chen (2004c) found that in spite of the near-perfect homology between *miR172a-1* and *miR172a-2*, and their *AP2-like* target — the silencing of the target was not done by degradation of the mRNAs but rather (mainly?) by suppression of translation. This latter kind of suppression was previously found to be typical for animal miRNAs rather than for plant miRNAs. However, it is too early to conclude if the suppression of target-gene expression in the miR172/*AP2* is really an exception.

I have mentioned above the role of AP2 with respect to its function in floral development (whorls 1 and 2). But there is a family of *AP2* genes and two of these act as suppressors of flowering (Aukerman and Sakai, 2003). The latter investigators also followed the impact of *miR172* on these AP2 targets. One of these target-genes is an *AP2-like* gene, having the rather long name *TARGET OF EAT1* or *TOE1*. It causes late flowering of *Arabidopsis*. A similar *AP2-like* gene, *TOE2*, could be down-regulated by *miR172* thus resulting in early flowering in *Arabidopsis*. In practice Aukerman and Sakai isolated two mutants for genes termed *EAT-D* and *TOE-1D*. The first mutant (*eat-D*) of these had very early flowering and the second (*toe1-1d*) showed both early flowering and floral defects. The cDNA sequences of these genes were amplified and employed in genetic transformations of w.t. *Arabidopsis*, using a strong constitutive promoter (35S of CaMV). Transformation with the 35S:EAT construct caused early flowering as in the respective *ap2-like* mutant and some flower defects, as occasionally seen in *EAT-D* were revealed among the transformants. The transcript of *EAT* was found to be a non-coding sequence that could fold and form 21 nt dsRNA that was identical to one of the *miR172* family members (*miR172a-2*). It was found that the 3' end of the coding sequence of *AP2* is nearly homologous to this *miR172*. The target site for *miR172* was revealed in several *AP2* family members of *Arabidopsis* (e.g. *TOE1, TOE2, TOE3*) as well as in *AP2* family members in other plant species including monocots (rice and wheat). In spite of the near-homology between the *miR172a-2* and its target the effect of this miRNA was by reduction of translation rather than by degradation of the respective mRNA. The authors came up with a model in which after germination of *Arabidopsis* there is a gradual build-up of a *miR172* level until at a certain level the proteins encoded by the AP2-like gene are reduced sufficiently to permit floral initiation. The promotion of flowering involves additional genes but the *miR172* gene and its target, the *AP2-like* gene, are obviously part of this system.

The patterning of leaves: Fundamental considerations and specific examples

The patterning of leaves in angiosperms is a rather old subject. It interested J.W. Goethe (1749–1832) who suggested 200 years ago that metamorphosis of leaves leads to the formation of flowers. Following the clarification of the functions of the *ABC* and *SEP* genes, vegetative leaves can indeed be converted into floral members — in support of Goethe's proposition. Experimental approaches to study the initiation of leaves in the shoot apical meristems (SAM) were undertaken already over 50 years ago (e.g. Sussex, 1954; Snow and Snow, 1959). The latter approaches indicated that the dorsoventrality of the leaf-initials is disrupted by a surgical incision between the dome of the SAM and the early leaf-initial. But specific genes that affect leaf-shape in dicots and in monocots were revealed only more recently (e.g. Waiters and Hudson, 1995; Waiters *et al.*, 1998). These and additional studies on the development of leaf shape were well reviewed by Rob Martienssen and associates (Byrne *et al.*, 2001) of the Cold Spring Harbor Laboratory, NY. Ample literature is supplied by this review.

The gene *PHANTASTICA (PHAN)* in snapdragon was one of the first genes that was revealed as involved in leaf-shape formation. In the mutant *phan*, patches of abaxial tissue form on the adaxial surface (the two surfaces can be easily identified by their palisade cells facing the adaxial surface and more abundant stomata on the abaxial epidermis). It was found that the PHAN protein is a MYB-domain transcription factor and can regulate other genes (e.g. *KNOX* genes). In *Arabidopsis* the gene *PHABULOSA* was revealed. The mutant *phb-1d* causes the formation of adaxialization of leaves with additional axillary meristems around their bases. The homozygous *phb-1d* mutants also have SAMs that are larger than w.t. SAMs.

John Bowman of the UC, in Davis, Kathryn Burton of the University of Wisconsin, in Madison and their associates (McConnell *et al.*, 2001) studied molecular aspects of the leaf-patterning affected by *PHABULOSA* and *PHAVOLUTA* genes. They based their study on *Arabidopsis* mutants in which dominant mutants of *phb* and *phv* cause a dramatic transformation of leaf symmetry: the abaxial side is converted into an adaxial leaf-fate causing radial symmetry. Also, one mutant *phb-1d* forms ectopic meristems at the adaxial leaf-base. The *PHB* gene was analyzed in detail. The genomic sequence has many exons and introns and the first base of one of the introns is typically changed in certain mutants (e.g. *phb-1d*, *phb-2d*). This change interferes with normal splicing, causing the addition of a short stretch of amino acids in a motif of the PHB protein that is essential for a normal distribution of PHB in the young leaf initials. The distribution of normal (w.t.) PHB is rather orderly-formed around the shoot apex meristem (SAM). The transcript level in the incipial leaf-primordium (PO) is low but it is equally distributed in this PO. In P1 (the next more advanced primordium) the transcript is more abundant but it is higher in the adaxial side of P1. The transcript is then clearly more abundant in the more advanced leaf initial, P2, where it is located at the adaxial side of this leaf primordium.

The aforementioned base change is in a region that code for a HD-ZIP domain that is followed by a START domain which has a regulatory role in animals (sterol/lipid binding). The additional amino acids in the proteins of mutants was correlated with the leaf phenotypes. The functions of PHB and PHV were not known in 2001 yet but the base-pair alterations that added a stretch of amino acids in mutations in both respective genes caused the adaxilation of leaves. Hence, the functions of PHB and PHV was suggested to be similar. There is also a typical change in occurrence and distribution of *PHB* and *PHV* transcripts in the early embryo of *Arabidopsis*. Also, this dynamic distribution is affected in *PHB* and *PHV* mutants. McConnell *et al.* (2001) proposed a model for the situation in which a ligand for

PHB was involved, causing a feedback loop of the regulation of the PHB transcript. The possibility of an active reduction of transcript and/or reduction of the protein by RNA silencing was not taken into account yet.

The subject of leaf symmetry and its control by the class III HD-ZIP and KANADI genes was further studied by the team of John Bowman in Davis (Emery *et al.*, 2003). This study was also conducted with *Arabidopsis* where the gain-of-function alleles of the genes *Phb* and *Phv* that are members of the class III HC-ZIP family and loss-of-function alleles of the *KANADI* genes cause adaxialization of lateral organs (e.g. leaves). While the Phb and Phv proteins tend to accumulate in the adaxial region of leaf primorid, the Kan protein accumulates in the abaxial region of these primordia. A gain-of-function allele of the *REVOLUTA* gene (*Rev-10d*), another Class III HC-ZIP family member causes an alteration of the radial pattern of the vascular bundles in the shoot (i.e. change in the relative location of the xylem and the phloem). A new approach was added to this symmetry study. As indicated above, Chen (2004c) changed base-pairs in mRNA involved in floral patterning. This change did not affect the derived protein. This approach was also utilized by Emery *et al.* (2003) but for the mRNA of *Rev*. The investigators were able to change certain bases in the mRNA. Two triplets GGT and CCG were changed to GGA and CCA, respectively. Hence, the respective encoded glycine and proline were retained. The changed mRNA, now termed *rev-δ miRNA*, was put behind the endogenous *REV* promoter and served in genetic transformation of *Arabidopsis*. The resulting transgenic plants showed the same phenotype, with respect to adaxilation of the vascular bundles, as the *rev-10d* mutant. Several controls verified that the altered phenotype was indeed caused by the two changes of bases. The *rev-10d* has also a change in one base: from CCG (proline) to CTG (leucine) but this slight change of one amino acid was not the cause of the altered phenotype. Clearly, the target of the miR165/166 was abolished so that these miRs no longer paired with the modified transcript of the *REV* gene. The differential accumulation of REV protein is actually initiated very early in the embryo and not only later in the apical meristem and flower meristem. The expression of *PHV* commences a little later than the expression of *REV* and *PHB* and these genes seem to interact: triple mutations show a very severe phenotype. The results of Emery *et al.* (2003) also indicated that *KANADI* is required in leaf-initials for the proper specification of adaxial cell types: loss of *KANADI* activity causes loss of adaxial cell types while ectopic supply of *KANADI* will promote the differentiation of abaxial cells with a concomitant loss of adaxial cell type in the leaves. The reactions of the stem cells to *Ken* mutants are different from the response of leaf cells to these mutants.

A further study on leaf development and miRNA was conducted by seven experienced investigators from several laboratories in the USA (Mallory *et al.*, 2004a; 2004b). These investigators from the Whitehead Institute in Cambridge, MA, the Carnegie Institution in Stanford, CA, the MIT in Cambridge, MA, and from the Massachusetts Medical School in Worcester, focused on the pairing between the transcript of *Phb* and the miRs that cause the silencing of this gene in *Arabidopsis* (i.e. miR165/166). In a way this study was a continuation of the Emery *et al.*'s (2003) study and provided interesting additional information on miRNAs and patterning in plants.

Mallory *et al.* (2004a, 2004b) asked what kinds of changes in the bases of the mRNA of *Phb* would cause the silencing of *Phb* and lead the adaxialized (i.e. radially symmetric) leaves. There was a mutant *phb-1d* that when used as a transgene in genetic transformation, caused adaxialization of leaves. In *phb-1d* there is a basic-alteration that caused a failure of splicing of one of the introns that added amino acids to the protein. The investigators used additional mutated genes of PHB for genetic transformation. Here, I shall demonstrate the approach with only one such mutation (PHB G202G) as shown in Fig. 42. In addition to the wild-type *PHB* the "natural" mutant *phb-1d* is shown. The third mRNA is a "synthetic" mRNA that served as an example of a "silent" mutation. In each of the three mRNAs the upper line shows the encoded amino acids (e.g. glycine, methionine, lysine, proline, glycine, proline and aspartic acid). The next line provides the target sequence from the mRNA; and the third line shows a region of the miR165/166 that binds to the mRNA. The *phb-1d* mutant is impaired in splicing (as indicated above). The misspliced mRNA has 33 additional bases and consequently 11 additional amino acids between K and P. Obviously, this mRNA will not serve as a target of pairing with

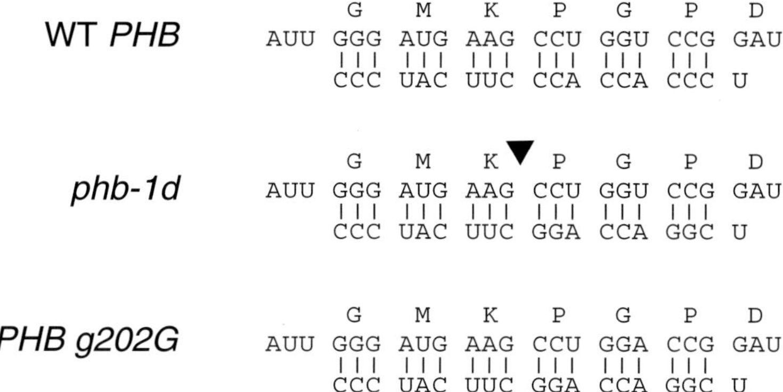

Fig. 42. Point mutations within the 3′ region of the *PHB* miR 165/166 complementary site confer stronger dominant phenotypes than those in central positions. Predicted base pairings of the *PHB* mRNA (top strand) and miR 165 (bottom strand) are shown below the amino acid sequence of the PHB protein. (From Mallory *et al.*, 2004b.)

miR165/166. The mRNA of *PHB-G202G* represents a *silent* mutation because in it a GGU was changed to GGA. Both these triplets code for glycine. Thus, no change in amino acids is expected in the resulting protein. But this mutation does have a severe phenotypic effect: there is an adaxialization of leaves in half of the transgenic plants in which this mutated mRNA is expressed. Other transgenic plants were also produced and the results lead the authors to the conclusion that the disruption of binding between the mRNA and the miR 165/166 is the cause of the mutated phenotype. Moreover, previous results of *in vitro* studies by Tang *et al.* (2003) support the notion that mismatch between the 3′ end of the mRNA of a *PHB* mutant and endogenous miRs prevents cleavage of the mRNA by miRs and will thus prevent normal leaf patterning. There were further indications that the miR165/166 normally cleaves all five III HD-ZIP mRNAs in plants. Also, the site of the base-mismatch between mRNA and miR165/166 is decisive. When this mismatch occurs more upstream rather than close to the 3′ end of the mRNA, there will be no adaxilization of the leaf-initials. As for the region of the miRNA that is crucial for an effective binding with its (mRNA) target, there are clear indications that the crucial region is at the 5′ end of the miRNA. The requirement of complementarity between miRNA and the target, at the 5′ end of the miRNA and at the 3′ end of the mRNA appears to be the norm in mRNA/miRNA interactions that lead to gene silencing.

Support for the miRNA/target relation that can lead to suppression of cleavage of a START domain (of III HD-ZIP encoding genes) came from the study by Zhong and Ye (2004) of the University of Georgia, in Athens, GA. This study discussed mainly the architecture of the vascular tissue in the stem (stele) of *Arabidopsis*. The normal pattern of the xylem and phloem tissues in *Arabidopsis* is a central xylem surrounded by phloem in a collateral pattern. A specific semidominant mutation, *amphivasal vascular bundle 1* (*avb1*), transforms the collateral vascular bundles into amphivasal bundles, disrupting the regular ring-like arrangement in the stele. The *avb1* mutant has several additional effects on patterning such as ectopic growth of carpel-like structures and changes in polarity. When the cDNA of *avb1* is used as a transgene in genetic transformation of w.t. *Arabidopsis*, to over-express *avb-1* the phenotype of the mutant was not only phenocopied but additional abberation in patterning were observed. The genomic sequence of the *avb1* was determined. It was found that in the mutant there is a C to T base change from w.t. to the mutant. This is in the region that codes for the START domain of this III HD-ZIP protein. The mRNA that encodes this region is a miR165 target. The change in the single base abolished the *avb1* transcript's ability to serve as a target for degradation by miR165: the mRNA of *avb1* was not cleaved while the w.t. mRNA (IFLI/REV) was cleaved, and the cleavage was at the site of the expected miR165 binding. It should also be noted that a loss of function in the null-mutation of the IFLI/REV causes a very different phenotype than the gain-of-function *avb1*. The former mutation causes reduced polar auxin flow, secondary xylem, reduced lateral branches etc. but not the formation of amphivasal vascular bundles. Could the change in vascular bundles architecture be a result of the disruption of normal auxin flow? The authors did not exclude this possibility. As for the change in miR165 target, there is one approach that was missing in the study by Zhong and Ye (2004). The transformation of *Arabidopsis* with several mutated IFLI/REV genes that have the same phenotype as *avb-1* but one silent mutation (change in bases without changes in the encoded protein) could strengthen the proposition that an altered target failed to reduce the protein where such a reduction is required for normal patterning.

The study by McHale and Koning (2004) from the Connecticut Agricultural Experiment Station and the Eastern Connecticut State University turned their attention to *Nicotiana sylvestris* to investigate the role of miRNA in spatial and temporal changes mediated by III HD-ZIP proteins. They based their study on the previously accumulated information that indicated that the adaxial development of leaf-initials radiated from the SAM is affected by the SAM and mediated by III HD-ZIP proteins. Moreover, the semidominant mutations of *PHB, PHV and REV* in *Arabidopsis* extended (as noted above) the adaxial domain in these leaf-initials. These investigators focused a *PHAVOLUTA*-like gene in *N. sylvestris* (NsPHAV). In a somewhat confusing manner these investigators termed a mutant in this gene *phv1*. So there is an *Arabidopsis phv1* and a *N. sylvestris phv1*. The two mutants have at the transcription level very similar but not identical changes relative to the wild type transcript.

Wild type: ...GGG ATG AAG CCT GGT CCG...
N. sylvestris phv1:   ...GGG ATG AAA CCT GGT CCG...
Arabidopsis phv1:   ...GGG ATG AAG CCT GAT CCG...

The above listed lines of 18 bases are targets of miR165/166. The base-change AAG to AAA (that does not change the code for lysine) in the *N. sylvestris phv1* is apparently disrupting the regular target of miRNA165/166, abolishing the role of these miRNAs in patterning. The phenotype caused by the *N. sylvestris phv1* is not identical to the phenotype of the *Arabidopsis phv1*. The former phenotype is typified by defects in the orientation of the cambium that causes a misdirection of lateral growth of leaf-midveins and stem vasculature, away from the shoot and consequently disrupts vascular connections in the stem nodes.

Two additional cases of involvement of miRNA in leaf development will be noted. The control of a III HD-ZIP protein in maize leaves was revealed by Juarez *et al.* (2004). The miR166 seems to act in maize leaves by regulating the shape of the leaves. The miR166 was revealed in leaf primordia in a defined dynamic pattern. This suggested the possibility that miR166 is able to move between cells in the leaf-primordia.

Another emerging research direction that involves miRNA and leaf morphogenesis was undertaken in *Arabidopsis* by a team that included James Carrington of the Oregon State University and Detlef Weigel of the MPI in Tübingen, Germany and the Salk Institute in La Jolla, California (Palatnik *et al.*, 2003). These investigators found that the *JAW* locus in *Arabidopsis* encodes a miRNA that can silence TCP genes. The TCP genes are coding for proteins that consist of a family of transcription factors termed by the acronym TCP for the first three factors that belong to this family (as noted above). Of the various TCP proteins of *Arabidopsis* TPC4 is required to prevent leaf abnormalities. Cleavage of the *TPC4* transcript by the *JAW* derived miRNA regulates the level of *TPC4* mRNA and thus seem to maintain normal leaf shape.

The study on the involvement of miRNAs in plant patterning is an emerging endeavor. I have described a few investigations on this involvement but as there are already many defined miRNAs in plants and several additional miRNAs of plants, with defined targets are expected to be reported in the future, further interesting information on miRNA and plant patterning will be provided in the coming years.

In "Pirkey Avot" of the Jewish Mishna, coded at about 200AD there is a collection of analects that is rather similar to the Analects of Kung Fu-Tze (Confucius). Among the former analects there is one attributed to Akavia ben Mehalelel: (who lived at about 30BC) — meaning "know from where you came and to where you are going...."

This wisdom (i.e. to know from where you came or where you are now and toward where you are going) was probably adopted many hundred millions of years ago by cells of multicellular organisms during their differentiation. What is left for us is to learn how the cells do it: how they perceive their location and how they regulate their participation in patterning. For our theme, one thing already became clear: miRNAs have a major role in the patterning.

The availability of miRNA (and siRNA) does not merely add "another tool" to solve questions on patterning in plants. It provides a very refined and versatile tool. There are several reasons for this versatility and precision:

1. miRNAs of plants (and synthetic siRNA) commonly guide the RISC to homologous sequences in the coding region of transcription. Thus, a given miRNA will cause the degradation of a very specific transcript.
2. The specificity is assured because a sequence of 19 or more nt is statistically unique; it is not expected to occur in different genes unless these genes encode proteins with identical amino acid motifs.
3. To assure the specificity of silencing a transgene expressing the target transcript can be introduced into the plant and one or more mismatches will be included in the sequence that has homology to the miRNA. These mismatches should suppress the silencing and can be in bases of codons, in a way that the mismatched codons still specify the same amino acids.
4. The impact of endogenous miRNA on patterning can be analyzed by introducing into a plant a miRNA that will silence one or more of the genes that are required for the activity of the endogenous miRNA (e.g. *DCL1*). When the transgene causing this silencing (e.g. miR162) is fused to a promoter that is activating this miR spatially, the impact of endogenous miRNA can be analyzed in a controlled manner.

Epilogue

Sontheimer and Carthew (2004) bring the Sophocles' drama on *Oedipus, the King* as an introduction to the recent results of investigating the protein complex termed *Argonaut(e)*. These investigations identified the long-sought catalytic subunit of this complex. This subunit executes the RNA interference and is therefore a prime component of RNA silencing. These authors claim that just as the ruling King of Thebes, Oedipus, was prominent in Thebes but was not identified as the murderer of his father, Laius, the previous King of Thebes, likewise Argonaute's presence was known for many years but its catalytic component was not characterized. The original *Argonauts* of the Greek mythology were the sailors (*nautes*) of the ship *Argo* who accompanied Jason in his complicated adventures during the pursuit after the Golden Fleece that also involved *Medea*. The results of the investigations that revealed the catalytic subunit of the Argonaute complex which will be summarized below. There was a good reason for the citizen of Thebes to identify the murderer of their previous King Laius because unless revealed Thebes was cursed with a plague. There was also a good reason for Jason and his Argonauts to pursue the Golden Fleece but why should an enormous amount of effort be dedicated to revealing an elusive scientific phenomenon? Since the time of the classical Greek philosophers (e.g. Aristotle) the acquisition of knowledge was claimed to provide utmost happiness. The English philosopher/politician/lawyer, Francis Bacon (1561–1626), provided a more practical role for knowledge: *knowledge is power* (ironically the "knowledge" of Bacon was not sufficient to keep him in power as Elizabeth's Lord Chancellor; he was removed from power by an unfair charge of corruption). Learning and acquisition of knowledge of various kinds including the knowledge of God was also claimed by old Jewish scholars (e.g. in the Mishna and the Talmud) to be of prime importance. But can we then deduce that present-day scholars will dedicate themselves to the demanding efforts of solving scientific riddles? Note that since the discovery by Fire *et al.* (1998) on the role of dsRNA fragments in RNA silencing, about 1000 investigations were published on RNA silencing. Were philosophical considerations or the hope for useful results the forces that prompted the scientists to be engaged in their efforts? As for philosophical considerations, the answer seems to be negative. Probably, only a minority among the scientists bother to follow philosophical arguments. I assume that the driving force behind the efforts of investigating RNA silencing (as of other scientific endeavors) is in our genes: the instinct of *curiosity*. This instinct preceded the evolution of man. Relative to other human instinct, as the instincts of territoriality, property and xenophobia, the instinct of curiosity is rather favorable; it *has no direct adverse effects* on other human beings.

Let us now turn back to the Argonaut(e)-proteins mentioned at the beginning of this Epilogue. As detailed in various chapters of this book, there are two complexes that cut specifically RNA sequences. One is the *Dicer* that cuts out ~21 nt fragments (siRNA) from longer fragments of dsRNA. The other is the RNA-induced silencing complex, *RISC* that cuts the single stranded mRNA at sites to where *RISC* is guided by the antisense strand of the ~21 nt fragment. The *RISC* thus harbors an RNA cutting capability in a subunit that was given the term "Slicer". It was revealed during the recent years that *RISCs* contain several subunits, one of which was a member of the family of Argonaute proteins. The Argonaute family was defined as proteins that have PAZ and PIWI domains. The PAZ domain is involved in binding the *RISC* to RNA. While it was clear that *RISC* has a "Slicer" function, the chemical-structural character of this "Slicer" remained elusive until two publications by Song *et al.* (2004) and Liu *et al.* (2004) appeared sequentially in the *Science* magazine. Both publications resulted from work performed at the Cold Spring Harbor Laboratory, NY, with Leemor Joshua–Tor and Gregory Hannon as corresponding authors. Various experimental approaches (Liu *et al.*, 2004) suggested that the ability of Argonaute 2 to assemble into catalytically active complexes may be critical for mouse development. This means that the active complexes can be guided by specific miRNAs to cleave specific mRNAs and consequently serve in the regulation of development. We should recall that in mammals it was considered that miRNAs act by suppression of translation (Chap. 10) rather than by cleavage of the mRNA (as is the "norm" in angiosperms). But an exception was already reported in which miRNA guided *RISC* in mammals to cause cleavage of mRNA rather than suppress translation (Yekta *et al.*, 2004). The experimental work of Liu *et al.* (2004) suggested that in mammals, mRNA cleavage guided by miRNA is more prevalent than previously assumed. As for the Argonaute proteins, it appeared most likely that Arg 2 itself provides the catalytic ("Slicer") activity of *RISC*. Strong support for this was furnished by the preceding publication (Song *et al.*, 2004) in which structural studies were conducted of a full-length Argonaute protein from the archebacterium *Pyrococcus furiosus*. These studies led to a structural model for siRNA-guided mRNA cleavage. In this model the siRNA binds to the PAZ domain that has a cleft into which the 3′ end of the siRNA enters. The mRNA is located between the PAZ domain and the PIWI domain, parallel to the siRNA. The cleavage of the mRNA is performed by the active site of the PIWI domain, opposite the middle of the guiding siRNA.

I began this book (in the Preface) with a simple cutting devise, the *sword* that was developed by humans for defence or attack against foes and ended this book with an elaborated "Slicer" that evolved in eukaryotes as an essential device for the control of gene activity. The evolution of the "Slicer" as well as other components of the RNA silencing mechanism started probably about one billion years ago but it took biologists only about 6 years to perceive the overall picture of this rather elaborate mechanism.

References

1. Abel, P. P., Nelson, R. S., De, B., Hoffmann, N., Rogers, S. G., Fraley, R. T. & Beachy, R. N. (1986). Delay of disease development in transgenic plants that express the tobacco mosaic-virus coat protein gene. *Science* **232**, 738–743.

2. Abrahante, J. E., Daul, A. L., Li, M., Volk, M. L., Tennessen, J. M., Miller, E. A. & Rougvie, A. E. (2003). The *Caenorhabditis elegans* hunchback-like gene *lin-57/hbl-1* controls developmental time and is regulated by microRNAs. *Developmental Cell* **4**, 625–637.

3. Agata, K. & Watanabe, K. (1999). Molecular and cellular aspects of planarian regeneration. *Cell & Developmental Biol.* **10**, 377–383.

4. Al-Anouti, F., Quach, T. & Ananvoranich, S. (2003). Double-stranded RNA can mediate the suppression of uracil phosphoribosyltransferase expression in *Toxoplasma gondii*. *Biochem. Biophys. Res. Comm.* **302**, 316–323.

5. Alexopoulou, L., Holt, A. C., Medzhitov, R. & Flavell, R. A. (2001). Recognition of double-stranded RNA and activation of NF-kappa B by Toll-like receptor 3. *Nature* **413**, 732–738.

6. Alvarado, A. S. & Newmark, P. A. (1999). Double-stranded RNA specifically disrupts gene expression during planarian regeneration. *Proc. Natl. Acad. Sci. USA* **96**, 5049–5054.

7. Alvarado, A. S., Newmark, P. A., Robb, S. M. C. & Juste, R. (2002). The Schmidtea mediterranea database as a molecular resource for studying platyhelminthes, stem cells and regeneration. *Development* **129**, 5659–5665.

8. Amarzguioui, M. & Prydz, H. (2004). An algorithm for selection of functional siRNA sequences. *Biochem. Biophys. Res. Comm.* **316**, 1050–1058.

9. Ambros, V. (2000). Control of developmental timing in *Caenorhabditis elegans*. *Curr. Opin. Genet. Develop.* **10**, 428–433.

10. Ambros, V., Bartel, B., Bartel, D. P., Burge, C. B., Carrington, J. C., Chen, X., Dreyfuss, G., Eddy, S. R., Griffiths-Jones, S., Marshall, M., Matzke, M., Ruvkun, G. & Tuschl, T. (2003). A uniform system for microRNA annotation. *RNA* **9**, 277–279.

11. Ambros, V. & Horvitz, H. R. (1984). Heterochronic mutants of the nematode *Caenorhabditis elegans*. *Science* **226**, 409–416.

12. Anandalakshmi, R., Marathe, R., Ge, X., Herr, J. M., Mau, C., Mallory, A., Pruss, G., Bowman, L. & Vance, V. B. (2000). A calmodulin-related protein that suppresses posttranscriptional gene silencing in plants. *Science* **290**, 142–144.

13. Anandalakshmi, R., Pruss, G. J., Ge, X., Marathe, R., Mallory, A. C., Smith, T. H. & Vance, V. B. (1998). A viral suppressor of gene silencing in plants. *Proc. Natl. Acad. Sci. USA* **95**, 13079–13084.

14. Angell, S. M. & Baulcombe, D. C. (1997). Consistent gene silencing in transgenic plants expressing a replicating potato virus X RNA. *EMBO J.* **16**, 3675–3684.

15. Angell, S. M. & Baulcombe, D. C. (1999). Potato virus X amplicon-mediated silencing of nuclear genes. *Plant J.* **20**, 357–362.

16. Aravin, A. A., Naumova, N. M., Tulin, A. V., Vagin, V. V., Rozovsky, Y. M. & Gvozdev, V. A. (2001). Double-stranded RNA-mediated silencing of genomic tandem repeats and transposable elements in the *D. melanogaster* germline. *Current Biology* **11**, 1017–1027.

17. Aravind, L., Iyer, L. M., Wellems, T. E. & Miller, L. H. (2003). Plasmodium biology: Genomic gleanings. *Cell* **115**, 771–785.

18. Ashrafl, K., Chang, F. Y., Watts, J. L., Fraser, A. G., Kamath, R. S., Ahringer, J. & Ruvkun, G. (2003). Genome-wide RNAi analysis of Caenorhabditis elegans fat regulatory genes. *Nature* **421**, 268–272.

19. Aufsatz, W., Mette, M. F., van der Winden, J., Matzke, A. J. M. & Matzke, M. (2002a). RNA-directed DNA methylation in Arabidopsis. *Proc. Natl. Acad. Sci. USA* **99**, 16499–16506.

20. Aufsatz, W., Mette, M. F., van der Winden, J., Matzke, M. & Matzke, A. J. M. (2002b). HDA6, a putative histone deacetylase needed to enhance DNA methylation induced by double-stranded RNA. *EMBO J.* **21**, 6832–6841.

21. Aukerman, M. J. & Sakai, H. (2003). Regulation of flowering time and floral organ identity by a microRNA and its *APETALA2*-like target genes. *Plant Cell* **15**, 2730–2741.

22. Aviv, D., Arzee-Gonen, P., Bleichman, S. & Galun, E. (1984). Novel alloplasmic Nicotiana plants by "donor-recipient" protoplast fusion: Cybrids having N. tabacum or N. sylvestris nuclear genomes and either or both plastomes and chondriomes from alien species. *Molecular and General Genetics* **196**, 244–253.

23. Aza-Blanc, P., Cooper, C. L., Wagner, K., Batalov, S., Deveraux, Q. L. & Cooke, M. P. (2003). Identification of modulators of TRAIL-induced apoptosis via RNAi-based phenotypic screening. *Molecular Cell* **12**, 627–637.

24. Barstead, R. (2001). Genome-wide RNAi. *Curr. Opin. Chemical Biology* **5**, 63–66.

25. Bartel, B. & Bartel, D. P. (2003). MicroRNAs: At the root of plant development. *Plant Physiology* **132**, 709–717.

26. Bartel, D. P. (2004). MicroRNAs: Genomics, biogenesis, mechanism, and function. *Cell* **116**, 281–297.

27. Bashirullah, A., Pasquinelli, A. E., Kiger, A. A., Perrimon, N., Ruvkun, G. & Thummel, C. S. (2003). Coordinate regulation of small temporal RNAs. *Developmental Biology* **259**, 1–8.

28. Bass, B. L. (2000). Double-stranded RNA as a template for gene silencing. *Cell* **101**, 235–238.

29. Bastin, P., Ellis, K., Kohl, L. & Gull, K. (2000). Flagellum ontogeny in trypanosomes studied via an inherited and regulated RNA interference system. *J. Cell Science* **113**, 3321–3328.

30. Baulcombe, D. C. (1999). Fast forward genetics based on virus-induced gene silencing. *Curr. Opin. Plant Biol.* **2**, 109–113.

31. Baulcombe, D. C. & English, J. J. (1996). Ectopic pairing of homologous DNA and post-transcriptional gene silencing in transgenic plants. *Curr. Opin. Biotech.* **7**, 173–180.

32. Baulcombe, D. C., Saunders, G. R., Bevan, M. W., Mayo, M. A. & Harrison, B. D. (1986). Expression of biologically active viral satellite RNA from the nuclear genome of transformed plants. *Nature* **321**, 446–449.

33. Beachy, R. N., Loesch-Fries, S. & Tumer, N. E. (1990). Coat proteins mediated resistance against virus infection. *Annu. Rev. Phytopathol.* **28**, 451–474.

34. Beclin, C., Bouet, S., Waterhouse, P. & Vaucheret, H. (2002). A branched pathway for transgene-induced RNA silencing in plants. *Current Biology* **12**, 684–688.

35. Beetham, P. R., Kipp, P. B., Sawycky, X. L., Arntzen, C. J. & May, G. D. (1999). A tool for functional plant genomics: Chimeric RNA/DNA oligonucleotides cause *in vivo* gene-specific mutations. *Proc. Natl. Acad. Sci. USA* **96**, 8774–8778.

36. Berg, P. & Singer, M. (2003). *George Beadle, an Uncommon Farmer.* Cold Spring Harbor Laboratory Press, Cold Spring Harbor, NY.

37. Bergmann, D. C., Lukowitz, W. & Somerville, C. (2004). Stomatal development and pattern controlled by a MAPKK kinase. *Science* **304**, 1494–1497.

38. Bernstein, E., Caudy, A. A., Hammond, S. M. & Hannon, G. J. (2001). Role for a bidentate ribonuclease in the initiation step of RNA interference. *Nature* **409**, 363–366.

39. Bernstein, E., Kim, S. Y., Carmell, M. A., Murchison, E. P., Alcorn, H., Li, M. Z., Mills, A. A., Elledge, S. J., Anderson, K. V. & Hannon, G. J. (2003). Dicer is essential for mouse development. *Nature Genetics* **35**, 215–217.

40. Beverley, S. M. (2003). Protozomics: Trypanosomatid parasite genetics comes of age. *Nature Reviews Genetics* **4**, 11–19.

41. Bezanilla, M., Pan, A. & Quatrano, R. S. (2003). RNA interference in the moss Physcomitrella patens. *Plant Physiology* **133**, 470–474.

42. Billy, E., Brondani, V., Zhang, H., Muller, U. & Filipowicz, W. (2001). Specific interference with gene expression induced by long, double-stranded RNA in mouse embryonal teratocarcinoma cell lines. *Proc. Natl. Acad. Sci. USA* **98**, 14428–14433.

43. Boden, D., Pusch, O., Lee, F., Tucker, L. & Ramratnam, B. (2003). Human immunodeficiency virus type 1 escape from RNA interference. *J. Virology* **77**, 11531–11535.

44. Bollman, K. M., Aukerman, M. J., Park, M.-Y., Hunter, C., Berardini, T. Z. & Poethig, R. S. (2003). HASTY, the Arabidopsis ortholog of exportin 5/MSN5, regulates phase change and morphogenesis. *Development* **130**, 1493–1504.

45. Boothroyd, J. C., Black, M., Bonnefoy, S., Hehl, A., Knoll, L. J., Manger, I. D., Ortega-Barria, E. & Tomavo, S. (1997). Genetic and biochemical analysis of development in *Toxoplasma gondii*. *Phil. Trans R. Soc. Lond. B.* **352**, 1347–1354.

46. Bosch, T. C. C. (1998). Hydra. In *Cellular and Molecualr Basis of Regeneration: From Invertebrates to Humans* (Ferreti, R. & Gerandie, J., eds.), pp. 111–134. John Wiley, Sussex, UK.

47. Bossinger, G. & Smyth, D. R. (1996). Initiation patterns of flower and floral organ development in Arabidopsis thaliana. *Development* **122**, 1093–1102.

48. Bourikas, D. & Stoeckli, E. T. (2003). New tools for gene manipulation in chicken embryos. *Oligonucleotides* **13**, 411–419.

49. Boutla, A., Delidakis, C., Livadaras, I., Tsagris, M. & Tabler, M. (2001). Short 5′-phosphorylated double-stranded RNAs induce RNA interference in *Drosophila*. *Current Biology* **11**, 1776–1780.

50. Bowman, J. L., Drews, G. N. & Meyerowitz, E. M. (1991). Expression of the Arabidopsis floral homeotic gene agamous is restricted to specific cell-types late in flower development. *Plant Cell* **3**, 749–758.

51. Bowman, J. L., Smyth, D. R. & Meyerowitz, E. M. (1989). Genes directing flower development in Arabidopsis. *Plant Cell* **1**, 37–52.

52. Bracha, R., Nuchamowitz, Y. & Mirelman, D. (2003). Transcriptional silencing of an amoebapore gene in *Entamoeba histolytica*: Molecular analysis and effect on pathogenicity. *Eukaryotic Cell* **2**, 295–305.

53. Brennecke, J., Hipfner, D. R., Stark, A., Russell, R. B. & Cohen, S. M. (2003). *Bantam* encodes a developmental regulated microRNA that controls cell proliferation and regulates the proapoptotic gene *hid* in *Drosophila*. *Cell* **113**, 25–36.

54. Brenner, S. (1974). The genetics oc *Caenorhabditis elegans*. *Genetics* **77**, 71–94.

55. Brigneti, G., Voinnet, O., Li, W.-X., Ji, L.-H., Ding, S.-W. & Baulcombe, D. C. (1998). Viral pathogenicity determinants are suppressors of transgene silencing in *Nicotiana benthamiana*. *EMBO J.* **17**, 6739–6746.

56. Brummelkamp, T. R., Bernards, R. & Agami, R. (2002). A system for stable expression of short interfering RNAs in mammalian cells. *Science* **296**, 550–553.

57. Bucher, E., Sijen, T., de Haan, P., Goldbach, R. & Prins, M. (2003). Negative-strand tospoviruses and tenuiviruses carry a gene for a suppressor of gene silencing at analogous genomic positions. *J. Virology* **77**, 1329–1336.

58. Byrne, M., Timmermans, M., Kidner, C. & Martienssen, R. (2001). Development of leaf shape. *Curr. Opin. Plant Biol.* **4**, 38–43.

59. Cadigan, K. M. & Nusse, R. (1997). Wnt signaling: A common theme in animal development. *Genes & Development* **11**, 3286–3305.

60. Calegari, F., Haubensak, W., Yang, D., Huttner, W. B. & Buchholz, F. (2002). Tissue-specific RNA interference in postimplantation mouse embryos with endoribonuclease-prepared short interfering RNA. *Proc. Natl. Acad. Sci. USA* **99**, 14236–14240.

61. Calin, G. A., Dumitru, C. D., Shimizu, M., Bichi, R., Zupo, S., Noch, E., Aldler, H., Rattan, S., Keating, M., Rai, K. L., Rassenti, L., Kipps, T., Negrini, M., Bullrich, F. & Croce, C. M. (2002). Frequent deletions and down-regulation of micro-RNA genes miR15 and miR16 at 13q14 in chronic lympocytic leukenia. *Proc. Natl. Acad. Sci. USA* **99**, 15524–15529.

62. Calin, G. A., Liu, C.-M., Sevignani, C., Ferracin, M., Felli, N., Dumitru, C. D., Shimizu, M., Cimmino, A., Zupo, S., Dono, M., Dell'Aquila, M. L., Alder, H., Rassenti, L., Kipps, T. J., Bullrich, F., Negrini, M. & Croce, C. M. (2004a). MicroRNA profiling reveals distinct signatures in B cell chronic lymphocytic leukemias. *Proc. Natl. Acad. Sci. USA* **101**, 11755–11760.

63. Calin, G. A., Sevignani, C., Dan Dumitru, C., Hyslop, T., Noch, E., Yendamuri, S., Shimizu, M., Rattan, S., Bullrich, F., Negrini, M. & Croce, C. M. (2004b). Human microRNA genes are frequently located at fragile sites and genomic regions involved in cancers. *Proc. Natl. Acad. Sci. USA* **101**, 2999–3004.

64. Cambareri, E. B., Jensen, B. C., Schabtach, E. & Selker, E. U. (1989). Repeat-Induced G-C to a-T mutations in Neurospora. *Science* **244**, 1571–1575.

65. Caplen, N. J., Parrish, S., Imani, F., Fire, A. & Morgan, R. A. (2001). Specific inhibition of gene expression by small double-stranded RNAs in invertebrate and vertebrate systems. *Proc. Natl. Acad. Sci. USA* **98**, 9742–9747.

66. Caplen, N. J., Zheng, Z., Falgout, B. & Morgan, R. A. (2002). Inhibition of viral gene expression and replication in mosquito cells by dsRNA-triggered RNA interference. *Molecular Therapy* **6**, 243–838.

67. Cardenas, M. M. & Salgado, L. M. (2003). STK, the src homologue, is responsible for the initial commitment to develop head structures in Hydra. *Developmental Biology* **264**, 495–505.

68. Carpenter, R. & Coen, E. S. (1990). Floral homeotic mutations produced by transposon-mutagenesis in *Antirrhinum majus*. *Genes & Development* **4**, 1483–1493.

69. Catalanotto, C., Azzalin, G., Macino, G. & Cogoni, C. (2002). Involvement of small RNAs and role of the qde genes in the gene silencing pathway in *Neurospora*. *Genes & Development* **16**, 790–795.

70. Caudy, A. A., Myers, M., Hannon, G. J. & Hammond, S. M. (2002). Fragile X-related protein and VIG associate with the RNA interference machinery. *Genes & Development* **16**, 2491–2496.

71. Causier, B., Kieffer, M. & Davies, B. (2002). MADS-box genes reach maturity. *Science* **296**, 275–276.

72. Cebria, F., Kobayashi, C., Umesono, Y., Nakazawa, M., Mineta, K., Ikeo, K., Gojobori, T., Itoh, M., Taira, M., Alvarado, A. S. & Agata, K. (2002). FGFR-related gene nou-darake restricts brain tissues to the head region of planarians. *Nature* **419**, 620–624.

73. Celotto, A. M. & Graveley, B. R. (2002). Exon-speciic RNAi: A tool for dissecting the functional relevance of alternative splicing. *RNA* **8**, 718–724.

74. Chalfie, M., Horvitz, H. R. & Sulston, J. E. (1981). Mutations that lead to reiterations in the cell lineages of *C. elegans*. *Cell* **24**, 59–69.

75. Chandler, V. L. & Walbot, V. (1986). DNA modification of a maize transposable element correlates with loss of activity. *Proc. Natl. Acad. Sci. USA* **83**, 1767–1771.

76. Chang, J., Provost, P. & Taylor, J. M. (2003). Resistance of human hepatitis delta virus RNAs to dicer activity. *J. Virology* **77**, 11910–11917.

77. Chen, C.-Z., Li, L., Lodish, H. F. & Bartel, D. P. (2004a). MicroRNAs modulate hematopoietic lineage differentiation. *Science* **303**, 83–86.

78. Chen, J., Li, W. X., Xie, D., Peng, J. R. & Ding, S. W. (2004b). Viral virulence protein suppresses RNA silencing-mediated defense but upregulates the role of microRNA in host gene expression. *Plant Cell* **16**, 1302–1313.

79. Chen, S., Hofius, D., Sonnewald, U. & Bornke, F. (2003). Temporal and spatial control of gene silencing in transgenic plants by inducible expression of double-stranded RNA. *Plant J.* **36**, 731–740.

80. Chen, X. (2004c). A microRNA as a translational repressor of APETALA2 in *Arabidopsis* flower development. *Science* **303**, 2022–2025.

81. Chi, J.-T., Chang, H. Y., Wang, N. N., Chang, D. S., Dunphy, N. & Brown, P. O. (2003). Genome wide view of gene silencing by small interfering RNAs. *Proc. Natl. Acad. Sci. USA* **100**, 6343–6346.

82. Chicas, A. & Macino, G. (2001). Characteristics of post-transcriptional gene silencing. *EMBO Reports* **21**, 992–996.

83. Clemens, M. J. (1997). PKR — A protein kinase regulated by double-stranded RNA. *Intern. J. Biochemi. Cell Biology* **29**, 945–949.

84. Coen, E. (1999). *The Art of the Genes*, Oxford University Press, Oxford, UK.

85. Coen, E. S. & Meyerowitz, E. M. (1991). The war of the whorls — Genetic interactions controlling flower development. *Nature* **353**, 31–37.

86. Coen, E. S., Romero, J. M., Doyle, S., Elliott, R., Murphy, G. & Carpenter, R. (1990). Floricaula — A homeotic gene required for flower development in *Antirrhinum majus. Cell* **63**, 1311–1322.

87. Cogoni, C., Irelan, J. T., Schumacher, M., Schmidhauser, T. J., Selker, E. U. & Macino, G. (1996). Transgene silencing of the al-1 gene in vegetative cells of Neurospora is mediated by a cytoplasmic effector and does not depend on DNA-DNA interactions or DNA methylation. *EMBO J.* **15**, 3153–3163.

88. Cogoni, C. & Macino, G. (1997). Isolation of quelling-defective (qde) mutants impaired in posttranscriptional transgene-induced gene silencing in *Neurospora crassa. Proc. Natl. Acad. Sci. USA* **94**, 10233–10238.

89. Cubas, P., Lauter, N., Doebley, J. & Coen, E. (1999). The TCP domain: A motif found in proteins regulating plant growth and development. *Plant J.* **18**, 215–222.

90. Dalmay, T., Hamilton, A., Mueller, E. & Baulcombe, D. C. (2000a). Potato *Virus X* amplicons in Arabidopsis mediate genetic and epigenetic gene silencing. *Plant Cell* **12**, 369–379.

91. Dalmay, T., Hamilton, A., Rudd, S., Angell, S. & Baulcombe, D. C. (2000b). An RNA-dependent RNA polymerase gene in *Arabidopsis* is required for posttranscriptional gene silencing mediated by a transgene but not by a virus. *Cell* **101**, 543–553.

92. Dalmay, T., Horsefield, Braunstein, T. H. & Baulcombe, D. C. (2001). *SDE3* encodes an RNA helicase required for post-transcriptional gene silencing in *Arabidopsis. EMBO J.* **20**, 2069–2077.

93. Dangl, J. L. & Jones, J. D. G. (2001). Plant pathogens and integrated defence responses to infection. *Nature* **411**, 826–833.

94. Danin-Kreiselman, M., Lee, C. Y. & Chanfreau, G. (2003). RNAse III-mediated degradation of unspliced pre-mRNAs and lariat introns. *Molecular Cell* **11**, 1279–1289.

95. DaRocha, W. D., Otsu, K., Teixeira, S. M. R. & Donelson, J. E. (2004). Tests of cytoplasmic RNA interference (RNAi) and construction of a tetracycline-inducible T7 promoter system in *Trypanosoma cruzi. Molecular & Biochemical Parasitology* **133**, 175–186.

96. Davidson, E. H. & Britten, R. J. (1979). Regulation of gene expression: Possible role of repetitive sequences. *Science* **204**, 1052–1059.

97. De-Haan, P., Gielen, J. J. L., Prins, M., Wijkamp, I. G., Vanschepen, A., Peters, D., Vangrinsven, M. Q. J. M. & Goldbach, R. (1992). Characterization of RNA-mediated resistance to tomato spotted wilt virus in transgenic tobacco plants. *Bio-Technology* **10**, 1133–1137.

98. Depicker, A. & Van Montagu, M. (1997). Post-transcriptional gene silencing in plants. *Curr. Opin. Cell Biology* **9**, 9373–9383.

99. Ding, S. W., Li, H. W., Lu, R., Li, F. & Li, W. X. (2004). RNA silencing: A conserved antiviral immunity of plants and animals. *Virus Research* **102**, 109–115.

100. Ding, S. W., Li, W. X. & Symons, R. H. (1995). A novel naturally-occurring hybrid gene encoded by a plant RNA virus facilitates long-distance virus movement. *EMBO J.* **14**, 5762–5772.

101. Djikeng, A., Shi, H., Tschudi, C., Shen, S. & Ullu, E. (2003). An siRNA ribonucleoprotein is found associated with polyribosomes in *Trypanosoma brucei. RNA* **9**, 802–808.

102. Djikeng, A., Shi, H., Tschudi, C. & Ullu, E. (2001). RNA interference in *Trypanosoma brucei*: Cloning of small interfering RNAs provides evidence for retroposon-derived 24-26 nucleotide RNAs. *RNA* **7**, 1522–1530.

103. Doench, J. G., Petersen, C. P. & Sharp, P. A. (2003). siRNAs can function as miRNAs. *Genes & Development* **17**, 438–442.

104. Doench, J. G. & Sharp, P. A. (2004). Specificity of microRNA target selection in translational repression. *Genes & Development* **18**, 504–511.

105. Doi, N., Zenno, S., Ueda, R., Ohki-Hamazaki, H., Ui-Tei, K. & Saigo, K. (2003). Short-interfering-RNA-mediated gene silencing in mammalian cells requires dicer and elF2C translation initiation factors. *Current Biology* **13**, 41–46.

106. Dong, X. L., van Wezel, R., Stanley, J. & Hong, Y. G. (2003). Functional characterization of the nuclear localization signal for a suppressor of posttranscriptional gene silencing. *J. Virology* **77**, 7026–7033.

107. Donze, O. & Picard, D. (2002). RNA interference in mammalian cells using siRNAs synthesized with T7 RNA polymerase. *Nucleic Acids Research* **30**, 1–4.

108. Doolan, D. L. & Hoffman, S. L. (1997). Pre-erythrocytic-stage immune effector mechanisms in *Plasmodium* spp. infections. *Phil. Trans. R. Soc. Lond. B.* **352**, 1361–1367.

109. Dorsett, Y. & Tuschl, T. (2004). siRNA: Application in functional genomics and potential therapeutics. *Nature Rev./Drug Discovery* **3**, 318–329.

110. Dostie, J., Mourelatos, Z., Yang, M., Sharma, A. & Dryfuss, G. (2003). Numerous microRNPs in neuronal cells containing novel microRNAs. *RNA* **9**, 180–186.

111. Drews, G. N., Bowman, J. L. & Meyerowitz, E. M. (1991). Negative regulation of the Arabidopsis homeotic gene agamous by the Apetala2 product. *Cell* **65**, 991–1002.

112. Driever, W., Solnica Krezel, L., Schier, A. F., Neuhauss, S. C. F., Malicki, J., Stemple, D. L., Stainier, D. Y. R., Zwartkruis, F., Abdelilah, S., Rangini, Z., Belak, J. & Boggs, C. (1996). A genetic screen for mutations affecting embryogenesis in zebrafish. *Development* **123**, 37–46.

113. Dudley, N. R., Labbe, J.-C. & Goldstein, B. (2002). Using RNA interference to identify genes required for RNA interference. *Proc. Natl. Acad. Sci. USA* **99**, 4191–4196.

114. Dunoyer, P., Pfeffer, S., Fritsch, C., Hemmer, O., Voinnet, O. & Richards, K. E. (2002). Identification, subcellular localization and some properties of a cysteine-rich supressor of gene silencing encoded by peanut clump virus. *Plant J.* **29**, 555–567.

115. Elbashir, S. M., Harborth, J., Lendeckel, W., Yalcin, A., Weber, K. & Tuschl, T. (2001b). Duplexes of 21-nucleotide RNAs mediate RNA interference in cultured mammalian cells. *Nature* **411**, 494–498.

116. Elbashir, S. M., Lendeckel, W. & Tuschl, T. (2001a). RNA interference is mediated by 21- and 22-nucleotide RNAs. *Genes & Development* **15**, 188–200.

117. Elbashir, S. M., Martinez, J., Patkaniowska, A., Lendeckel, W. & Tuschl, T. (2001c). Functional anatomy of siRNAs for mediating efficient RNAi in *Drosophila melanogaster* embryo lysate. *EMBO J.* **20**, 6877–6888.

118. Ellstrand, N. C. (2004). Dangerous liaisons — When cultivated plants mate with their wild relatives. *Plant Science* **167**, 187–188.

119. Elmayan, T., Balzergue, S., Beon, F., Bourdon, V., Daubremet, J., Guenet, Y., Mourrain, P., Palauqui, J.-C., Vernhettes, S., Vialle, T., Wostrikoff, K. & Vaucheret, H. (1998). Arabidopsis mutants impaired in cosuppression. *Plant Cell* **10**, 1747–1757.

120. Elmayan, T. & Vaucheret, H. (1996). Expression of single copies of a strongly expressed 35S transgene can be silenced post-transcriptionally. *Plant J.* **9**, 787–797.

121. Emery, J. F., Floyd, S. K., Alvarez, J., Eshed, Y., Hawker, N. P., Izhaki, A., Baum, S. F. & Bowman, J. L. (2003). Radial patterning of *Arabidopsis* shoots by class III HD-ZIP and KANADI Genes. *Current Biology* **13**, 1768–1774.

122. English, J. J., Mueller, E. & Baulcombe, D. C. (1996). Suppression of virus accumulation in transgenic plants exhibiting silencing of nuclear genes. *Plant Cell* **8**, 179–188.

123. Fagard, M., Boutet, S., Morel, J.-B., Bellini, C. & Vaucheret, H. (2000). AGO1, QDE-2, RDE-1 are related proteins required for post-transcriptional gene silencing in plants, quelling in fungi, and RNA interference in animals. *Proc. Natl. Acad. Sci. USA* **97**, 11650–11654.

124. Fedoroff, N. V. (1995). DNA methylation and activity of the maize *Spm* transposable element. In *Gene Silencing in Higher Plants and Related Phenomena in Other Eukaryotes* (Meyer, P., ed.), Vol. 174, pp. 144–164. *Curr. Topics Microbiol. Immuno.*

125. Feinbaum, R. & Ambros, V. (1999). The timing of *lin-4* RNA accumulation controls the timing of postembryonic developmental events in *Caenorhabditis elegans*. *Developmental Biology* **210**, 87–95.

126. Feinberg, E. H. & Hunter, C. P. (2003). Transport of dsRNA into cells by the transmembrane protein SID-1. *Science* **301**, 1545–1547.

127. Finnegan, E. J., Martgis, R. & Waterhouse, P. M. (2003). Posttranscriptional gene silencing is not compromised in the *Arabidopsis* CARPEL FACTORY (DICER-LIKE1) mutant, a homolog of dicer-1 from Drosophila. *Current Biology* **13**, 236–240.

128. Fire, A., Albertson, D., Harrison, S. W. & Moerman, D. G. (1991). Production of antisense RNA leads to effective and specific-inhibition of gene-expression in *C. elegans* muscle. *Development* **113**, 503–514.

129. Fire, A. & Waterston, R. H. (1989). Proper expression of myosin genes in transgenic nematodes. *EMBO J.* **8**, 3419–3428.

130. Fire, A., Xu, S., Montgomery, M. K., Kostas, S. A., Driver, S. E. & Mello, C. C. (1998). Potent and specific genetic interference by double-stranded RNA in *Caenorhabditis elegans*. *Nature* **391**, 806–811.

131. Firtel, R. A. (1995). Integration of signaling information in controlling cell-fate decisions in Dictyostelium. *Genes & Development* **9**, 1427–1444.

132. Flavell, R. B. (1994). Inactivation of gene-expression in plants as a consequence of specific sequence duplication. *Proc. Natl. Acad. Sci. USA* **91**, 3490–3496.

133. Floyd, S. K. & Bowman, J. L. (2004). Ancient microRNA target sequences in plants. *Nature* **428**, 485–486.

134. Frankel, R. (1956). Graft-induced transmission to progeny of cytoplasmic male sterility in petunia. *Science* **124**, 684–685.

135. Frankel, R. & Galun, E. (1977). *Pollination Mechanisms, Reproduction and Plant Breeding*, Springer-Verlag, Berlin.

136. Fraser, A. G., Kamath, R. S., Zipperlen, P., Martinez-Campos, M., Sohrmann, M. & Ahringer, J. (2000). Functional genomic analysis of C. *elegans* chromosome I by systematic RNA interference. *Nature* **408**, 325–330.

137. Fritsch, L., Martinez, L. A., Sekhri, R., Naguibneva, I., Gerard, M., Vandromme, M., Schaeffer, L. & Harel-Bellan, A. (2004). Conditional gene knock-down by CRE-dependent short interfering RNAs. *EMBO Reports* **5**, 178–182.

138. Galili, G. & Hofgen, R. (2002). Metabolic engineering of amino acids and storage proteins in plants. *Metabolic Engineering* **4**, 3–11.

139. Galun, E. (2003). *Transposable Elements — A Guide to the Perplexed and the Novice*, Kluwer Academic Publishers, Dordrecht.

140. Galun, E. & Breiman, A. (1997). *Transgenic Plants*, Imperial College Press, London.

141. Galun, E. & Galun, E. (2001). *Manufacture of Medical and Health Products by Transgenic Plants*, Imperial College Press, London.

142. Galun, E. & Torrey, J. G. (1969). Initiation and suppression of apical hairs of Fucus embryos. *Develop. Biol.* **19**, 447–459.

143. Galvani, A. & Sperling, L. (2001). Transgene-mediated post-transcriptional gene silencing is inhibited by 3′ non-coding sequences in Paramecium. *Nucleic Acids Research* **29**, 4387–4394.

144. Galvani, A. & Sperling, L. (2002). RNA interference by feeding in Paramecium. *Trends Genetics* **18**, 11–12.

145. Garsin, D. A., Villanueva, J. M., Begun, J., Kim, D. H., Sifri, C. D., Calderwood, S. B., Ruvkun, G. & Ausubel, F. M. (2003). Long-lived C. *elegans daf-2* mutants are resistant to bacterial pathogens. *Science* **300**, 1921–1921.

146. Gaudilliere, B., Shi, Y. & Bonni, A. (2002). RNA interference reveals a requirement for myocyte enhancer factor 2A in activity-dependent neuronal survival. *J. Biol. Chem.* **277**, 46442–46446.

147. Glazov, E., Phillips, K., Budziszewski, G. J., Meins, J. F. & Levin, J. Z. (2003). A gene encoding an RNase D exonuclease-like protein is required for post-transcriptional silencing in Arabidopsis. *Plant J.* **35**, 342–349.

148. Golemboski, D. B., Lomonossoff, G. P. & Zaitlin, M. (1990). Plants transformed with a tobacco mosaic-virus nonstructural gene sequence are resistant to the virus. *Proc. Natl. Acad. Sci. USA* **87**, 6311–6315.

149. Gonczy, P., Echeverri, C., Oegema, K., Coulson, A., Jones, S. J. M., Copley, R. R., Dupeeron, J., Oegema, J., Brehm, M., Cassin, E., Hannak, E., Kirkham, M., Pichler, S., Flohrs, K., Goessen, A., Leidel, S., Alleaume, A.-M., Martin, C., Oztu, N., Bork, P. & Hyman, A. A. (2000). Functional genomic analysis of cell division in C. *elegans* using RNAi of genes on chromosome III. *Nature* **408**, 331–336.

150. Goto, K., Kyozuka, J. & Bowman, J. L. (2001). Turning floral organs into leaves, leaves into floral organs. *Curr. Opin. Genetics Development* **11**, 449–456.

151. Gottesman, S. (2004a). The small RNA regulators of *Escherichia coli*: Roles and mechanisms. *Annu. Rev. Microbiology* **58**, 303–328.
152. Gottesman, S. (2004b). Small RNAs shed some light. *Cell* **118**, 1–2.
153. Goyon, C. & Faugeron, G. (1989). Targeted transformation of *Ascobolus immersus* and de novo methylation of the resulting duplicated DNA-sequences. *Molecular Cellular Biology* **9**, 2818–2827.
154. Grad, Y., Aach, J., Hayes, G. D., Reinhart, B. J., Church, G. M., Ruvkun, G. & Kim, J. (2003). Computational and experimental identification of *C. elegans* microRNAs. *Molecular Cell* **11**, 1253–1263.
155. Graveley, B. R. (2001). Alternative splicing: Increasing diversity in the proteomic world. *Trends Genetics* **17**, 100–107.
156. Grewal, S. I. S. (2000). Transcriptional silencing in fission yeast. *J. Cell Physiol.* **184**, 311–318.
157. Griffiths-Jones, S. (2004). The microRNA registry. *Nucleic Acids Research* **32**, D109–D111.
158. Grishok, A., Pasquinelli, A. E., Conte, D., Li, N., Parrish, S., Ha, I., Baillie, D. L., Fire, A., Ruvkun, G. & Mello, C. C. (2001). Genes and mechanisms related to RNA interference regulate expression of the small temporal RNAs that control *C. elegans* developmental timing. *Cell* **106**, 23–34.
159. Guo, H.-S., Fei, J.-F., Xie, Q. & Chua, N.-H. (2003). A chemical-regulated inducible RNAi system in plants. *Plant J.* **34**, 383–392.
160. Guo, H. S. & Ding, S. W. (2002). A viral protein inhibits the long range signaling activity of the gene silencing signal. *EMBO J.* **21**, 398–407.
161. Guo, S. & Kemphues, K. J. (1995). par-1, a gene required for establishing polarity in *C. elegans* embryos, encodes a putative ser/thr kinase that is asymmetrically distributed. *Cell* **81**, 611–620.
162. Gupta, S., Schoer, R. A., Egan, J. E., Hannon, G. J. & Mittal, V. (2004). Inducible, reversible, and stable RNA interference in mammalian cells. *Proc. Natl. Acad. Sci. USA* **101**, 1927–1932.
163. Ha, I., Wightman, B. & Ruvkun, G. (1996). A bulged *lin-4/lin-14* RNA duplex is sufficient for *Caenorhabditis elegans lin-14* temporal gradient formation. *Genes & Development* **10**, 3041–3050.
164. Haffter, P., Granato, M., Brand, M., Mullins, M. C., Hammerschmidt, M., Kane, D. A., Odenthal, J., vanEeden, F. J. M., Jiang, Y. J., Heisenberg, C. P., Kelsh, R. N., FurutaniSeiki, M., Vogelsang, E., Beuchle, D., Schach, U., Fabian, C. & Nusslein-Volhard, C. (1996). The identification of genes with unique and essential functions in the development of the zebrafish, *Danio rerio. Development* **123**, 1–36.
165. Hamilton, A., Voinnet, O., Chappell, L. & Baulcombe, D. C. (2002). Two classes of short interfering RNA in RNA silencing. *EMBO J.* **21**, 4671–4679.
166. Hamilton, A. J. & Baulcombe, D. C. (1999). A species of small antisense RNA in posttranscriptional gene silencing in plants. *Science* **286**, 950–952.
167. Hammond, S. M., Bernstein, B. E., Beach, D. & Hannon, G. J. (2000). An RNA-directed nuclease mediates post-transcriptional gene silencing in *Drosophila* cells. *Nature* **404**, 293–296.
168. Hammond, S. M., Boettcher, S. & Caudy, A. A. (2001a). Argonaute2, a link between genetic and biochemical analyses of RNAi. *Science* **293**, 1146–1150.

169. Hammond, S. M., Caudy, A. A. & Hannon, G. J. (2001b). Post-transcriptional gene silencing by double-standed RNA. *Nature Reviews Genetics* **2**, 110–119.

170. Han, M. H., Goud, S., Song, L. & Fedoroff, N. (2004). The Arabidopsis double-stranded RNA-binding protein HYL1 plays a role in microRNA-mediated gene regulation. *Proc. Natl. Acad. Sci. USA* **101**, 1093–1098.

171. Hannon, G. J. (2002). RNA interference. *Nature* **418**, 244–251.

172. Hannon, G. J. (2003). *RNAi: A Guide to Gene Silencing* (Hannon, G. J., ed.), Cold Spring Harbor Laboratory Press, Cold Spring Harbor, NY.

173. Harrison, B. D., Mayo, M. A. & Baulcombe, D. C. (1987). Virus resistance in transgenic plants that express cucumber mosaic virus satellite RNA. *Nature* **328**, 799–802.

174. Hasuwa, H., Kaseda, K., Einarsdottir, T. & Okabe, M. (2002). Small interfering RNA and gene silencing in transgenic mice and rats. *FEBS Letters* **532**, 227–230.

175. He, L. & Hannon, G. J. (2004). MicroRNAs: Small RNAs with a big role in gene regulation. *Nature Reviews Genetics* **5**, 522–531.

176. Heggestad, A. D., Notterpek, L. & Fletcher, B. S. (2004). Transposon-based RNAi delivery system for generating knockdown cell lines. *Biochem. Biophys. Res. Comm.* **316**, 643–650.

177. Helliwell, C. & Waterhouse, P. (2003). Constructs and methods for high-throughput gene silencing in plants. *Methods* **30**, 289–295.

178. Hemann, M. T., Fridman, J. S., Zilfou, J. T., Hernando, E., Paddison, P. J., Cordon-Cardo, C., Hannon, G. J. & Lowe, S. W. (2003). An epi-allelic series of p53 hypomorphs created by stable RNAi produces distinct tumor phenotypes *in vivo*. *Nature Genetics* **33**, 396–400.

179. Herwaldt, B. L. (1999). Leishmaniasis. *The Lancet* **354**, 1191–1199.

180. Himber, C., Dunoyer, P., Moissiard, G., Ritzenthaler, C. & Voinnet, O. (2003). Transitivity-dependent and independent cell-to-cell movement of RNA silencing. *EMBO J.* **22**, 4523–4533.

181. Hobmayer, B., Rentzsch, F., Kuhn, K., Happel, C. M., von Laue, C. C., Snyder, P., Rothbacher, U. & Holstein, T. W. (2000). WNT signalling molecules act in axis formation in the diploblastic metazoan Hydra. *Nature* **407**, 186–189.

182. Hoffmann, L., Besseau, S., Geoffroy, O., Ritzenthaler, C., Meyer, D., Lapierre, C., Pollet, B. & Legrand, M. (2004). Silencing of hydroxycinnamoyl-Coenzyme A Shikimate/Quinate hydroxycinnamoyltransferase affects phenylpropanoid biosynthesis. *Plant Cell* **16**, 1446–1465.

183. Horiguchi, G. (2004). RNA silencing in plants: A shortcut to functional analysis. *Differentiation* **72**, 65–73.

184. Houbaviy, H. B., Murray, M. F. & Sharp, P. A. (2003). Embryonic stem cell-specific microRNAs. *Developmental Cell* **5**, 351–358.

185. Hutvagner, G., McLachlan, J., Pasquinelli, A., Balint, E., Tuschl, T. & Zamore, P. D. (2001). A cellular function for the RNA-interference enzyme dicer in the maturation of the *let-7* small temporal RNA. *Science* **293**, 834–838.

186. Hutvagner, G., Simard, M. J., Mello, C. C. & Zamore, P. D. (2004). Sequence-specific inhibition of small RNA function. *PloS Biology* **2**(4), 1–11.

187. Hutvagner, G. & Zamore, P. D. (2002a). A microRNA in a multiple-turnover RNAi enzyme complex. *Science* **297**, 2056–2060.

188. Hutvagner, G. & Zamore, P. D. (2002b). RNAi: Nature abhors a double-strand. *Curr. Opin. Genetics Development* **12**, 225–232.

189. Hynes, M. J. & Todd, R. B. (2003). Detection of unpaired DNA at meiosis results in RNA-mediated silencing. *BioEssays* **25**, 99–103.

190. Inoue, N., Osu, K., Ferraro, D. M. & Donelson, J. E. (2002). Tetracycline-regulated RNA interference in *Trypanosoma congolense*. *Molec. Biochem. Parasitology* **120**, 309–313.

191. Irish, V. F. (2003). The evolution of floral homeotic gene function. *BioEssays* **25**, 637–646.

192. Ishizuka, A., Siomi, M. C. & Siomi, H. (2002). A *Drosophila* fragile X protein inter-acts with components of RNAi and ribosomal proteins. *Genes & Development* **16**, 2497–2508.

193. Izsvak, Z. & Ivics, Z. (2004). Sleeping Beauty transposition: Biology and applica-tions for molecular therapy. *Molecular Therapy* **9**, 147–156.

194. Jack, T. (2001a). Plant development going MADS. *Plant Molecular Biology* **46**, 515–520.

195. Jack, T. (2001b). Relearning our ABCs: New twists on an old model. *Trends Plant Science* **6**, 310–316.

196. Jack, T. (2004). Molecular and genetic mechanisms of floral control. *Plant Cell* **16**, S1–S17.

197. Jackson, A. L., Bartz, S. R., Schelter, J., Kobayashi, S. V., Burchard, J., Mao, M., Li, B., Cavet, G. & Linsley, P. S. (2003). Expression profiling reveals off-target gene regulation by RNAi. *Nature Biotechnology* **21**, 635–637.

198. Jensen, S., Gassama, M.-P. & Heidmann, T. (1999a). Taming of transposable ele-ments by homology-dependent gene silencing. *Nature Genetics* **21**, 209–212.

199. Jensen, S., Gassama, M.-P. & Heidmann, T. (1999b). Cosuppression of I trans-poson activity in drosophila by I-containing sense and antisense transgenes. *Genetics* **153**, 1767–1774.

200. Joliot, F. & Curie, I. (1934). Artificial production of a new kind of radio-element. *Nature* **138**, 201–202.

201. Jones-Rhoades, M. W. & Bartel, D. P. (2004). Computational identification of plant MicroRNAs and their targets, including a stress-induced miRNA. *Molecular Cell* **14**, 787–799.

202. Jorgensen, R., Atkinson, R. G., Forster, R. L. S. & Lucas, W. J. (1998). An RNA-based information superhighway in plants. *Science* **279**, 1486–1487.

203. Jorgensen, R. A. (2002). RNA traffics information systemically in plants. *Proc. Natl. Acad. Sci. USA* **99**, 11561–11563.

204. Jorgensen, R. A., Cluster, P. D., English, J., Que, Q. & Napoli, C. A. (1996). Chal-cone synthase cosuppression phenotypes in petunia flowers: Comparison of sense vs. antisense constructs and single-copy vs. complex T-DNA sequences. *Plant Molecular Biology* **31**, 957–973.

205. Juarez, M. T., Kui, J. S., Thomas, J., Heller, B. A. & Timmermans, M. C. P. (2004). microRNA-mediated repression of rolled leaf1 specifies maize leaf polar-ity. *Nature* **428**, 84–88.

206. Kaempfer, R. (2003). RNA sensors: Novel regulators of gene expression. *EMBO Reports* **4**, 1043–1047.

207. Kalidas, S. & Smith, D. P. (2002). Novel genomic cDNA hybrids produce effective RNA interference in adult *Drosophila*. *Neuron* **33**, 177–184.

208. Kamath, R. S., Fraser, A. G., Dong, Y., Poulin, G., Durbin, R., Gotta, M., Kanapin, A., Le Bot, N., Moreno, S., Soyhrmann, M., Welchman, D. P., Zipperlen, P. & Ahringer, J. (2003). Systematic functional analysis of the *Caenorhabditis elegans* genome using RNAi. *Nature* **421**, 231–237.

209. Kariko, K., Bhuyan, P., Capodici, J. & Weissman, D. (2004). Small interfering RNAs mediate sequence-independent gene suppression and induce immune activation by signaling through toll-like receptor 3. *J. Immunology* **172**, 6545–6549.

210. Kasschau, K., Xie, Z., Allen, E., Llave, C., Chapman, E. J., Krizan, K. A. & Carrington, J. C. (2003). P1/HC-Pro, a viral suppressor of RNA silencing, interferes with *Arabidopsis* development and miRNA function. *Developmental Cell* **4**, 205–217.

211. Kasschau, K. D. & Carrington, J. C. (1998). A counterdefensive strategy of plant viruses. *Cell* **95**, 461–470.

212. Kawasaki, H., Suyama, E., Iyo, M. & Taira, K. (2003). SiRNAs generated by recombinant human dicer induce specific and significant but target site-independent gene silencing in human cells. *Nucleic Acids Research* **31**, 981–987.

213. Kawasaki, H. & Taira, K. (2003a). Hes1 is a target of microRNA-23 during retinoic-acid-induced neuronal differentiation of NT2 cells. *Nature* **423**, 838–842.

214. Kawasaki, H. & Taira, K. (2003b). Short hairpin type of dsRNAs that are controlled by tRNA^{Val} promoter significantly induce RNAi-mediated gene silencing in the cytoplasm of human cells. *Nucleic Acids Research* **31**, 700–707.

215. Kawasaki, H., Wadhwa, R. & Taira, K. (2004). World of small RNAs: From ribozymes to siRNA and miRNA. *Differentiation* **72**, 58–64.

216. Keating, C. D., Kriek, N., Daniels, M., Ashcroft, N. R., Hopper, N. A., Siney, E. J., Holden-Dye, L. & Burke, J. F. (2003). Whole-genome analysis of 60 G protein-coupled receptors in Caenorhabditis elegans by gene knockout with RNAi. *Current Biology* **13**, 1715–1720.

217. Kennedy, S., Wang, D. & Ruvkun, G. (2004). A conserved siRNA-degrading RNase negatively regulates RNA interference in *C. elegans*. *Nature* **427**, 645–649.

218. Kennerdell, J. R. & Carthew, R. W. (1998). Use of dsRNA-mediated genetic interference to demonstrate that frizzled and frizzled 2 act in the wingless pathway. *Cell* **95**, 1017–1026.

219. Kennerdell, J. R. & Carthew, R. W. (2000). Heritable gene silencing in *Drosophila* using double-stranded RNA. *Nature Biotechnology* **17**, 896–898.

220. Ketting, R. F., Fischer, S. E. J., Bernstein, E., Sijen, T., Hannon, G. J. & Plasterk, R. H. A. (2001). Dicer functions in RNA interference and in synthesis of small RNA involved in developmental timing in *C. elegans*. *Genes & Development* **15**, 2654–2659.

221. Ketting, R. F., Haverkamp, T. H. A., van Luenen, H. G. A. M. & Plasterk, R. H. A. (1999). mut-7 of *C. elegans*, required for transposon silencing and RNA interference, is a homolog of Werner syndrome helicase and RNaseD. *Cell* **99**, 133–141.

222. Ketting, R. F. & Plasterk, R. H. A. (2000). A genetic link between co-suppression and RNA interference in C-elegans. *Nature* **404**, 296–298.

223. Ketting, R. F., Tijsterman, M. & Plasterk, R. H. A. (2003). RNAi in *Caenorhabditis elegans*. In *RNAi- A Guide to Gene Silencing* (Hannon, G. J., ed.), pp. 65–85. Cold Spring Harbor Laboratory Press, Cold Spring Harbor, N.Y.

224. Khvorova, A., Reynolds, A. & Jayasena, S. D. (2003). Functional siRNAs and miRNAs exhibit strand bias. *Cell* **115**, 209–216.

225. Kidner, C. A. & Martienssen, R. A. (2003). Macro effects of microRNAs in plants. *Trends Genetics* **19**, 13–16.

226. Kidner, C. A. & Martienssen, R. A. (2004). Spatially restricted microRNA directs leaf polarity through ARGONAUTE1. *Nature* **428**, 81–84.

227. Kim, J., Krichevsky, A. M., Grad, Y., Hayes, G. D., Kosik, K. S., Church, G. M. & Ruvkun, G. (2004). Identification of many microRNAs that copurify with polyribosomes in mammalian neurons. *Proc. Natl. Acad. Sci. USA* **101**, 360–365.

228. Kim, M., Canio, W., Kessler, S. & Sinha, N. (2001). Developmental changes due to long-distance movement of a homeobox fusion transcript in tomato. *Science* **293**, 287–289.

229. Kinney, A. J. (1996). Designer oils for better nutrition. *Nature Biotechnology* **14**, 946–946.

230. Kjemtrup, S., Sampson, K. S., Peele, C. G., Nguyen, L. V., Conkling, M. A., Thompson, W. F. & Robertson, D. (1998). Gene silencing from plant DNA carried by a Geminivirus. *Plant J.* **14**, 91–100.

231. Klahre, U., Crete, P., Leuenberger, S. A., Iglesias, V. A. & Mains, F. (2002). High molecular weight RNAs and small interfering RNAs induce systemic posttranscriptional gene silencing in plants. *Proc. Natl. Acad. Sci. USA* **99**, 11981–11986.

232. Knight, S. W. & Bass, B. L. (2001). A role for the RNase III enzyme DCR-1 in RNA interference and germ line development in *Caenorhabditis elegans*. *Science* **293**, 2269–2271.

233. Knutzon, D. S., Thompson, G. A., Radke, S. E., Johnson, W. B., Knauf, V. C. & Kridl, J. C. (1992). Modification of Brassica seed oil by antisense expression of a stearoyl-acyl carrier protein desaturase gene. *Proc. Natl. Acad. Sci. USA* **89**, 2624–2628.

234. Kooter, J. M., Matzke, M. A. & Meyer, P. (1999). Listening to the silent genes: Transgene silencing, gene regulation and pathogen control. *Trends Plant Science* **4**, 1360–1385.

235. Krichevsky, A. M., King, K. S., Donahue, C. P., Khrapko, K. & Kosik, K. S. (2003). A microRNA array reveals extensive regulation of microRNAs during brain development. *RNA* **9**, 1274–1281.

236. Krichevsky, A. M. & Kosik, K. S. (2002). RNAi functions in cultured mammalian neurons. *Proc. Natl. Acad. Sci. USA* **99**, 11926–11929.

237. Kumar, R., Conklin, D. S. & Mittal, V. (2003). High-throughput selection of effective RNAi probes for gene silencing. *Genome Research* **13**, 2333–2340.

238. Kusaba, M., Miyahara, K., Lida, S., Fukuoka, H., Takano, T., Sassa, H., Nishimura, M. & Nishio, T. (2003). Low glutelin content1: A dominant mutation that suppresses the glutelin multigene family via RNA silencing in rice. *Plant Cell* **15**, 1455–1467.

239. Lagos-Quintana, M., Rauhut, R., Lendeckel, W. & Tuschi, T. (2001). Identification of novel genes coding for small expressed RNAs. *Science* **294**, 853–858.

240. Lagos-Quintana, M., Rauhut, R., Meyer, J., Borkhardt, A. & Tuschl, T. (2003). New microRNAs from mouse and human. *RNA* **9**, 175–179.

241. Lagos-Quintana, M., Rauhut, R., Yalcin, A., Meyer, J., Lendeckel, W. & Tuschl, T. (2002). Identification of tissue-specific microRNAs from mouse. *Current Biology* **12**, 735–739.

242. Lai, E. C. (2003). MicroRNAs: Runts of the genome assert themselves. *Current Biology* **13**, R925–R936.

243. Lakatos, L., Szittya, G., Silhavy, D. & Burgyan, J. (2004). Molecular mechanism of RNA silencing suppression mediated by p19 protein of tombusviruses. *EMBO J.* **23**, 876–884.

244. Langland, J. O., Jin, S., Jacobs, B. L. & Roth, D. A. (1995). Identification of a plant-encoded analog of PKR, the mammalian double-stranded RNA-dependent protein kinase. *Plant Physiology* **108**, 1259–1267.

245. Lau, N. C., Lim, L. P., Weinstein, E. G. & Bartel, D. P. (2001). An abundant class of tiny RNAs with probable regulatory roles in *Caenorhabditis elegans*. *Science* **294**, 858–862.

246. Lechtenberg, B., Schubert, D., Forsbach, A., Gils, M. & Schmidt, R. (2003). Neither inverted repeat T-DNA configurations nor arrangmenets of tandemly repeated transgenes are sufficient to trigger transgene silencing. *Plant J.* **34**, 507–517.

247. Lee, R. C. & Ambros, V. (2001). An extensive class of small RNAs in *Caenorhabditis elegans*. *Science* **294**, 862–864.

248. Lee, R. C., Feinbaum, R. L. & Ambros, V. (1993). The *C. elegans* heterochronic gene *lin-4* encodes small RNAs with antisense complementarity to *lin-14*. *Cell* **75**, 843–854.

249. Lee, Y., Ahn, C., Han, J. J., Choi, H., Kim, J., Yim, J., Lee, J., Provost, P., Radmark, O., Kim, S. & Kim, V. N. (2003). The nuclear RNase III Drosha initiates microRNA processing. *Nature* **425**, 415–419.

250. Lee, Y., Jeon, K., Lee, J.-T., Kim, S. & Kim, V. N. (2002). MicroRNA maturation: Stepwise processing and subcellular localization. *EMBO J.* **21**, 4663–4670.

251. Lewis, B. P., Shih, I. H., Jones-Rhoades, M. W., Bartel, D. P. & Burge, C. B. (2003). Prediction of mammalian microRNA targets. *Cell* **115**, 787–798.

252. Lewis, D. L., Hagstrom, J. E., Loomis, A. G., Wolff, J. A. & Herweijer, H. (2002). Efficient delivery of siRNA for inhibition of gene expression in postnatal mice. *Nature Genetics* **32**, 107–108.

253. Li, H., Li, W. X. & Ding, S. W. (2002). Induction and suppression of RNA silencing by an animal virus. *Science* **296**, 1319–1321.

254. Li, H.-W., Lucy, A. P., Guo, H.-S., Li, W.-X., Ji, L.-H., Wong, S.-M. & Ding, S.-W. (1999). Strong host resistance targeted against a viral suppressor of the plant gene silencing defence mechanism. *EMBO J.* **18**, 2683–2691.

255. Li, Y.-X., Farrell, M. J., Liu, R., Mohanty, N. & Kirby, M. L. (2000). Double-stranded RNA injection produces null phenotypes in zebrafish. *Developmental Biology* **217**, 394–405.

256. Liang, X.-H., Liu, Q. & Michaeli, S. (2003a). Small nucleolar RNA interference induced by antisense or double-stranded RNA in trypanosomatids. *Proc. Natl. Acad. Sci. USA* **100**, 7521–7526.

257. Liang, X. H., Haritan, A., Uliel, S. & Michaeli, S. (2003b). trans and cis splicing in trypanosomatids: Mechanism, factors, and regulation. *Eukaryotic Cell* **2**, 830–840.

258. Lim, L. P., Lau, N. C., Weinstein, E. G., Abdelhakim, A., Yekta, S., Rhoades, M. W., Burge, C. B. & Bartel, D. P. (2003b). The microRNAs of *Caenorhabditis elegans*. *Genes Development* **17**, 991–1008.

259. Lin, S.-Y., Johnson, M. B., Abraham, M., Vella, M. C., Pasquinelli, A., Gamberi, C., Gottlieb, E. & Slack, F. J. (2003). The *C. elegans* hunchback homolog, hb1-1, controls temporal patterning and is a probable microRNA target. *Developmental Cell* **4**, 639–650.

260. Lindbo, J. A., Silva-Rosales, L., Proebsting, W. M. & Dougherty, W. G. (1993). Induction of a highly specific antiviral state in transgenic plants: Implications for regulation of gene expression and virus resistance. *Plant Cell* **5**, 1749–1759.

261. Lipardi, C., Wei, Q. & Paterson, B. M. (2001). RNAi as random degradative PCR: siRNA primers convert mRNA into dsRNAs that are degraded to generate new siRNAs. *Cell* **107**, 297–307.

262. Liu, J., Carmell, M. A., Rivas, F. V., Marsden, C. G., Thomson, J. M., Song, J.-J., Hammond, S. M., Joshua-Tor, L. & Hannon, G. J. (2004). Argonaute2 is the catalytic engine of mammalian RNAi. *Science* **305**, 1437–1441.

263. Liu, Q., Singh, S. P. & Green, A. G. (2002). High-stearic and high-oleic cottonseed oils produced by hairpin RNA-mediated post-transcriptional gene silencing. *Plant Physiology* **129**, 1732–1743.

264. Llave, C., Kasschau, K., Rector, M. A. & Carrington, J. C. (2002a). Endogenous and silencing-associated small RNAs in plants. *Plant Cell* **14**, 1605–1619.

265. Llave, C., Kasschau, K. D. & Carrington, J. C. (2000). Virus-encoded suppressor of posttranscriptional gene silencing targets a maintenance step in the silencing pathway. *Proc. Natl. Acad. Sci. USA* **97**, 13401–13406.

266. Llave, C., Xie, Z. X., Kasschau, K. D. & Carrington, J. C. (2002). Cleavage of Scarecrow-like mRNA targets directed by a class of Arabidopsis miRNA. *Science* **297**, 2053–2056.

267. Lohmann, J. U., Endl, I. & Bosch, T. C. G. (1999). Silencing of developmental genes in *Hydra*. *Developmental Biology* **214**, 211–214.

268. Lomonossoff, G. P. (1995). Pathogen-derived resistance to plant viruses. *Annu. Rev. Phytopathol.* **33**, 323–343.

269. Longstaff, M., Brigneti, G., Boccard, F., Chapman, S. & Baulcombe, D. (1993). Extreme resistance to potato virus-X infection in plants expressing a modified component of the putative viral replicase. *EMBO J.* **12**, 379–386.

270. Loomis, W. F. (1993). Lateral inhibition and pattern-formation in Dictyostelium. *Curr. Topics Developmental Biology*, **28**, 1–46.

271. Lu, R., Martin-Hernandez, A. M., Peart, J. R., Malcuit, I. & Baulcombe, D. C. (2003). Virus-induced gene silencing in plants. *Methods* **30**, 296–303.

272. Lund, E., Guttinger, S., Calado, A., Dahlberg, J. E. & Kutay, U. (2004). Nuclear export of microRNA precursors. *Science* **303**, 95–98.

273. Luo, B., Heard, A. D. & Lodish, H. F. (2004). Small interfering RNA production by enzymatic engineering of DNA (Speed). *Proc. Natl. Acad. Sci. USA* **101**, 5494–5499.

274. Maeda, I., Kohara, Y., Yamamoto, M. & Sugimoto, A. (2001). Large-scale analysis of gene function in *Caenorhabditis elegans* by high-throughput RNAi. *Current Biology* **11**, 171–176.

275. Malhotra, P., Dasaradhi, P. V. N., Kumar, A., Mohmmed, A., Agrawal, N., Bhatnagar, R. K. & Chauhan, V. S. (2002). Double-stranded RNA-mediated gene silencing of cysteine proteases (falcipain-1 and -2) of *Plasmodium falciparum*. *Molecular Microbiology* **45**, 1245–1254.

276. Mallory, A., Dugas, D. V., Bartel, D. P. & Bartel, B. (2004a). MicroRNA regulation of NAC-Domain targets is required for proper formation and separation of adjacent embryonic, vegetative, and floral organs. *Current Biology* **14**, 1035–1046.

277. Mallory, A. C., Ely, L., Smith, T. H., Marathe, R., Anandalakshmi, R., Fagard, M., Vaucheret, H., Pruss, G., Bowman, L. & Vance, V. B. (2001). HC-Pro suppression of transgene silencing eliminates the small RNAs but not transgene methylation or the mobile signal. *Plant Cell* **13**, 571–583.

278. Mallory, A. C., Mlotshwa, S., Bowman, L. H. & Vance, V. B. (2003). The capacity of transgenic tobacco to send a systemic RNA silencing signal depends on the nature of the inducing transgene locus. *Plant J.* **35**, 82–92.

279. Mallory, A. C., Parks, G., Endres, M. W., Baulcombe, D., Bowman, L. H., Pruss, G. J. & Vance, V. B. (2002a). The amplicon-plus system for high-level expression of transgenes in plants. *Nature Biotechnology* **20**, 622–625.

280. Mallory, A. C., Reinhart, B. J., Bartel, D., Vance, V. B. & Bowman, L. H. (2002b). A viral suppressor of RNA silencing differentially regulates the accumulation of short interfering RNAs and micro-RNAs in tobacco. *Proc. Natl. Acad. Sci. USA* **99**, 15228–15233.

281. Mallory, A. C., Reinhart, B. J., Jones-Rohoads, M. W., Tang, G., Zamore, P. D., Barton, M. K. & Bartel, D. P. (2004b). MicroRNA control of *PHABULOSA* in leaf-development: Importance of pairing to microRNA 5' region. *EMBO J.* **23**, 3356–3364.

282. Mandelboim, M., Barth, S., Biton, M., Liang, X. H. & Michaeli, S. (2003). Silencing of Sm proteins in *Trypanosoma brucei* by RNA interference captured a novel cytoplasmic intermediate in spliced leader RNA biogenesis. *J. Biolog. Chem.* **278**, 51469–51478.

283. Marano, M. R. & Baulcombe, D. (1998). Pathogen-derived resistance targeted against the negative-strand RNA of tobacco mosaic virus: RNA strand-specific gene silencing? *Plant J.* **13**, 537–546.

284. Martens, H., Novotny, J., Oberstrass, J., Steck, T. L., Postlethwait, P. & Nellen, W. (2002). RNAi in *Dictyostelium*: The role of RNA-directed RNA polymerases and double-stranded RNase. *Mol. Biol. Cell* **13**, 445–453.

285. Martinek, S. & Young, M. W. (2000). Specific genetic interference with behavioral rhythms in Drosophila by expression of inverted repeats. *Genetics* **156**, 1717–1725.

286. Martinez, D. E., Dirksen, M. L., Bode, P. M., Jamrich, M., Steele, R. E. & Bode, H. R. (1997). Budhead, a fork head HNF-3 homologue, is expressed during axis formation and head specification in hydra. *Developmental Biology* **192**, 523–536.

287. Matsuda, T. & Cepko, C. L. (2004). Electroporation and RNA interference in the rodent retina *in vivo* and *in vitro*. *Proc. Natl. Acad. Sci. USA* **101**, 16–22.

288. Matzke, M. A. & Matzke, A. J. M. (1995). How and why do plants inactivate homologous (trans)genes. *Plant Physiology* **107**, 679–685.

289. Matzke, M. A., Matzke, A. J. M. & Kooter, J. M. (2001a). RNA: Guiding gene silencing. *Science* **293**, 1080–1083.

290. Matzke, M. A., Matzke, A. J. M., Pruss, G. J. & Vance, V. B. (2001b). RNA-based silencing strategies in plants. *Curr. Opin. Genetics Development* **11**, 221–227.

291. Matzke, M. A., Primig, M., Trnovsky, J. & Matzke, A. J. M. (1989). Reversible methylation and inactivation of marker genes in sequentially transformed tobacco plants. *EMBO J.* **8**, 643–649.

292. McCaffrey, A. P., Meuse, L., Pham, T.-T. T., Conklin, D. S., Hannon, G. J. & Kay, M. A. (2002). RNA interference in adult mice. *Nature* **418**, 38–39.

293. McConnell, J. R., Emery, J., Eshed, Y., Bao, N., Bowman, J. & Barton, M. K. (2001). Role of PHABULOSA and PHAVOLUTA in determining radial patterning in shoots. *Nature* **411**, 709–713.

294. McHale, N. A. & Koning, R. E. (2004). PHANTASTICA regulates development of the adaxial mesophyll in Nicotiana leaves. *Plant Cell* **16**, 1251–1262.

295. McManus, M. T., Haines, B. B., Dillon, C. P., Whitehurst, C. E., Van Parijs, L., Chen, J. & Sharp, P. A. (2002). Small interfering RNA-mediated gene silencing in T lymphocytes. *J. Immunology* **169**, 5754–5760.

296. McManus, M. T. & Sharp, P. A. (2002). Gene silencing in mammals by small interfering RNAs. *Nature Reviews Genetics* **3**, 737–747.

297. McRobert, L. & McConkey, G. A. (2002). RNA interference (RNAi) inhibits growth of *Plasmodium falciparum. Molec. Biochem. Parasitology* **119**, 273–278.

298. Meister, G., Landthaler, M., Dorsett, Y. & Tuschl, T. (2004). Sequence-specific inhibition of microRNA- and siRNA-induced RNA silencing. *RNA* **10**, 544–550.

299. Mette, M. F., Matzke, A. J. M. & Matzke, M. A. (2001). Resistance of RNA-mediated TGS to HC-Pro, a viral suppressor of PTGS, suggests alternative pathways for dsRNA processing. *Current Biology* **11**, 1119–1123.

300. Mette, M. F., Van der Winden, J., Matzke, M. A. & Matzke, A. J. M. (2002). Short RNAs can identify new candidate transposable element families in Arabidopsis. *Plant Physiology* **130**, 6–9.

301. Metzlaff, M., O'Dell, M., Cluster, P. D. & Flavell, R. B. (1997). RNA-mediated RNA degradation and chalcone synthase A silencing in Petunia. *Cell* **88**, 845–854.

302. Michael, M. Z., O'Connor, S. M., Pellekaan, N. G. V., Young, G. P. & James, R. J. (2003). Reduced accumulation of specific microRNAs in colorectal neoplasia. *Molecular Cancer Research* **1**, 882–891.

303. Mirelman, D., Galun, E., Sharon, N. & Lotan, R. (1975). Inhibition of fungal growth by wheat germ agglutinin. *Nature* **256**, 414–416.

304. Misquitta, L. & Paterson, B. M. (1999). Targeted disruption of gene function in *Drosophila* by RNA interference (RNA-i): A role for nautilus in embryonic somatic muscle formation. *Proc. Natl. Acad. Sci. USA* **96**, 1451–1456.

305. Mittal, V. (2004). Improving the efficiency of RNA interference in mammals. *Nature Reviews Genetics* **5**, 355–365.

306. Miyagishi, M. & Taira, K. (2002). U6 promoter-driven siRNAs with four uridine 3′ overhangs efficiently suppress targeted gene expression in mammalian cells. *Nature Biotechnology* **19**, 497–500.

307. Moazed, D. (2001). Common themes in mechanisms of gene silencing. *Molecular Cell* **8**, 489–498.

308. Mochizuki, K., Fine, N. A., Fujisawa, T. & Gorovsky, M. A. (2002). Analysis of a piwi-related gene implicates small RNAs in genome rearrangement in *Tetrahymena. Cell* **110**, 689–699.

309. Montgomery, M. K., Xu, S. & Fire, A. (1998). RNA as a target of double-stranded RNA-mediated genetic interference in *Caenorhabditis elegans*. *Proc. Natl. Acad. Sci. USA* **95**, 15502–15507.

310. Mourelatos, Z., Dostie, J., Paushkin, S., Sharma, A., Charroux, B., Abel, L., Rappsilber, J., Mann, M. & Dreyfuss, G. (2002). miRNPs: A novel class of ribonucleoproteins containing numerous microRNAs. *Genes & Development* **16**, 720–728.

311. Mourrain, P., Beclin, C., Elmayan, T., Feuerbach, F., Godon, C., Morel, J.-B., Jouette, D., Lacombe, A.-M. & Nikic, S. (2000). *Arabidopsis* SGS2 and SGS3 genes are required for posttranscriptional gene silencing and natural virus resistance. *Cell* **101**, 533–542.

312. Mousses, S., Caplen, N. J., Cornelison, M., Weaver, D., Basik, M., Hautaniemi, S., Elkahloun, A. G., Lotufo, R. A., Choudary, A., Dougherty, E. R., Suh, E. & Kallioniemi, O. (2003). RNAi microarray analysis in cultured mammalian cells. *Genome Research* **13**, 2341–2347.

313. Mueller, E., Gilbert, J., Davenport, G., Brigneti, G. & Baulcombe, D. C. (1995). Homology-dependent resistance — Transgenic virus-resistance in plants related to homology-dependent gene silencing. *Plant J.* **7**, 1001–1013.

314. Myers, J. W., Jones, J. T., Meyer, T. & Ferrell, J. E. J. (2003). Recombinant dicer efficiently converts large dsRNAs into siRNAs suitable for gene silencing. *Nature Biotechnology* **21**, 324–328.

315. Nakano, H., Amemiya, S., Shiokawa, K. & Taira, M. (2000). RNA interference for the organizer-specific gene Xlim-1 in *Xenopus* embryos. *Biochem. Biophys. Res. Comm.* **274**, 434–439.

316. Nakayashiki, H., Ikeda, K., Hashimoto, Y., Tosa, Y. & Mayama, S. (2001). Methylation is not the main force repressing the retrotransposon MAGGY in *Magnaporthe grisea*. *Nucleic Acids Research* **29**, 1278–1284.

317. Napoli, C., Lemieux, C. & Jorgensen, R. (1990). Introduction of a chimeric chalcone synthase gene into petunia results in reversible co-suppression of homologous genes in *trans*. *Plant Cell* **2**, 279–289.

318. Nellen, W., Datta, S., Reymond, C., Sivertsen, A., Mann, S., Crowley, T. & Firtel, R. A. (1987). Molecular biology in Dictyostelium: Tools and applications. In *Methods in Cell biology*, Vol. 28, pp. 67–100. Academic Press Inc., San Diego, USA.

319. Nelson, P. T., Hatzigeorgiou, A. G. & Mourelatos, Z. (2004). miRNP : mRNA association in polyribosomes in a human neuronal cell line. *RNA* **10**, 387–394.

320. Newmark, P. A. & Alvarado, A. S. (2002). Not your father's planarian: A classic model enters the era of functional genomics. *Nature Reviews Genetics* **3**, 210–219.

321. Newmark, P. A., Reddien, P. W., Cebria, F. & Alvarado, A. S. (2003). Ingestion of bacterially expressed double-stranded RNA inhibits gene expression in planarians. *Proc. Natl. Acad. Sci. USA* **100**, 11861–11865.

322. Ngo, H., Tschudi, C., Gull, K. & Ullu, E. (1998). Double-stranded RNA induces mRNA degradation in *Trypanosoma brucei*. *Proc. Natl. Acad. Sci. USA* **95**, 14687–14692.

323. Niblett, C. L., Dickson, E., Fernow, K. H., Horst, R. K. & Zaitlin, M. (1978). Cross protection among four viroids. *Virology* **91**, 198–203.

324. Nicolas, F. E., Torres-Martinez, S. & Ruiz-Vazquez, R. M. (2003). Two classes of small antisense RNAs in fungal RNA silencing triggered by non-integrative transgenes. *EMBO J.* **22**, 3983–3991.

325. Novotny, J., Diegel, S., Schirmacher, H., Mohrle, A., Hildebrandt, M., Oberstrass, J. & Nellen, W. (2001). Dictyostelium double-stranded ribonuclease. *Methods Enzymology* **342**, 193–212.

326. Nusslein-Volhard, C. (1994). Of flies and fishes. *Science* **266**, 572–574.

327. Nykanen, A., Haley, B. & Zamore, P. D. (2001). ATP requirements and small interfering RNA structure in the RNA interference pathway. *Cell* **107**, 309–321.

328. Oates, A. C., Bruce, A. E. E. & Ho, R. K. (2000). Too much interference: Injection of double stranded RNA does not have specific effects in the zebrafish embryo. *Developmental Biology* **222**, 227–227.

329. Ogawa, K., Kobayashi, C., Hayashi, T., Orii, H., Watanabe, K. & Agata, K. (2002). Planarian fibroblast growth factor receptor homologs expressed in stem cells and cephalic ganglions. *Development Growth & Differentiation* **44**, 191–204.

330. Ogita, S., Uefuji, H., Yamaguchi, Y., Koizumi, N. & Sano, H. (2003). RNA interference — Producing decaffeinated coffee plants. *Nature* **423**, 823–823.

331. Okamoto, H. & Hirochika, H. (2001). Silencing of transposable elements in plants. *Trends Plant Sciences* **16**, 527–534.

332. Ori, N., Eshed, Y., Chuck, G., Bowman, J. L. & Hake, S. (2000). Mechanisms that control knox gene expression in the Arabidopsis shoot. *Development* **127**, 5523–5532.

333. Orii, H., Mochii, M. & Watanabe, K. (2003). A simple "soaking method" for RNA interference in the planarian *Dugesia japonica*. *Dev. Genes Evol.* **213**, 138–141.

334. Paddison, P. J., Caudy, A. A., Bernstein, E., Hannon, G. J. & Conklin, D. S. (2002b). Short hairpin RNAs (shRNAs) induce sequence-specific silencing in mammalian cells. *Genes Development* **16**, 948–958.

335. Paddison, P. J., Caudy, A. A. & Hannon, G. J. (2002a). Stable suppression of gene expression by RNAi in mammalian cells. *Proc. Natl. Acad. Sci. USA* **99**, 1443–1448.

336. Paddison, P. J. & Hannon, G. J. (2002b). RNA interference: The new somatic cell genetics? *Cancer Cell* **2**, 17–23.

337. Paddison, P. J., Silva, J. M., Conklin, D. S., Schlabach, M., Li, M. M., Aruleba, S., Balija, V., O'Shaughnessy, A., Gnoj, L., Scobie, K., Chang, K., Westbrook, T., Cleary, M., Sachidanandam, R., McCombie, W. R., Elledge, S. J. & Hannon, G. J. (2004). A resource for large-scale RNA-interference-based screens in mammals. *Nature* **428**, 427–431.

338. Pal-Bhadra, M., Bhadra, U. & Birchler, J. A. (2002). RNAi related mechanisms affect both transcriptional and posttranscriptional transgene silencing in Drosophila. *Molecular Cell* **9**, 315–327.

339. Palatnik, J. F., Allen, E., Wu, X. L., Schommer, C., Schwab, R., Carrington, J. C. & Weigel, D. (2003). Control of leaf morphogenesis by microRNAs. *Nature* **425**, 257–263.

340. Palauqui, J.-C., Elmayan, T., Pollien, J.-M. & Vaucheret, H. (1997). Systemic acquired silencing: Transgene-specific post-transcriptional silencing is transmitted by grafting from silenced stocks to non-silenced scions. *EMBO J.* **16**, 4738–4745.

341. Palauqui, J.-C. & Vaucheret, H. (1998). Transgenes are dispensable for the RNA degradation step of cosuppression. *Proc. Natl. Acad. Sci. USA* **95**, 9675–9680.

342. Papaefthimiou, J., Hamilton, A. J., Denti, M. A., Baulcombe, D. C., Tsagris, M. & Tabler, M. (2001). Replicating potato spindle tuber viroid RNA is accompanied by short RNA fragments that are characteristic of post-transcriptional gene silencing. *Nucleic Acids Research* **29**, 2395–2400.

343. Papp, I., Mette, M. F., Aufsatz, W., Daxinger, L., Schauer, S. E., Ray, A., van der Winden, J., Matzke, M. & Matzke, A. J. M. (2003). Evidence for nuclear processing of plant microRNA and short interfering RNA precursors. *Plant Physiology* **132**, 1382–1390.

344. Park, W., Li, J., Song, R., Messing, J. & Chen, X. (2002). CARPEL FACTORY, a dicer homolog and HEN1, a novel protein, act in microRNA metabolism in *Arabidopsis thaliana*. *Current Biology* **12**, 1481–1495.

345. Parrish, S. & Fire, A. (2001). Distinct roles for RDE-1 and RDE-4 during RNA interference in *Caenorhabditis elegans*. *RNA* **7**, 1397–1402.

346. Parrish, S., Fleenor, J., Xu, S., Mello, C. & Fire, A. (2000). Functional anatomy of a dsRNA trigger: Differential requirement for the two trigger strands in RNA interference. *Molecular Cell* **6**, 1077–1087.

347. Pasquinelli, A., Reinhart, B. J., Slack, F., Martindale, M. Q., Kuruda, M. I., Maller, B., Hayward, D. C., Ball, E. E., Degnan, B., Muller, P., Spring, J., Srinivasan, A., Fishman, M., Finnerty, J., Coro, J., Levine, M., Leahy, P., Davidson, E. & Ruvkun, G. (2000). Conservation of the sequence and temporal expression of *let-7* heterochronic regulatory RNA. *Nature* **408**, 86–89.

348. Peele, C., Jordan, C. V., Muangsan, N., Turnage, M., Egelkrout, E., Eagle, P., Hanley-Bowdoin, L. & Robertson, D. (2001). Silencing of a meristematic gene using geminivirus-derived vectors. *Plant J.* **27**, 357–366.

349. Pekarik, V., Bourikas, D., Miglino, N., Joset, P., Preiswerk, S. & Stoeckli, E. T. (2003). Screening for gene function in chicken embryo using RNAi and electroporation. *Nature Biotechnology* **21**, 93–96.

350. Pfeffer, S., Dunoyer, P., Heim, F., Richards, K. E., Jonard, G. & Ziegler-Graff, V. (2002). P0 of beet western yellows virus is a suppressor of posttranscriptional gene silencing. *J. Virology* **76**, 6815–6824.

351. Pfeifer, A., Ikawa, M., Dayn, Y. & Verma, I. M. (2002). Transgenesis by lentiviral vectors: Lack of gene silencing in mammalian embryonic stem cells and preimplantation embryos. *Proc. Natl. Acad. Sci. USA* **99**, 2140–2145.

352. Pickford, A. S., Catalanotto, C., Cogoni, C. & Macino, G. (2002). Quelling in *Neurospora crassa*. *Advances in Genetics* **46**, 277–303.

353. Pineda, D., Gonzalez, J., Callaerts, P., Ikeo, K., Gehring, W. J. & Salo, E. (2000). Searching for the prototypic eye genetic network: Sine oculis is essential for eye regeneration in planarians. *Proc. Natl. Acad. Sci. USA* **97**, 4525–4529.

354. Plasterk, R. H. A. (2002). RNA silencing: The genome's immune system. *Science* **296**, 1263–1265.

355. Plasterk, R. H. A. & Ketting, R. F. (2000). The silence of the genes. *Curr Opin. Genet. Development* **10**, 562–567.

356. Prescott, D. M. (1994). The DNA of Ciliated Protozoa. *Microbiological Reviews* **58**, 233–267.

357. Provost, P., Dishart, D., Doucet, J., Frendewey, D., Samuelsson, B. & Radmark, O. (2002). Ribonuclease activity and RNA binding of recombinant human Dicer. *EMBO J.* **21**, 5864–5874.

358. Pruss, G., Ge, X., Shi, X. M., Carrington, J. C. & Vance, V. B. (1997). Plant viral synergism: The potyviral genome encodes a broad-range pathogenicity enhancer that transactivates replication of heterologous viruses. *Plant Cell* **9**, 859–868.

359. Pruss, G. J., Lawrence, C. B., Bass, T., Li, Q. Q., Bowman, L. H. & Vance, V. (2004). The potyviral suppressor of RNA silencing confers enhanced resistance to multiple pathogens. *Virology* **320**, 107–120.

360. Qu, F. & Morris, T. J. (2002). Efficient infection of *Nicotiana benthamiana* by tomato bushy stunt virus is facilitated by the coat protein and maintained by p19 through suppression of gene silencing. *Molecular Plant-Microbe Interactions* **15**, 193–202.

361. Qu, F., Ren, T. & Morris, T. J. (2003). The coat protein of turnip crinkle virus suppresses posttranscriptional gene silencing at an early initiation step. *J. Virology* **77**, 511–522.

362. Que, Q. & Jorgensen, R. A. (1998). Homology-based control of gene expression patterns in transgenic petunia flowers. *Develop. Genetics* **22**, 100–109.

363. Ratcliff, F., Harrison, B. D. & Baulcome, D. C. (1997). A similarity between viral defense and gene silencing in plants. *Science* **276**, 1558–1560.

364. Ratcliff, F. G., MacFarlane, S. A. & Baulcombe, D. C. (1999). Gene silencing without DNA: RNA-mediated cross-protection between viruses. *Plant Cell* **11**, 1207–1215.

365. Raven, P. H. & Johnson, G. B. (1996). *Biology*, Wm. C. Brown Publishers, Dubuque, IA, USA.

366. Reed, J. C., Kasschau, K. D., Prokhnevsky, A. I., Gopinath, K., Pogue, G. P., Carrington, J. C. & Dolja, V. V. (2003). Suppressor of RNA silencing encoded by beet yellows virus. *Virology* **306**, 203–209.

367. Reinhart, B. J., Slack, F., Basson, M., Pasquinelli, A. E., Bettinger, J. C., Rougvie, A. E., Horvitz, H. R. & Ruvkun, G. (2000). The 21-nucleotide *let-7* RNA regulates developmental timing in *Caenorhabditis elegans*. *Nature* **403**, 901–906.

368. Reinhart, B. J., Weinstein, E. G., Rhoades, M. W., Bartel, B. & Bartel, D. P. (2002). MicroRNAs in plants. *Genes & Development* **16**, 1616–1626.

369. Rhoades, M. W., Reinhart, B. J., Lim, L. P., Burge, C. B., Bartel, B. & Bartel, D. P. (2002). Prediction of plant microRNA targets. *Cell* **110**, 513–520.

370. Robinson, K. A. & Beverley, S. M. (2003). Improvements in transfection efficiency and tests of RNA interference (RNAi) approaches in the protozoan parasite *Leishmania*. *Molecular and Biochemical Parasitology* **128**, 217–228.

371. Romano, N. & Macino, G. (1992). Quelling: Transient inactivation of gene expression in *Neurospora crassa* by transformation with homologous sequences. *Molecular Microbiology* **6**, 3343–3353.

372. Ron, M. & Avni, A. (2004). The receptor for the fungal elicitor ethylene-inducing xylanase is a member of a resistance-like gene family in tomato. *Plant Cell* **16**, 1604–1615.

373. Ross, J. M. & Zarkower, D. (2003). Polycomb group regulation of Hox gene expression in C-elegans. *Developmental Cell* **4**, 891–901.

374. Roth, B. M., Pruss, G. J. & Vance, V. B. (2004). Plant viral suppressors of RNA silencing. *Virus Research* **102**, 97–108.

375. Rubinson, D. A., Dillon, C. P., Kwiatkowski, A. V., Sieversm, C., Yang, L., Kopinja, J., Zhang, M., McManus, M. T., Gertler, F. B., Scott, M. L. & Van Parijs, L. (2003). A lentivirus-based system to functionally silence genes in primary mammalian cells, stem cells and transgenic mice by RNA interference. *Nature Genetics* **33**, 401–406.

376. Rudel, D. & Sommer, R. J. (2003). The evolution of developmental mechanisms. *Developmental Biology* **264**, 15–37.

377. Ruiz, F., Vayssie, L., Klotz, C., Sperling, L. & Madeddu, L. (1998a). Homology-dependent gene silencing in *Paramecium*. *Mol. Biol. Cell* **9**, 931–943.

378. Ruiz, M. T., Voinnet, O. & Baulcombe, D. C. (1998b). Initiation and maintenance of virus-induced gene silencing. *Plant Cell* **10**, 937–946.

379. Ruvkun, G. (2001). Glimpses of a tiny RNA world. *Science* **294**, 797–799.

380. Ruvkun, G., Wightman, B. & Ha, I. (2004). The 20 years it took to recognize the importance of tiny RNAs. *Cell* S**116**, S93–S96.

381. Sanchez Alvarado, A. & Newmark, P. A. (1999). Double-stranded RNA specifically disrupts gene expression during planarian regeneration. *Proc. Natl. Acad. Sci. USA* **96**, 5049–5054.

382. Savenkov, E. I. & Valkonen, J. P. T. (2002). Silencing of a viral RNA silencing suppressor in transgenic plants. *J. General Virology* **83**, 2325–2335.

383. Saxena, S., Jonsson, Z. O. & Dutta, A. (2003). Small RNAs with imperfect match to endogenous mRNA repress translation — Implications for off-target activity of small inhibitory RNA in mammalian cells. *J. Biol. Chem.* **278**, 44312–44319.

384. Scacheri, P. C., Rozenblatt-Rosen, O., Caplen, N. J., Wolfsberg, T. G., Umayam, L., Lee, J. C., Hughes, C. M., Shanmugam, K. S., Bhattacharjee, A., Meyerson, M. & Collins, F. S. (2004). Short interfering RNAs can induce unexpected and divergent changes in the levels of untargeted proteins in mammalian cells. *Proc. Natl. Acad. Sci. USA* **101**, 1892–1897.

385. Scheid, O. M., Paszkowski, J. & Potrykus, I. (1991). Reversible inactivation of a transgene in *Arabidopsis thaliana*. *Molecular & General Genetics* **228**, 104–112.

386. Schiebel, W., Pelissier, T., Riedel, L., Thalmeir, S., Schiebel, R., Kempe, D., Lottspeich, F., Sager, H. L. & Wassenegger, M. (1998). Isolation of an RNA-directed RNA polymerase-specific cDNA clone from tomato leaf tissue mRNA. *Plant Cell* **10**, 2087–2101.

387. Schmid, A., Schindelholz, B. & Zinn, K. (2002). Combinatorial RNAi: A method for evaluating the functions of gene families in *Drosophila*. *Trends Neurosciences* **25**, 71–74.

388. Schwarz, D. S., Hutvagner, G., Du, T., Xu, Z. S., Aronin, N. & Zamore, P. D. (2003). Asymmetry in the assembly of the RNAi enzyme complex. *Cell* **115**, 199–208.

389. Schwarz, D. S., Hutvagner, G., Haley, B. & Zamore, P. D. (2002). Evidence that siRNAs function as guides, not primers, in the *Drosophila* and human RNAi pathways. *Molecular Cell* **10**, 537–548.

390. Schwarz-Sommer, Z., Huijser, P., Nacken, W., Saedler, H. & Sommer, H. (1990). Genetic-control of flower development by homeotic genes in *Antirrhinum majus*. *Science* **250**, 931–936.

391. Schweizer, P., Pokorny, J., Schulze-Lefert, P. & Dudler, R. (2000). Double-stranded RNA interferes with gene function at the single-cell level in cereals. *Plant J.* **24**, 895–903.

392. Segal, G., Song, R. T. & Messing, J. (2003). A new opaque variant of maize by a single dominant RNA-interference-inducing transgene. *Genetics* **165**, 387–397.

393. Selker, E. U. (2002). Repeat-induced gene silencing in fungi. *Advances in Genetics* **46**, 439–450.

394. Selker, E. U., Cambareri, E. B., Jensen, B. C. & Haack, K. R. (1987). Rearrangement of duplicated DNA in specialized cells of *Neurospora*. *Cell* **51**, 741–752.

395. Selker, E. U. & Garrett, P. W. (1988). DNA sequence duplicatons trigger gene inactivation in *Neurospora crassa*. *Proc. Natl. Acad. Sci. USA* **85**, 6870–6874.

396. Semizarov, D., Frost, L., Sarthy, A., Kroeger, P., Halbert, D. N. & Fesik, S. W. (2003). Specificity of short interfering RNA determined through gene expression signatures. *Proc. Natl. Acad. Sci. USA* **100**, 6347–6352.

397. Sempere, L. F., Sokol, N. S., Dubrovsky, E. B., Berger, E. M. & Ambros, V. (2003). Temporal regulation of microRNA expression in *Drosophila melanogaster* mediated by hormonal signals and broad-complex gene activity. *Developmental Biology* **259**, 9–18.

398. Sen, G., Wehrman, T. S., Myers, J. W. & Blau, H. M. (2004). Restriction enzyme-generated siRNA (REGS) vectors and libraries. *Nature Genetics* **36**, 183–189.

399. Sharp, P. A. (1999). RNAi and double-stranded RNA. *Genes & Development* **13**, 139–141.

400. Sharp, P. A. (2001). RNA interference — 2001. *Genes & Development* **15**, 485–490.

401. Sherwood, J. L. (1987). Mechanisms of cross-protection between plant virus strains. *Ciba Foundation Symp.* **133**, 136–150.

402. Shi, H., Dijkeng, A., Tschudi, C. & Ullu, E. (2004). Argonaute protein in the early divergent eukaryote *Trypanosoma brucei*: Control of small interfering RNA accumulation and retroposon retrocirpt abundance. *Molec. Cell. Biology* **24**, 420–427.

403. Shi, H., Djikeng, A., Mark, T., Wirtz, E., Tschudi, C. & Ullu, E. (2000). Genetic interference in *Trypanosoma brucei* by heritable and inducible double-stranded RNA. *RNA* **6**, 1069–1076.

404. Shi, X. M., Miller, H., Verchot, J., Carrington, J. C. & Vance, V. B. (1997). Mutations in the region encoding the central domain of helper component-proteinase (HC-Pro) eliminate potato virus X/potyviral synergism. *Virology* **231**, 35–42.

405. Shi, Y. (2003). Mammalian RNAi for the masses. *Trends Genetics* **19**, 9–12.

406. Shinagawa, T. & Ishii, S. (2003). Generation of Ski-knockdown mice by expressing a long double-strand RNA from an RNA polymerase II promoter. *Genes & Development* **17**, 1340–1345.

407. Shirane, D., Sugao, K., Namiki, S., Tanabe, M., Iino, M. & Hirose, K. (2004). Enzymatic production of RNAi libraries form cDNAs. *Nature Genetics* **36**, 190–196.

408. Shiu, P. K. T. & Metzenberg, R. L. (2002). Meiotic silencing by unpaired DNA: Properties, regulation and suppression. *Genetics* **161**, 1483–1495.

409. Shiu, P. K. T., Raju, N. B., Zickler, D. & Metzenberg, R. L. (2001). Meiotic silencing by unpaired DNA. *Cell* **107**, 905–916.

410. Sijen, T., Fleenor, J., Simmer, F., Thijssen, K. L., Parrish, S., Timmons, L. & Plasterk, R. H. A. (2001). On the role of RNA amplification in dsRNA-triggered gene silencing. *Cell* **107**, 465–476.

411. Sijen, T. & Plasterk, R. H. A. (2003). Transposon silencing in *Caenohabditis elegans* germ line by natural RNAi. *Nature* **426**, 310–314.

412. Silhavy, D., Molnar, A., Lucioli, A., Szittya, G., Hornyik, C., Tavazza, M. & Burgyan, J. (2002). A viral protein suppresses RNA silencing and binds silencing-generated, 21- to 25-nucleotide double-stranded RNAs. *EMBO J.* **21**, 3070–3080.

413. Silva, J. M., Mizuno, H., Brady, A., Lucito, R. & Hannon, G. J. (2004). RNA interference microarrays: High-throughput loss-of-function genetics in mammalian cells. *Proc. Natl. Acad. Sci. USA* **101**, 6548–6552.

414. Silvarolla, M. B., Mazzafera, P. & Fazuoli, L. C. (2004). Plant biochemistry — A naturally decaffeinated arabica coffee. *Nature* **429**, 826–826.

415. Simmer, F., Tijsterman, M., Parrish, S., Koushika, S. P., Nonet, M. L., Fire, A., Ahringer, J. & Plasterk, R. H. A. (2002). Loss of the putative RNA-directed RNA polymerase RRF-3 makes *C. elegans* hypersensitive to RNAi. *Current Biology* **12**, 1317–1319.

416. Simon, J. A. & Tamkun, J. W. (2002). Programming off and on states in chromatin: Mechanisms of polycomb and trithorax group complexes. *Curr. Opin. Genet. Dev.* **12**, 210–218.

417. Singer, O., Yanai, A. & Verma, I. M. (2004). Silence of the genes. *Proc. Natl. Acad. Sci. USA* **101**, 5313–5314.

418. Slack, F., Basson, M., Liu, Z., Ambros, V., Horvitz, H. R. & Ruvkun, G. (2000). The *lin-41* RBCC gene acts in the *C. elegans* heterochronic pathway between the *let-7* regulatory RNA and the LIN-29 transcription factor. *Molecular Cell* **5**, 659–669.

419. Slack, F. & Ruvkun, G. (1997). Temporal pattern formation by heterochronic genes. *Annu. Rev. Genet.* **31**, 611–634.

420. Smardon, A., Spoerke, J. M., Stacey, S. C., Klein, M. E., Mackin, N. & Maine, E. M. (2000). EGO-1 is related to RNA-directed RNA polymerase and functions in germ-line development and RNA interference in *C. elegans*. *Current Biology* **10**, 169–178.

421. Smith, C. J. S., Watson, C. F., Bird, C. R., Ray, J., Schuch, W. & Grierson, D. (1990). Expression of a truncated tomato polygalacturonase gene inhibits expression of the endogenous gene in transgenic plants. *Molecular Gen. Genet.* **224**, 477–481.

422. Smith, H. A., Swaney, S. L., Parks, T. D., Wernsman, E. A. & Dougherty, W. G. (1994). Transgenic plant-virus resistance mediated by untranslatable sense RNAs — Expression, regulation, and fate of nonessential RNAs. *Plant Cell* **6**, 1441–1453.

423. Smith, N. A., Singh, S. P., Wang, M.-B., Stoutjesdijk, P. A., Green, A. G. & Waterhouse, P. M. (2000). Total silencing by intron-spliced hairpin RNAs. *Nature* **407**, 319–320.

424. Snow, M. & Snow, R. (1959). The dorsoventrality of leaf primordia. *New Phytology* **58**, 188–207.

425. Sogin, M. L., Gunderson, J. H., Elwood, H. J., Alonso, R. A. & Peattie, D. A. (1989). Phylogenetic meaning of the kingdom concept — An unusual ribosomal-RNA from *Giardia lamblia*. *Science* **243**, 75–77.

426. Somma, M. P., Fasulo, B., Cenci, G., Cundari, E. & Gatti, M. (2002). Molecular dissection of cytokinesis by RNA interference in Drosophila cultured cells. *Molecular Biol. Cell* **13**, 2448–2460.

427. Song, E., Lee, S.-K., Wang, J., Ince, N., Ouyang, N., Min, J., Chen, J., Shankar, P. & Lieberman, J. (2003). RNA interference targeting Fas protects mice from fulminant hepatitis. *Nature Medicine* **9**, 347–351.

428. Song, J.-J., Smith, S. K., Hannon, G. J. & Joshua-Tor, L. (2004). Crystal structure of Argonaute and its implications for RISC slicer activity. *Science* **305**, 1434–1437.

429. Sontheimer, E. J. & Carthew, R. W. (2004). Argonaute journeys into the heart of RISC. *Science* **305**, 1409–1410.

430. Stein, P., Svoboda, P., Anger, M. & Schultz, R. M. (2003a). RNAi: Mammalian oocytes do it without RNA-dependent RNA polymerase. *RNA* **9**, 187–192.

431. Stein, P., Svoboda, P. & Schultz, R. M. (2003b). Transgenic RNAi in mouse oocytes: A simple and fast approach to study gene function. *Developmental Biology* **256**, 187–193.

432. Stein, P., Svoboda, P., Stumpo, D. J., Blackshear, P. J., Lombard, D. B., Johnson, B. & Schultz, R. M. (2002). Analysis of the role of RecQ helicases in RNAi in mammals. *Biochem. Biophys. Res. Comm.* **291**, 1119–1122.

433. Stoeckli, E. T. & Landmesser, L. T. (1998). Axon guidance at choice points. *Curr. Opin. Neurobiology* **8**, 73–79.

434. Storz, G., Opdyke, J. A. & Zhang, A. X. (2004). Controlling mRNA stability and translation with small, noncoding RNAs. *Curr. Opin. Microbiology* **7**, 140–144.

435. Stoutjesdijk, P. A., Singh, S. P., Liu, Q., Hurlstone, C. J., Waterhouse, P. A. & Green, A. G. (2002). hpRNA-mediated targeting of the Arabidopsis FAD2 gene gives highly efficient and stable silencing. *Plant Physiology* **129**, 1723–1731.

436. Streisinger, G., Walker, C., Dower, N., Knauber, D. & Singer, F. (1981). Production of clones of homozygous diploid zebra fish (*Brachydanio rerio*). *Nature* **291**, 293–296.

437. Sui, G., Soohoo, C., Affar, E. B., Frederique, G., Shi, Y., Forrester, W. C. & Shi, Y. (2002). A DNA vector-based RNAi technology to suppress gene expression in mammalian cells. *Proc. Natl. Acad. Sci. USA* **99**, 5515–5520.

438. Sullivan, B. A., Blower, M. D. & Karpen, G. H. (2001). Determining centromere identity: Cyclical stories and forking paths. *Nature Reviews Genetics* **2**, 584–596.

439. Sulston, J. E., Schierenberg, E., White, J. G. & Thomson, J. N. (1983). The embryonic-cell lineage of the nematode *Caenorhabditis elegans*. *Developmental Biology* **100**, 64–119.

440. Sussex, I. M. (1954). Experiments on the cause of dorsoventrality in leaves. *Nature* **174**, 351–352.

441. Svoboda, P., Stein, P., Hayashi, H. & Schultz, R. M. (2000). Selective reduction of dormant maternal mRNAs in mouse oocytes by RNA interference. *Development* **127**, 4147–4156.

442. Svoboda, P., Stein, P. & Schultz, R. M. (2001). RNAi in mouse oocytes and preimplantation embryos: Effectiveness of hairpin dsRNA. *Biochem. Biophys. Res. Comm.* **287**, 1099–1104.

443. Tabara, H., Grishok, A. & Mello, C. C. (1998). RNAi in *C. elegans*: Soaking in the genome sequence. *Science* **282**, 430–431.

444. Tabara, H., Sarkissian, M., Kelly, W. G., Fleenor, J., Grishok, A., Timmons, L., Fire, A. & Mello, C. C. (1999). The rde-1 gene, RNA interference, and transposon silencing in *C. elegans*. *Cell* **99**, 123–132.

445. Tabara, H., Yuigit, E., Siomi, H. & Mello, C. C. (2002). The dsRNA binding protein RDE-4 interacts with RDE-1, DCR-1, and a DExH-bod helicase to direct RNAi in *C. elegans*. *Cell* **109**, 861–871.

446. Takeda, A., Sugiyama, K., Nagano, H., Mori, M., Kaido, M., Mise, K., Tsuda, S. & Okuno, T. (2002). Identification of a novel RNA silencing suppressor, NSs protein of tomato spotted wilt virus. *FEBS Letters* **532**, 75–79.

447. Tang, G. & Galili, G. (2004). Using RNAi to improve plant nutritional value: From mechanism to application. *Trends Biotechnology* **22**, 463–469.

448. Tang, G., Reinhart, B. J., Bartel, D. P. & Zamore, P. D. (2003). A biochemical framework for RNA silencing in plants. *Genes & Development* **17**, 49–63.

449. Tanzer, M. M., Thompson, W. F., Law, M. D., Wernsman, E. A. & Uknes, S. (1997). Characterization of post-transcriptionally suppressed transgene expression that confers resistance to tobacco etch virus infection in tobacco. *Plant Cell* **9**, 1411–1423.

450. Taverna, S. D., Coyne, R. S. & Allis, C. D. (2002). Methylation of histone H3 at lysine 9 targets programmed DNA elimination in *Tetrahymena*. *Cell* **110**, 701–711.

451. Taylor, J. M. (2003). Replication of human hepatitis delta virus: Recent developments. *Trends Microbiology* **11**, 185–190.

452. Tazaki, A., Kato, K., Orii, H., Agata, K. & Watanabe, K. (2002). The body margin of the planarian *Dugesia japonica*: Characterization by the expression of an intermediate filament gene. *Develop. Genes & Evolution* **212**, 365–373.

453. Technau, U. & Bode, H. R. (1999). HyBra1, a Brachyury homologue, acts during head formation in Hydra. *Development* **126**, 999–1010.

454. Tenllado, F., Llave, C. & Diaz-Ruiz, J. R. (2004). RNA interference as a new biotechnological tool for the control of virus diseases in plants. *Virus Research* **102**, 85–96.

455. Theissen, G., Becker, A., Di Rosa, A., Kanno, A., Kim, J. T., Munster, T., Winter, K.-U. & Saedler, H. (2000). A short history of MADS-box genes in plants. *Plant Molecular Biology* **42**, 115–149.

456. Thomas, C. L., Leh, V., Lederer, C. & Maule, A. J. (2003). Turnip crinkle virus coat protein mediates suppression of RNA silencing in *Nicotiana benthamiana*. *Virology* **306**, 33–41.

457. Tijsterman, M., Ketting, R. F. & Plasterk, R. H. A. (2002). The genetics of RNA silencing. *Annu. Rev. of Genetics* **36**, 489–519.

458. Tijsterman, M., May, R. C., Simmer, F., Okihara, K. L. & Plasterk, R. H. A. (2004). Genes required for systemic RNA interference in *Caenorhabditis elegans*. *Current Biology* **14**, 111–116.

459. Timmons, L., Court, D. L. & Fire, A. (2001). Ingestion of bacterially expressed dsRNAs can produce specific and potent genetic interference in *Caenorhabditis elegans*. *Gene* **263**, 103–112.

460. Timmons, L. & Fire, A. (1998). Specific interference by ingested dsRNA. *Nature* **395**, 854–854.

461. Tiscornia, G., Singer, O., KIkawa, M. & Verma, I. M. (2003). A general method for gene knockdown in mice by using lentiviral vectors expressing small interfering RNA. *Proc. Natl. Acad. Sci. USA* **100**, 1844–1848.

462. Torrey, J. G. & Galun, E. (1970). Apolar embryos of Fucus resulting from osmotic and chemical treatment. *American J. Botany* **57**, 111–119.

463. Trewavas, A. (2003). Aspects of plant intelligence. *Annals Botany* **92**, 1–20.
464. Trompouki, E., Hatzivassiliou, E., Tsichritzis, T., Farmer, H., Ashworth, A. & Mosialos, G. (2003). CYLD is a deubiquitinating enzyme that negatively regulates NF-kappa B activation by TNFR family members. *Nature* **424**, 793–796.
465. Turing, A. (1952). The chemical basis of differentiation. *Phil. Trans. R. Soc. B.* **237**, 37–72.
466. Tuschl, T., Zamore, P. D., Lehmann, R., Bartel, D. P. & Sharp, P. A. (1999). Targeted mRNA degradation by double-stranded RNA *in vitro*. *Genes & Development* **13**, 3191–3197.
467. Ui-Tei, K., Naito, Y., Takahashi, F., Haraguchi, T., Ohki-Hamazaki, H., Juni, A., Ueda, R. & Saigo, K. (2004). Guidelines for the selection of highly effective siRNA sequences for mammalian and chick RNA interference. *Nucleic Acids Research* **32**, 936–948.
468. Ui-Tei, K., Zenno, S. H., Miyata, Y. & Saigo, K. (2000). Sensitive assay of RNA interference in Drosophila and Chinese hamster cultured cells using firefly luciferase gene as target. *FEBS Letters* **479**, 79–82.
469. Ullu, E. & Tschudi, C. (2003). RNA interference in *Trypanosoma brucei* and other non-classical model organisms. In *RNAi — A Guide to Gene Silencing* (Hannon, G. J., ed.), pp. 401–412, Cold Spring Harbor Laboratory Press, Cold Spring Harbor, NY.
470. Vaistij, F. E., Jones, L. & Baulcombe, D. C. (2002). Spreading of RNA targeting and DNA methylation in RNA silencing requires transcription of the target gene and a putative RNA-dependent RNA polymerase. *Plant Cell* **14**, 857–867.
471. Van Blokland, R., Van der Geest, N., Mol, J. N. M. & Kooter, J. M. (1994). Transgene-mediated suppression of chalcone synthase expression in *Petunia hybrida* results from an increase in RNA turnover. *Plant J.* **6**, 861–877.
472. van der Krol, A., Mur, L. A., Beld, M., Mol, J. N. M. & Stuitje, A. R. (1990). Flavonoid genes in petunia: Addition of a limited number of gene copies may lead to a suppression of gene expression. *Plant Cell* **2**, 291–299.
473. Van West, P., Kamoun, S., Van Klooster, J. W. & Govers, F. (1999). Internuclear gene silencing in *Phytophthora infestans*. *Molecular Cell* **3**, 339–348.
474. van Wezel, R., Dong, X. L., Liu, H. T., Tien, P., Stanley, J. & Hong, Y. G. (2002). Mutation of three cysteine residues in tomato yellow leaf curl virus-China C2 protein causes dysfunction in pathogenesis and posttranscriptional gene-silencing suppression. *Molecular Plant-Microbe Interactions* **15**, 203–208.
475. Vance, V. & Vaucheret, H. (2001). RNA silencing in plants — Defense and counterdefense. *Science* **292**, 2277–2280.
476. Vance, V. B. (1991). Replication of potato virus-X RNA is altered in coinfections with potato virus-Y. *Virology* **182**, 486–494.
477. Vance, V. B., Berger, P. H., Carrington, J. C., Hunt, A. G. & Shi, X. M. (1995). 5′-proximal potyviral sequences mediate potato-virus-X potyviral synergistic disease in transgenic tobacco. *Virology* **206**, 583–590.
478. Vanitharani, R., Chellappan, P. & Fauquet, C. M. (2003). Short interfering RNA-mediated interference of gene expression and viral DNA accumulation in cultured plant cells. *Proc. Natl. Acad. Sci. USA* **100**, 9632–9636.
479. Vargason, J. M., Szittya, G., Burgyan, J. & Hall, T. M. T. (2003). Size selective recognition of siRNA by an RNA silencing suppressor. *Cell* **115**, 799–811.

480. Vastenhouw, N. L., Fischer, S. E. J., Robert, V. J. P., Thijssen, K. L., Fraser, A. G., Kamath, R. S. & Ahringer, J. (2003). A genome-wide screen identifies 27 genes involved in transposon silencing in *C. elegans*. *Current Biology* **13**, 1311–1316.

481. Vaucheret, H., Beclin, C. & Farard, M. (2001). Post-transcriptional gene silencing in plants. *J. Cell Science* **114**, 3083–3091.

482. Vaucheret, H. & Fagard, M. (2001). Transcriptional gene silencing in plants: Targets, inducers and regulators. *Trends Genetics* **17**, 29–35.

483. Voinnet, O. (2001). RNA silencing as a plant immune system against viruses. *Trends Genetics* **17**, 449–459.

484. Voinnet, O. (2002). RNA silencing: Small RNAs as ubiquitous regulators of gene expression. *Curr. Opin. Plant Biology* **5**, 444–451.

485. Voinnet, O. & Baulcombe, D. C. (1997). Systemic signalling in gene silencing. *Nature* **389**, 553–553.

486. Voinnet, O., Lederer, C. & Baulcombe, D. C. (2000). A viral movement protein prevents spread of the gene silencing signal in *Nicotiana benthamiana*. *Cell* **103**, 157–167.

487. Voinnet, O., Pinto, Y. M. & Baulcombe, D. C. (1999). Suppression of gene silencing: A general strategy used by diverse DNA and RNA viruses of plants. *Proc. Natl. Acad. Sci. USA* **96**, 14147–14152.

488. Voinnet, O., Rivas, S., Mestre, P. & Baulcombe, D. (2003). An enhanced transient expression system in plants based on suppression of gene silencing by the p19 protein of tomato bushy stunt virus. *Plant J.* **33**, 949–956.

489. Voinnet, O., Vain, P., Angell, S. & Baulcombe, D. C. (1998). Systemic spread of sequence-specific transgene RNA degradation in plants is initiated by localized introduction of ectopic promoterless DNA. *Cell* **95**, 177–187.

490. Volpe, T. A., Kidner, C., Hall, I. M., Teng, G., Grewal, S. I. S. & Martienssen, R. A. (2002). Regulation of heterochromatic silencing and histone H3 lysine-9 methylation by RNAi. *Science* **297**, 1833–1837.

491. Waites, R. & Hudson, A. (1995). Phantastica — A gene required for dorsoventrality of leaves in *Antirrhinum majus*. *Development* **121**, 2143–2154.

492. Waites, R., Selvadurai, H. R. N., Oliver, I. R. & Hudson, A. (1998). The PHANTASTICA gene encodes a MYB transcription factor involved in growth and dorsoventrality of lateral organs in *Antirrhinum*. *Cell* **93**, 779–789.

493. Wang, J., Tekle, E., Oubrahim, H., Mieyal, J. J., Stadtman, E. R. & Chock, P. B. (2003). Stable and controllable RNA interference: Investigating the physiological function of glutathionylated actin. *Proc. Natl. Acad. Sci. USA* **100**, 5103–5106.

494. Wang, M.-B., Abbott, D. C. & Waterhouse, P. M. (2000a). A single copy of a virus-derived transgene encoding hairpin RNA gives immunity to barley yellow dwarf virus. *Molecular Plant Pathology* **1**, 347–356.

495. Wang, M.-B. & Waterhouse, P. M. (2001). Application of gene silencing in plants. *Curr. Opin. Plant Biology* **5**, 146–150.

496. Wang, Z., Morrris, C., Drew, M. & Englund, P. T. (2000b). Inhibition of *Trypanosoma brucei* gene expression by RNA interference using an integratable vector with opposing T7 promoters. *J. Biol. Chem.* **275**, 40174–40179.

497. Wargelius, A., Ellingsen, S. & Fjose, A. (1999). Double-stranded RNA induces specific developmental defects in zebrafish embryos. *Biochem. Biophys. Res. Comm.* **263**, 156–161.

498. Wassenegger, M., Heimes, S., Riedel, L. & Sanger, H. L. (1994). RNA-directed de-novo methylation of genomic sequences in plants. *Cell* **76**, 567–576.

499. Wassenegger, M. & Pelissier, T. (1998). A model for RNA-mediated gene silencing in higher plants. *Plant Molecular Biology* **37**, 349–362.

500. Waterhouse, P. M., Graham, M. W. & Wang, M.-B. (1998). Virus resistance and gene silencing in plants can be induced by simultaneous expression of sense and antisense RNA. *Proc. Natl. Acad. Sci. USA* **95**, 13959–13964.

501. Waterhouse, P. M., Wang, M.-B. & Lough, T. (2001). Gene silencing as an adaptive defence against viruses. *Nature* **411**, 834–842.

502. Watson, J. D. & Crick, H. C. (1953). Molecular structure of nucleic acids: A structure for deoxynucleic acids. *Nature* **171**, 737–738.

503. Weigel, D. & Jurgens, G. (2002). Stem cells that make stems. *Nature* **415**, 751–754.

504. Wellmer, F., Riechmann, J. L., Alves-Ferreira, M. & Meyerowitz, E. M. (2004). Genome-wide analysis of spatial gene expression in *Arabidopsis* flowers. *Plant Cell* **16**, 1314–1326.

505. Wesley, S. V., Helliwell, C. A., Smith, N. A., Wang, M., Rouse, D. T., Liu, Q., Gooding, P. S., Singh, S. P., Abbott, D., Stoutjesdijk, P. A., Robinson, S. P., Gleave, A. P., Green, A. G. & Waterhouse, P. M. (2001). Construct design for efficient, effective and high-throughput gene silencing in plants. *Plant J.* **27**, 581–590.

506. Wianny, F. & Zernicka-Goetz, M. (2000). Specific interference with gene function by double-stranded RNA in early mouse development. *Nature Cell Biology* **2**, 70–75.

507. Wienholds, E., Koudijs, M. J., Van Eeden, F. J. M., Cuppen, E. & Plasterk, R. H. A. (2003). The microRNA-producing enzyme Dicer1 is essential for zebrafish development. *Nature Genetics* **35**, 217–218.

508. Wightman, B., Ha, I. & Ruvkun, G. (1993). Posttranscriptional regulation of the heterochronic gene *lin-14* by *lin-4* mediates temporal pattern formation in *C. elegans*. *Cell* **75**, 855–862.

509. Williams, B. R. G. (1999). PKR; a sentinel kinase for cellular stress. *Oncogene* **18**, 6112–6120.

510. Williams, R. W. & Rubin, G. M. (2002). ARGONAUTE1 is required for efficient RNA interference in *Drosophila* embryos. *Proc. Natl. Acad. Sci. USA* **99**, 6889–6894.

511. Winston, W. M., Molodowitch, C. & Hunter, C. P. (2002). Systemic RNAi in *C. elegans* requires the putative transmembrane protein SID-1. *Science* **295**, 2456–2459.

512. Wirtz, E., Leal, S., Ochatt, C. & Cross, G. A. M. (1999). A tightly regulated inducible expression system for conditional gene knock-outs and dominant-negative genetics in *Trypanosoma brucei*. *Molec. Biochem. Parasitology* **99**, 89–101.

513. Wu, W. L., Hodges, E., Redelius, J. & Hoog, C. (2004). A novel approach for evaluating the efficiency of siRNAs on protein levels in cultured cells. *Nucleic Acids Research* **32**, e17.

514. Wu-Scharf, D., Jeong, B.-R., Zhang, C. & Cerutti, H. (2000). Transgene and transposon silencing in *Chlamydomonas reinhardtii* by a DEAH-box RNA helicase. *Science* **290**, 1159–1162.

515. Xie, Z., Kasschau, K. D. & Carrington, J. C. (2003). Negative feedback regulation of *Dicer-Like1* in *Arabidopsis* by microRNA-guided mRNA degradation. *Current Biology* **13**, 784–789.

516. Xu, P., Vernooy, S. Y., Guo, M. & Hay, B. A. (2003). The *Drosophila* microRNA miR-14 suppresses cell death and is required for normal fat metabolism. *Current Biology* **13**, 790–795.

517. Xu, Y. H., Linde, A., Larsson, A., Thormeyer, D., Elmen, J., Wahlestedt, C. & Liang, Z. C. (2004). Functional comparison of single- and double-stranded siRNAs in mammalian cells. *Biochem. Biophys. Res. Comm.* **316**, 680–687.

518. Yang, D., Buchholz, F., Huang, Z., Goga, A., Chen, C.-Y., Brodsky, F. M. & Bishop, J. M. (2002). Short RNA duplexes produced by hydrolysis with *Escherichia coli* RNase III mediate effective RNA interference in mammalian cells. *Proc. Natl. Acad. Sci. USA* **99**, 9942–9947.

519. Yang, D., Lu, H. & Erickson, J. W. (2000). Evidence that processed small dsRNAs may mediate sequence-specific mRNA degradation during RNAi in *Drosophila* embryos. *Current Biology* **10**, 1191–1200.

520. Yang, S., Tutton, S., Pierce, E. & Yoon, K. (2001). Specific double-stranded RNA interference in undifferentiated mouse embryonic stem cells. *Molec. Cell. Biol.* **21**, 7807–7816.

521. Yang, S. J., Carter, S. A., Cole, A. B., Cheng, N. H. & Nelson, R. S. (2004). A natural variant of a host RNA-dependent RNA polymerase is associated with increased susceptibility to viruses by *Nicotiana benthamiana*. *Proc. Natl. Acad. Sci. USA* **101**, 6297–6302.

522. Yanofsky, M. F., Ma, H., Bowman, J. L., Drews, G. N., Feldmann, K. A. & Meyerowitz, E. M. (1990). The protein encoded by the Arabidopsis homeotic gene *Agamous* resembles transcription factors. *Nature* **346**, 35–39.

523. Ye, M. K., Malinina, L. & Patel, D. J. (2003). Recognition of small interfering RNA by a viral suppressor of RNA silencing. *Nature* **426**, 874–878.

524. Yekta, S., Shih, I. H. & Bartel, D. P. (2004). MicroRNA-directed cleavage of HOXB8 mRNA. *Science* **304**, 594–596.

525. Yelina, N. E., Savenkov, E. I., Solovyev, A. G., Morozov, S. Y. & Valkonen, J. P. T. (2002). Long-distance movement, virulence, and RNA silencing suppression controlled by a single protein in hordei- and potyviruses: Complementary functions between virus families. *J. Virology* **76**, 12981–12991.

526. Yi, R., Qin, Y., Macara, I. G. & Cullen, B. R. (2003). Exportin-5 mediates the nuclear export of pre-microRNAs and short hairpin RNAs. *Genes & Development* **17**, 3011–3016.

527. Yu, J.-Y., DeRuiter, S. L. & Turner, D. L. (2002). RNA interference by expression of short-interfering RNAs and hairpin RNAs in mammalian cells. *Proc. Natl. Acad. Sci. USA* **99**, 6047–6052.

528. Zamore, P. D. (2004). Plant RNAi: How a viral silencing suppressor inactivates siRNA. *Current Biology* **14**, R198–R200.

529. Zamore, P. D., Tuschl, T., Sharp, P. A. & Bartel, B. (2000). RNAi: Double-stranded RNA directs the ATP-dependent cleavage of mRNA at 21 to 23 nucleotide intervals. *Cell* **101**, 25–33.

530. Zender, L., Hutker, S., Liedtke, C., Tillmann, H. L., Zender, S., Mundt, B., Waltemathe, M., Osling, T., Flemming, P., Malek, N. P., Trautwein, C., Manns, M. P., Kuhnel, F. & Kubicka, S. (2003). Caspase 8 small interfering RNA prevents acute liver failure in mice. *Proc. Natl. Acad. Sci. USA* **100**, 7797–7802.

531. Zeng, Y. & Cullen, B. R. (2003). Sequence requirements for microRNA processing and function in human cells. *RNA* **9**, 112–123.

532. Zeng, Y., Wagner, E. J. & Cullen, B. R. (2002). Both natural and designed microRNAs can inhibit the expression of cognate mRNAs when expressed in human cells. *Molecular Cell* **9**, 1327–1333.

533. Zeng, Y., Yi, R. & Cullen, B. R. (2003). MicroRNAs and small interfering RNAs can inhibit mRNA expression by similar mechanisms. *Proc. Natl. Acad. Sci. USA* **100**, 9779–9784.

534. Zhang, H., Azevedo, R. B. R., Lints, R., Doyle, C., Teng, Y., Haber, D. & Emmons, S. W. (2003). Global regulation of Hox gene expression in *C. elegans* by a SAM domain protein. *Developmental Cell* **4**, 903–915.

535. Zhang, H., Kolb, F. A., Brondani, V., Billy, E. & Filipowicz, W. (2002). Human dicer preferentially cleaves dsRNAs at their termini without a requirement for ATP. *EMBO J.* **21**, 5875–5885.

536. Zhang, J. & Carthew, R. W. (1998). Interactions between Wingless and DFz2 during *Drosophila* wing development. *Development* **125**, 3075–3085.

537. Zhao, Z., Cao, Y., Li, M.-J. & Meng, A. (2001). Double-stranded RNA injection produces nonspecific defects in zebrafish. *Developmental Biology* **229**, 215–223.

538. Zheng, L., Liu, J., Batalov, S., Zhou, D., Orth, A., Ding, S. & Schultz, P. G. (2004). An approach to genomewide screens of expressed small interfering RNAs in mammalian cells. *Proc. Natl. Acad. Sci. USA* **101**, 135–140.

539. Zheng, L., Zhang, J. & Carthew, R. W. (1995). Frizzled regulates mirror-symmetric pattern formation in the *Drosophila* eye. *Development* **121**, 3045–3055.

540. Zhong, R. Q. & Ye, Z. H. (2004). Amphivasal vascular bundle 1, a gain-of-function mutation of the IFL1/REV gene, is associated with alterations in the polarity of leaves, stems and carpels. *Plant Cell Physiology* **45**, 369–385.

541. Zhou, Y., Ching, Y.-P., Kok, K.-H., Kung, H.-F. & Jin, D.-Y. (2002). Post-transcriptional suppression of gene expression in *Xenopus* embryos by small interfering RNA. *Nucleic Acids Research* **30**, 1664–1669.

542. Ziauddin, J. & Sabatini, D. M. (2001). Microarrays of cells expressing defined cDNAs. *Nature* **411**, 107–110.

 The Use of RNAi for Gene Therapy

EITHAN GALUN

Hadassah University Hospital, Israel

Introduction

Although the concept of gene therapy is not new in medical sciences, the studies aimed at specifically assessing this approach were initiated just over 15 years ago. The emergence of this new medical discipline coincided with the Human Genome Project. Initially, these "original" studies questioned a very basic thought — whether this new treatment paradigm was in any way feasible. Preliminary results carried out in tissue culture and in specific animal models suggested that this therapeutic modality, in its most general sense, indeed held significant potential. Throughout these "pioneering" investigations, the "art" of gene therapy has expanded exponentially. In fact, innovative therapeutic modalities, currently investigated, include numerous potent drugs such as anti-sense, ribozymes, group I or II introns or chimeraplasts, as well as the use of complete ORF and regulatory elements for the correction of monogenetic maladies or reversing a malignant phenotype. A recent report by the Alan Fischer group from the Necker Hospital in Paris presents a complete genetic and phenotypic correction of X-linked severe combined immunodeficiency (SCID-X1, also called gamma chain or *gamma*(c) deficiency) in nine out of 10 patients by retrovirus-mediated gamma(c) gene transfer into autologous CD34 bone marrow cells. However, 3 years later, uncontrolled exponential clonal proliferation of mature T cells (with $\gamma\delta+$ or $\alpha\beta+$ T cell receptors) occurred in three patients. Clinically, these patients had developed a leukemia-like disease that was successfully treated with conventional approaches. Clones from two patients showed retrovirus vector integration in proximity to the LMO2 proto-oncogene promoter, leading to aberrant transcription and expression of LMO2. This suggests that the retrovirus vector insertion can trigger deregulated premalignant cell proliferation with unexpected frequency, most likely driven by retrovirus enhancer activity on the LMO2 gene promoter (Hacein–Bey–Abina *et al.*, 2002, 2003). This Appendix highlights and symbolizes the current state of gene therapy: a potential efficacy for specific cases, for example, ex-vivo transduction, with an unexpected frequency of major side-effects such as oncogenic activation. Presently, the most significant barrier in gene therapy is the development of simple, efficient and low-cost gene delivery methods with minute side-effects. Currently, these are still unmet objectives for most gene therapy targets. Overall, *in vitro* and *in vivo* gene therapy experiments using animal models have proven its efficacy and in some cases in the most profound way. Until now, the majority of clinical studies treating over 7000 patients have applied viral vectors. These vehicles encounter numerous disadvantages, including the augmentation of the immune response against the transgene or the activation of oncogenes in transduced cells, resulting in short-term expression or the development of unwarranted maladies. Furthermore, the production of replication-free batches of viral vectors carrying the therapeutic gene or the genetic therapy payload could be complicated and expensive.

Gene Therapy: Selecting a Delivery System is a Trade-Off Business

Non-viral delivery systems

A prerequisite for translating the promise of gene therapy from the laboratory, where ongoing *in vitro* and *in vivo* studies are conducted to the clinical arena, is to devise successful and efficient gene therapy delivery systems. Currently, overcoming this obstacle remains a major challenge. However, this is not the only barrier. While the simplest method of delivery is the use of naked DNA, the effectiveness of this method was only proven in specific tissues such as the dermis and the muscle. In both cases, the longevity of expression did not meet expectations in spite of the fact that muscle cells are terminally differentiated and enter into the cell cycle very rarely; thus, they are perceived as an ideal site for a "cell factory" and the production of secreted proteins like in the case of coagulopathies, for example, Hemophilia A. The introduction of naked DNA into these tissues could be done by physical and chemical methods or a combination of both. One of the robust methods for gene transduction into muscle tissue is electroporation (Tranchant *et al.*, 2004) or iontophoresis (Sakamoto *et al.*, 2004). The former could result in massive muscle injury and the latter has not yet been proven to be effective. However, both approaches will soon meet clinical studies and their applicability in translating these methods into proof-of-concept would probably be reported in the very near future. To improve gene transduction into cells, various chemical delivery vehicles have been developed. Investigators engaged in gene therapy use these reagents for *in vitro* transduction into cells, for example, lipofectamine™. The use of cationic liposomes *in vitro* and in some animal models have been shown to be effective in gene delivery. Most liposomes are complexes formed from cationic liposomes (or cationic polymers) and DNA, that is, lipoplexes (or polyplexes). While many lipoplex formulations have been studied, *in vivo* activity is generally low compared to that of viral systems. In addition, specific side-effects such as lipoplex-induced hemagglutination (Eliyahu *et al.*, 2002) have encouraged investigators to develop other non-viral delivery systems as an alternative to liposomes. Recent reports have also suggested the use of polysaccharides (Hosseinkhani *et al.*, 2004) or cell-penetrating peptides to improve the efficiency of delivery into cells. All of these approaches are still waiting for in-depth investigations to assess their applicability in animal studies and later in human studies. Overall, methods of gene delivery, whether using naked DNA or carrier delivery-like liposomes, have the potential of enhancing gene delivery systems. However, since their effectiveness have not been verified yet, further developments of these and other types of carriers are expected.

Viral vectors for gene delivery

Viral vectors are efficient delivery systems and several types of these vectors have been developed in recent years. These include adenovirus and adenovirus–associated virus (AAV), herpesvirus, the oncoretroviruses and lentiviruses. In addition to these viral vectors, others have also been developed for specific therapeutic targets such as oncolytic viruses, replicative competent vectors and SV40 based delivery systems (Arad *et al.*, 2004; Rund *et al.*, 1998). These are not the only types of delivery agents developed and the reader should be aware that additional viral delivery systems are available and new ones would probably emerge in the near future. Each of these delivery systems encounters specific advantages and disadvantages. In this section, I will attempt to highlight the specific issues related to the properties of the currently most used vectors for siRNA expression and delivery, and in a later section, some specific examples will be discussed.

The first signs of converting the adenovirus to an **adenovector** (Ad) could be tracked back to the 1980s. A number of prominent investigators such as Frank Graham and Marshall Horwitz as well as others have contributed through their studies to the development of the first generation of adenovirus vectors. Since the early days of assessment, a wealth of information has been generated and it is currently apparent that this vector has some significant advantages: (1) A simple cloning strategy; (2) High titers could be produced in the laboratory; (3) The virus is non-biohazard and most adults who have been exposed to the common serotypes currently used for gene therapy (Ad5 and Ad2) have developed neutralizing antibodies; (4) The Ad transduces most cell lineages due to the fact that the viral receptor is expressed in most human cells, the CAR (coxsackie adenovirus receptor); (5) This vector enables the transportation of its genome into the host cell nucleus to reside as an episome and its genome rarely integrates. Although this vector has also encountered some significant disadvantages, some of these have been recently challenged with moderate success. These disadvantages include: (1) An immune response to the vector and its transgene which has led to short-term expression *in vivo* (10–30 days, depending on the animal used for the experiment). To overcome this hurdle, the Ad-gutless vector has been developed in recent years. Upon its introduction to animal models, the Ad-gutless vector supports transgene expression for many months (Morral *et al.*, 1999); (2) One of the major challenges in recent years for the development of Ad is to eliminate the generation of replication competent viruses (RCV) due to homologous recombination between the transfer plasmids and the cell lines supporting the generation of the vectors. Although there are currently cell lines and protocols that enable the reduction of the level of RCV to a very minute population, this issue should always be kept in the minds of those using first generations of Ad. In the case of the Ad-gutless system, the production of high titers without a replication competent contaminant is complicated and improved protocols to overcome this problem are currently under development. For those interested in assessing Ad-gutless for clinical translation, this could represent a major barrier with the current methodology. Until today, there has not been even a single human reported study that has used the Ad-gutless vector.

Many human studies and those that were most successful in terms of efficacy applied the **amphotropic retroviruses** (an oncoretrovirus) based vectors. In recent years, retroviruses were the most common vectors used for gene therapy in clinical studies (Edelstein *et al.*, 2004). The retroviral vectors harness beneficial properties to those interested in using them as gene delivery systems: (1) A simple cloning system enables the introduction of transgenes into these delivery vectors; (2) It provides a platform for prolonged expression due to its property to integrate into the host cell genome; (3) Low immunogenicity; (4) Simple production system of the viral vector. However, some of these vector properties could present major disadvantages for the gene therapist: (1) The integration of the transgene is not site-controlled and could induce a side-effect on neighboring genes; (2) The titers produced with the available methods are relatively low for systemic administration (up to $\sim 10^8$ particles/ml); (3) In specific cases, there is size limitation in these vectors although for the cloning of siRNA expression cassette this is not the case; (4) These vectors do not express their transgene in non-dividing cells since they do not travel through the nuclear pores into the nucleus of cells. To overcome this last major hurdle, the lentivirus vector has been developed in recent years. The advantage of the **lentiviral vectors** over the retroviruses stems from their potential to transduce non-dividing cells (Galimi and Verma, 2002). In addition, this vector could harbor a larger transgene although in the case of siRNA expression, this is probably a lesser important feature of these vectors. In spite of the major efforts conducted to incorporate safety measures into the HIV lentiviral vector, there are some concerns for its use in humans. Until this Appendix was written, there was no report of a clinical study in non-HIV infected patients that used the HIV-based vector. Furthermore, the production of a high titer vector for systemic administration also represents a major barrier in the translation of this vector into clinical studies. An alternative lentiviral vector is the feline immunodeficiency virus (FIV) based vector (Poeschla, 2003). To overcome the need to generate large titers for systemic administration, we have combined the use of this vector with the hydrodynamic administration method to target the vector to a specific organ, the liver (Condiotti *et al.*, 2004). However, in spite of all these developments, the use of lentiviruses in the clinical setting is still awaiting approval from the regulatory agencies.

The last viral vector, which I believe is relevant to this Appendix on RNAi gene therapy, is the **adenovirus–associated virus (AAV)**. For those who are interested in learning about other viral vectors such as the Vaccinia virus or the herpes simplex virus and others, I suggest, as an initial step, to search for an updated review individually. The AAV vector carries a number of important features for the gene therapists: (1) The expression from the AAV vector in many tissues was found to be relatively stable for a prolonged period of time; (2) Transduction *in vivo* with the AAV vector induces a minute immune response; (3) The capacity of this vector is ~4.5 kbp, sufficient for most targets; as for the siRNA expression vectors, this size is more than enough; (4) New methods have led to the generation of a relatively high titer of this vector, up to 10^{13} TU/ml; (5) It is possible to encapsidate one serotype of AAV with a capsid of a different one, leading to tissue specific AAVs; (6) The AAV expresses its transgene from episomal sequences as the integration events are not essential for gene expression. However, the issue of AAV integration is still under investigation. In spite of these major advantages, it must be remembered that there are still some disadvantages and unresolved issues. In a few recent reports, unexpected immune responses have been generated against the AAV transgene (Gao *et al.*, 2004). In addition, in a recent clinical study using AAV to treat hemophilia, the vector might have induced inflammation in the AAV infected tissue (Kaiser, 2004). This last study has been put on hold temporarily. It must also be remembered that AAV vectors do integrate and they do so preferentially at existing chromosomal breaks (Miller *et al.*, 2004a).

Although there has been significant progress in the design of viral vectors for gene delivery including our basic understanding of the biology of the vectors *in vitro* and *in vivo* as well as in the generation of safety measures, at this point in time, it is still required to improve, simplify and better understand the available vectors before translating them into clinical practice.

Delivery of RNAi

The major challenge in RNAi gene therapy is to transform the *in vitro* robust effects of siRNA into an *in vivo* gene silencing method. In other words, what would be the preferred delivery system to use in animals and at a later stage in humans? As for gene therapy in general and the specific aims of delivering RNAi platforms, we need to tailor the tools to be used to the sought objective. This includes targeting the tissue, adjusting the desired level of expression [high level of siRNA could induce non-specific silencing (Fish and Kruithof, 2004)], the longevity and the specific maladies that we wish to treat. This is a complex situation, especially since our major barrier is the lack of a simple non-immunogenic targeted delivery system without side-effects. However, in spite of all these hurdles, I would like to discuss the potential available methods to deliver siRNA.

The most straightforward method of using siRNA *in vivo* is by administering synthetic siRNA. Upon co-injecting siRNA and its target being expressed from a plasmid vector, we could achieve knock-down of expression in the liver following a hyperdynamic injection. In my laboratory we could show that HBV expression out of a HBV head-to-tail plasmid that supports viral replication in the liver of mice could effectively be silenced transiently with synthetic siRNA (Giladi *et al.*, 2003). This effect was dose-dependent. However, the kinetics of this effect revealed that the silencing was transient as was also shown by other groups using a similar approach (McCaffrey *et al.*, 2002; Lewis *et al.*, 2003; Braash *et al.*, 2003; Song *et al.*, 2003). The effect subsides after 48–72 hours and is probably completely lost after 7 days. The silencing effect of siRNA following systemic administration of duplex

RNA could have been hampered by various factors thus imposing some logistic hurdles upon translating this approach into the clinical setting (additional examples of RNAi against viral infections are depicted in Table 1). Although duplex RNA is quite stable in serum (Braash *et al.*, 2003) and more stable than ssDNA or ssRNA, a high serum concentration could reduce stability. The introduction of phosphorothioate linkages could enhance stability in the serum (Braash *et al.*, 2003). Recently, numerous studies have conducted chemical modifications of siRNA in order to enhance its stability and effect; a short review of these modifications is summarized in a report by Paroo and Corey (2004). However, although siRNA is relatively stable in the serum, there are disadvantages of using synthetic siRNA: (1) The effect of synthetic siRNA is transient; in order to impose a long-term effect, repeated administrations would be needed; (2) The production of synthetic siRNA is expensive, making repeated administrations for long-term effect very costly; (3) It is very complicated to target synthetic siRNA to a specific cell or tissue. However, in specific cases, the use of siRNA directly administered into the target tissue could encounter a significant effect. In a recent report by Dorn *et al.* (2004), they have used siRNA against the pain-related cation-channel $P2X_3$ by intrathecal injection of phosphorothioated (PS) siRNA in a rat model of neuropathic pain (Dorn *et al.*, 2004). Although they did not compare the non-PS modified to PS modified siRNA, they have clearly shown a significant effect of siRNA in relieving chronic pain. Furthermore, they were also able to show that the effect was superior to the comparable $P2X_3$ antisense oligonucleotides. One specific case where such an approach could be translated into a clinical applicable therapeutic modality is in post-herpetic neuralgia. However, since chronic pain is a condition that is generally expected to last for months, this type of treatment would need to be readministered several times.

In an effort to enhance and prolong the effect of siRNA, various approaches have been undertaken using non-viral reagents (Table 2). Tailoring the specific delivery tactic and method is essential while designing a specific therapeutic strategy. One example is the use of siRNA targeting the influenza virus genome. This malady can cause moderate to severe illnesses, affect millions of people each year and could be life-threatening. For the gene therapist this means that the

Table 1. The effect of RNA interference against viral infection

Virus	RNAi expression system	Delivery method	Model system	Effect	Ref.*
RNA viruses non-integrating:					
Severe acute respiratory syndrome — coronavirus (CoV- SARS)	pSUPER	Transfection	Infected Vero cells *in vitro*	Message and titer inhibition, knockdown to ~20%	61 100 110
Foot-and-mouth disease virus (FMDV)	U6 promoter	Transfection	*In vitro* and *in vivo*	Inhibits VP1 expression and FMDV replication in BHK-21 cells and suckling mice	17
Influenza A virus	Naked siRNA & U6 promoter	*In vivo* i.v. delivery by polyethyleneimine (PEI), a cationic polymer with siRNA mixture; and Lentivirus. Also intranasal delivery	Animal model	Preventive and therapeutic effects	30 65 96
Hepatitis A virus (HAV)	Naked siRNA	Transfection	*In vitro* in HuH7 cells	Inhibition of replication and gene expression	47
Hepatitis C virus (HCV)	H1 promoter	Retroviral transduction: MoMuLV-based vector pBABE/puro	*In vitro* against the HCV replicon	Endoribonuclease-prepared siRNAs (esiRNAs) to simultaneously target multiple sites. IRES Domain IV is the preferred target	55 87 107
Human rhinovirus (HRV)	Naked siRNA	Transfection	*In vitro* in HeLa cells	Suppression of HRV-16 replication	80

Table 1. (*Continued*)

Virus	RNAi expression system	Delivery method	Model system	Effect	Ref.*
Hepatitis delta virus (HDV)	pSilencer	Transfection	*In vitro* in HuH7 cells	siRNA against mRNA were also effective against HDV replication	14
Retroviruses:					
Porcine endogenous retroviruses (PERV)	Pol III vector	Transfection	*In vitro*	Significant suppression of replication	49
HIV	Utilization of the human miR-30 pre-microRNA	Transfection	*In vitro*	Effective reduction of HIV-1 p24 antigen	9 74 57 76 2 5 6
DNA viruses:					
JC virus	Naked siRNA	Transfection	*In vitro*: SV40-transformed human fetal glial cells	Significant inhibition of JCV production	75
EBV	pSUPER	Transfection	*In vitro*: NPC cell line	EBV lytic cycle effectively blocked	15
Hepatitis B virus (HBV)	Naked siRNA	Transfection *in vitro* and hydrodynamic *in vivo*	HepG2, 2215 cells and SCID mice	Significant short term anti-viral effect	31 54 106 63 88 37

*Refs. are numbers in the literature list.

Table 2.   RNAi expression systems, non-viral and viral delivery systems

Target genes	Delivery method & siRNA expression system	Description	Ref.*
Signal transduction and tumor suppressor genes	HIV based vector. Cre/lox controlled inducible U6 promoter	A lentiviral delivery of dsRNA tightly controlled through the expression of the CRE recombinase. The target genes knockdown were GFP, p53 and NF-κB transcription factor subunit p65	95 50
Marker genes in tumor tissue	Conditionally replicating adenoviruses (CRAd). U6 promoter	The CRAd vector was designed with the bidirectional human aldehyde reductase promoter driving firefly luciferase and Renilla luciferase. In the same vector under the U6 promoter siRNA targeted the luciferase gene. Gene silencing was effective in tumor cells.	13
Marker gene & Lamin A gene	Maleficent: Tc1-like transposon Sleeping Beauty	Plasmid-based system for the generation of long-term gene knockdown	38
HIV-1 replication	AAV2 vector MTD promoter	Suppression of HIV-1 replication in primary Lymphocytes and H9 infected cells	7
Marker gene, MyoD and p53	SIN MoMuLV vector	Ecdysone-inducible synthesis of shRNA in mammalian cells with a tight control effect	33
Marker gene: GFP	HIV-1 based vector with U6 promoter	Suppression of GFP expression in CHO and 293T cells and in mouse brain	98 1
Marker genes	Hydrodynamic tail vein injection for liver directed RNAi with U6 promoter	A study describing the kinetics and dose response effects of siRNA following hydrodynamic injection	53
Marker gene & Crx and Nrl genes	Electroporation into the rat retina with U6 promoter	Over 50 day expression following electroporation	62
p53 and VprBP/ KIAA0800	Adenovirus H1 promoter in HeLa cells	Specific reduction of target gene expression	4 111

*Refs. are numbers in the literature list.

genetic therapy effect could be designed for a short window period. In this case, the use of synthetic siRNA with an enhanced transduction is an appropriate approach. In a study by Ge *et al.* (2004) (see also Table 1), they were able to show that the systemic and intratracheal delivery of polyethyleneimine (PEI), a cationic polymer in promoting siRNA delivery in mice, is beneficial for prophylaxis and therapy of the influenza virus infection. PEI was developed for *in vitro* and *in vivo* local and systemic gene delivery and PEI-mediated gene delivery/transduction into the lung following systemic and intratracheal administrations. Tompkins *et al.* (2004) also assessed the effect of siRNA against the same pathogen, the influenza virus. This group used a different therapeutic regimen. They used a preventive measure by administering naked siRNA systemically, that is, intravenously, before infection, and at the time of infection they administered the siRNA/oligofectamine™ (a lipid carrier from Invitrogen) intranasally. In this model, they were able to show that their therapeutic approach prevented death of animals. However, this study was limited to asking the general question of siRNA effect against influenza virus infection rather than comparing different siRNA delivery systems. Thus, we cannot draw any conclusions that would suggest a preference for any specific delivery method in this disease model system. Future animal studies would be required to determine the preferred delivery method. In addition to the lipid carriers, traditionally developed for *in vitro* and *in vivo* delivery of DNA and now for siRNA, alternative approaches have also been developed. One specific interesting report by Minakuchi *et al.* (2004) describes the use of Atelocollagen (AT) which is prepared from type I collagen from calf dermis. This is a low immunogenic product that is already available in clinics for various indications like promoting wound healing. The same investigators showed that AT enhanced DNA delivery and supported prolonged expression. AT mediated the delivery of siRNA *in vitro* and was found *in vivo* to encounter a significant advantage over the siRNA/liposome complex in inhibiting tumor growth in mice. Recently, more sophisticated delivery methods of siRNA have also been developed. Zhang *et al.* (2004b) were interested in developing an siRNA delivery system to treat brain tumors. The model that they used *in vivo* to assess their delivery and RNAi effects was an immunodeficient mouse with an inoculated intracranial human U87 glioma tumor that was dependent on EGF signaling for growth. One of the major barriers for macromolecules to travel to diseased tissue in the brain is the blood brain barrier (BBB). Passing the BBB is a major challenge faced for the development of any pharmacological compound with high molecular weight. Pegylated (polyethylene glycol) immunoliposomes (85 nm size liposomes designed with monoclonal antibodies over its outer surface) were able to support transvascular delivery of plasmid DNA and to target and transduce specific cells in the brain. This group conjugated the PEGylated liposomes with two monoclonal antibodies, one against the mouse transferrin receptor to enable BBB crossing and the second against the insulin receptor to enhance cellular uptake. The generated PEGylated immunoliposomes, encapsidated a plasmid payload that was designed to express the siRNA against the EGF receptor in the transduced cell. The siRNA complex of PEGylated immunoliposomes showed an enhanced antitumor effect by prolonging survival of animals. The ability to treat brain tumors by a systemic approach rather than the need to perform stereotactic injection is an outmost achievement. Combining this unique delivery method with a sustained antitumor molecule is significant. It would be interesting to see if such an approach could be translated into the clinical setting. Additional examples of RNAi-mediated approaches against viral, cancer and metabolic diseases are summarized in Tables 1, 3 and 4, respectively.

The potential advantage of viral-mediated siRNA delivery encouraged numerous groups to clone expression cassettes in transgenes and to encapsidate these into viral particles (Table 2). As described in the previous section of this Appendix, each type of viral vector holds specific properties. These viral vector characteristics should be those that determine which viral delivery system to apply for the specific therapeutic target. The adenovector (Hosono *et al.*, 2004) and in particular the Ad-gutless vector hold a major promise for liver directed systemic delivery. In cases where short-term silencing effect is warranted, the non-gutless vectors could then be applied; however, for prolonged silencing effects, the gutless vector might be more beneficial. In the gutless vector, there are practically no restrictions as for the size of the sequences to be incorporated and these could include marker genes,

Table 3. RNAi as a therapeutic tool for cancer

Tumor (Tu:) & Target gene (Ta:)	RNAi expression system	Delivery method	Model	Effect	Ref.*
Tu: Hepatocellular Carcinoma (HCC) Ta: u-PA	U6 promoter plasmid with 9-bp hairpin spacer	Stable transfection	SKHep1C3 cells	Reduction of migration, invasion, and proliferation	86
Tu: Melanoma Ta: BRAF	3° generation SIN-HIV vector, expressing siRNA from U6	Lentivirus infection	*In vitro* and in NOD/SCID	*In vitro*: inhibition of growth *In vivo*: significantly reduced tumor growth	32 93
Tu: Leukemia Ta: TEL-PDGFßR	H1 promoter/Super	Oncoretrovirus	Ba/F3 cells *in vitro* and Nude mice	Stable siRNA expression inhibits TEL-PDGFβR *in vitro* and *in vivo*. Enhanced effect with small drugs	16
Tu: Cervical & Lung carcinoma Ta: Polo-like kinase 1 (PLK1)	U6 promoter	Transfection and i.v delivery of plasmid	HeLa S3 or A549 cells *in vitro* & *in vivo* in nude mice	Significant anti-tumor effects *in vitro* and *in vivo*. [ATA protects plasmid DNA degradation in mammalian blood]	92
Tu: Human Glioma Ta: Epidermal growth factor receptor (EGFR)	U6 promoter	Pegylated immunoliposome targeted to glioma with conjugated MAb insulin R or transferrin R	U87 glioma *in vitro* & *in vivo* in SCID mice with brain tumors	Anti tumor effect *in vitro* and increase in survival time *in vivo*	109

Table 3. (*Continued*)

Tumor (Tu:) & Target gene (Ta:)	RNAi expression system	Delivery method	Model	Effect	Ref.*
Tu: Prostate Ta: Vascular endothelial growth factor (VEGF)	Naked dsRNA	Transfection and direct injection	PC3 cells *in vitro* and in nude mice	Anti tumor effects	94
Tu: Cervical Cancer Ta: HPV-E6	pSUPER, H1 promoter	Transfection	HPV18-positive HeLa cervical carcinoma cells	siRNA against E6 induced massive apoptosis in HeLa cells	12 36
Tu: Breast cancer Ta: CXCR4	Tet inducible shRNA expression system with U6 promoter	Stable transfection	MDA-MB-231 breast cancer cells	Reduced breast cancer cell migration *in vitro* with potential implication to invasion	18
Tu: Naso-pharyngeal carcinoma (NPC) Ta: EBV — LMP-1	U6 promoter	Transfection	EBV-positive NPC cell line C666	Suppressing NPC motility, invasion ability and metastasis	60

*Refs. are numbers in the literature list.

Tu = Tumor type, and Ta = Target gene.

Table 4. RNAi as a therapeutic tool targeting human genes

Target gene	RNAi expression system	Delivery method	Model system	Effect	Ref.*
Multidrug resistance (MDR) P-glycoprotein	SUPER	Stable transfection	K562 leukaemic cells	Completely reversed the MDR phenotype	104
Pain-related cation-channel P2X$_3$	Naked dsRNA	Direct injection	Rats	Diminished pain responses and P2X$_3$ protein translocated into the dorsal horn	23
Alzheimer's disease: Tau and amyloid precursor protein (APP)	T7 polymerase	Transfection	COS7 and HeLa cells	siRNA effectively suppressed the expression of the targeted gene	67
Tyrosine hydroxylase (th)	U6 promoter	Stereotaxic injections of AAV	Mice model	th knockdown resulted in behavioral changes, motor performance deficit and reduced response to a psychostimulants	41
Amyotrophic lateral sclerosis (ALS) Superoxide dismutase (SOD1) gene	U6 promoter	Transfection and i.v. systemic injection	HeLa cells and mice model	A selectively silencing of two dominant mutant SOD1 genes	22
IL-12 in dendritic cell	Naked siRNA	Transfection	*In vitro* dendritic cells	Shifting Th1 to Th2 effect by reducing IL-12 and enhancing IL-10 effect	40
Caspase-8 in acute liver failure (ALF)	Naked siRNA	Hydrodynamic tail vein injection	*In vivo* in mice with ALF following Fas activation	Caspase-8 siRNA prevents liver damage in a model of AdFasL-induced ALF	108

*Refs. are numbers in the literature list.

regulatory controlled cassettes or matrix controlled regions for prolonged expression. Controlling the expression of a siRNA specifically in tumor cells could also be designed in a conditionally replicating adenovirus (CRAd) which are designed to replicate and kill tumor cells specifically without harming normal cells. Carette *et al.* (2004) applied CRAd, which is dependent on Rb deficiency for replication, to test its potential in silencing a marker gene *in vitro* in tumor cells. They showed that the silencing effect of a marker gene is dependent on CRAd replication. The combination of the CRAd antitumor effect with a siRNA against a tumor dependent growth gene should be assessed in the future and this was not considered by this group, neither *in vitro* nor *in vivo*. These issues would need further studies in an effort to assess and enhance their therapeutic potential.

The major advantage of the retroviral delivery method is their potential to incorporate the payload transgene they "carry" into the host cell genome although the integration site could not be specified and this could cause side-effects. In the past 2 years, numerous groups have reported on various retroviral (Gupta *et al.*, 2004), including lentiviral systems to express siRNA's against viral pathogens and tumor cells. Ralf Bartenschlager and associates from Heidelberg reported recently on the use of the Moloney murine leukemia virus (Mo-MuLV) based vector (pBABE) as a delivery system for siRNA targeting hepatitis C virus (HCV) (Kronke *et al.*, 2004). In the publication these investigators also assessed a unique RNAi approach against HCV infection. This was done in an effort to overcome the low fidelity of the viral polymerase, establishing a state of quasispecies by generating endoribonuclease-prepared siRNA to simultaneously target multiple sites of the viral genome in order to prevent escape. As for the retroviral delivery approach, this group designed their siRNA mainly against the viral IRES sequences. Their readout system to assess the silencing effect involved tissue culture cells transfected with the subgenomic HCV replicon. This replicon harbors the HCV IRES upstream of the luciferase gene and the neomycin resistance region and an additional IRES (non-HCV) upstream of the viral non-structural sequences. The presence of the luciferase gene enables the determination of HCV IRES activity *in vitro*. This system enables the assessment of the effect of siRNA against the HCV IRES as well as against the viral non-structural genome. The siRNA in the retrovirus was expressed out of the H1 promoter. They designed numerous siRNAs in the vector and found that a specific region in the HCV IRES, near the beginning of the viral coding nucleotides, was the most sensitive to RNAi effect. Although this is an interesting approach with clearcut results, significant development and modifications are required in order to translate this modality into the clinical setting. However, the results described in this report again suggest that the RNAi encounters a therapeutic potential for chronic viral infection. Similar observations were reported by other groups (Sen *et al.*, 2003; Yokota *et al.*, 2003, see: Table 1).

Presently, most groups who use retroviral vectors to express siRNA apply lentiviral vectors. Numerous papers were recently published describing differently designed lentiviral vectors to meet different needs of siRNA expression. These include systems which support the control (Wimerowicz and Trono, 2003) or conditioning (Ventura *et al.*, 2004; Tiscornia *et al.*, 2004) of siRNA expression. Veerle Baekelandt and associates have designed a study to assess the potential use of lentiviruses in delivering siRNA into brain tissue. In their recent report (Van den Haute *et al.*, 2003), they constructed a lentiviral vector with siRNA against the marker gene eGFP. Upon simultaneous administration into the brain tissue by stereotactic injection of the lentivirus expressing the eGFP and the lentiviral vector expressing the siRNA against the same gene, they were able to show almost complete knockdown of eGFP expression. In two additional experiments in which the siRNA lentivirus was administered before or after the marker gene, they were also able to show significant silencing of expression. Interestingly, they claimed, albeit without showing the data, that this effect persisted for 6 months. If others could reproduce these results, in particular the longevity of expression described in the study, this would be a very interesting finding. However, the issue of delivery is again a major barrier. Stereotactic administrations are possible but alternative approaches would be beneficial. Lentiviruses hold major promises in gene therapy. Once the issue of integration and production is overcome, I expect that the lentiviral-based vector would be integrated in the clinical setting.

The AAV vector is currently perceived as a relatively safe vector since it supports long-term expression in most tissues from an episome. Beverly Davidson and associates reported an interesting study on the suppression of polyglutamine-induced neurodegeneration in a mouse model of spinocerebellar ataxia (SCA1), a disease of the polyglutamine-expansion group which also includes Huntington chorea (Xia *et al.*, 2004). They had expressed the siRNA under a modified CMV promoter due to its enhanced expression/silencing effect compared to the pol III promoter in their hands. In addition, they have revealed that incorporating the miR23 loop into the siRNA expression cassette enhanced the silencing effect, resulting in improved suppression of the ataxin protein levels (Kawasaki and Taira, 2003). However, this effect was only apparent in the pol III expression vector. How to select the best loop in any specific case is still an open question, and presently, this is a matter of empiric assessment (additional tools for the designing of the specific siRNA are depicted in Table 5). In their mouse model, they injected the AAV vector expressing the shRNA against the mutant SCA1 message directly into the brain tissue. This treatment showed long-term therapeutic effect on motor coordination as well as a histological improvement by reducing intranuclear inclusions. Although this represents a step forward from previous studies with similar models that used antisense and ribozymes as therapeutic agents, we are still far from the clinic. The direct administration of a viral vector into brain tissue is a significant drawback for the current delivery systems. The potential side-effects of AAV administration into the brain may soon be revealed once results of clinical studies using the AAV for direct brain administrations in Parkinson disease and Alzheimer disease studies are available (Howard, 2003). One specific point of importance should be mentioned on the issue of designing loop sequences stated previously. A number of investigators have assessed this matter as related to the effectiveness of silencing. We must bear in mind that for each specific case there is a need to develop a specific structure to improve the silencing effect (Miyagishi *et al.*, 2004).

RNAi Gene Therapy Silencing Effects along the Genome

In theory, we could think of dividing this issue into three types of RNAi gene silencing "strategies": (1) siRNA targeting a message ORF

Table 5.   Gene therapy RNAi resource tool

Name of tool	Website	Properties	Ref.*
siDirect	http://design.RNAi.jp	Online software system for computing effective siRNA sequences with target-specificity for mammalian RNAi	73
T7 RNAi oligo Designer (TROD)	http://www.cellbio.unige.ch/RNAi.html.	Target sequences based on the constraints of the T7 RNA polymerase for RNA interference with siRNAs	24
DEQOR	http://cluster-1.mpi-cbg.de/Deqor/deqor.html.	Scoring system based on parameters for siRNA design to evaluate the inhibitory potency of siRNAs	39
OptiRNAi	http://bioit.dbi.udel.edu/rnai/.	An RNAi design tool	20
Ambion siRNA Target Finder and Design Tool	http://www.ambion.com/techlib/misc/siRNA_finder.html	Computer driven RNAi design tools	Ambion
Qiagen RNAi Design Tool	http://www.qiagen.com/siRNA/ordering.asp	RNAi design tool	Qiagen
Nature Review on RNAi	http://www.nature.com/focus/rnai/index.html	General information	NPG
General Information		Improving the efficiency of RNA interference in mammals	69 56
RNAi central	http://katahdin.cshl.org:9331/RNAi.web/scripts/main2.pl	Multiple tools for RNAi delivery and expression and target selection	Hannon Lab
Dharmacon	Through: www.cshl.org/public/SCIENCE/hannon.html	Rational siRNA design for RNA interference	82

*Refs. are either numbers in the literature list or sources indicated by the respective websites.

(siRNA-ORF). This is the most commonly used type until now by gene therapy investigators and for non-gene therapy avenues where siRNA is applied as a gene knockdown tool for the study of gene expression. In most cases, the investigators use one of the bioinformatic tools available in the web (see: Table 5 for specific sites) to design their potential siRNA payloads. The effectiveness and sensitivity is assessed *in vitro* in transfection experiments by determination of the level of knockdown at the lowest concentration; (2) In humans, there are ~250 known miRs. They usually target the 3′ UTRs. The targeting of the 3′ UTR by miRs also reduces expression. Future studies will determine if there are additional miRs in the human genome including those that not only reduce expression but also increase it. The use of one of the known miRs to control gene expression is most appealing and would probably be tested in the future. This approach is now in development. I suggest calling this type of gene therapy **miRx**. The use of miRx in gene therapy could enhance specificity; (3) Recent studies have suggested that siRNA could also control gene expression through induction of methylation of CpG islands in promoter regions of human cells (Kawasaki and Taira, 2004). I propose to call this type of gene therapy knockdown **siRNA-CpGx**. Although this type of siRNA biology is in its infancy, I expect that in the near future, we shall be informed about the use of these last two molecular platforms as gene therapy modalities: miRx and siRNA-CpGx.

Specific Points for Consideration upon Designing a Gene Therapy Approach Utilizing RNAi

Availability of the Dicer machinery

During the course of differentiation, the expression of proteins involved in RNA interference decreases (Sago *et al.*, 2004). The cellular level of *Dicer* could be crucial for gene therapy approaches while utilizing the RNA interference machinery in targeted cells. Recent data suggest that although the expression of *Dicer* and other proteins that participate in digesting long dsRNAs into 21–25 nucleotide, for example, *eIF2C1~4*, decreases in differentiated cells but retains sufficient amount of enzymatic activity to induce RNAi. However, upon designing and planning any specific approach applying siRNA for gene therapy, it is advisable to assess the *Dicer* activity in the targeted tissue if the expected siRNA is unmet.

Vector design

Recent reports have showed that the expression of shRNA from the H1 or the U6 pol III promoters in a HIV-based vector induced the expression of interferon-stimulated genes (ISGs). This effect was dependent on the presence of an AA di-nucleotide near the transcription start site. Preserving a C/G sequence at positions $-1/+1$ could prevent this effect as the authors have suggested (Pebernard and Iggo, 2004).

In some cases, the expression from the U6 promoter could be relatively low. The enhancer from the cytomegalovirus immediate-early promoter can enhance the U6 promoter activity (Xia *et al.*, 2003). CMV enhancer increased the inhibition of target gene expression in 293T cells transfected with $SOD1^{G93A}GFP$ fusion protein compared to unmodified promoter and had no effect on the $SOD1^{wt}GFP$ levels, suggesting specificity of the expression system. Others had also tested various promoters reporting some beneficial effects on the expression with the modified tRNA(met)-derived (MTD) promoter, upon expressing shRNA against HIV-1 compared to U6 or H1 promoters (Boden *et al.*, 2003b). It may happen that for each specific application, we would need to compare numerous regulatory elements to achieve the desired RNAi effect. In some systems, it may be important to tightly control the expression of the shRNA. Although most controlled systems do not reach the desired stringency *in vivo*, some reports have suggested the use of specific systems. One such method is the tet-on-off expression technology (Miyagishi and Taira, 2002). Again, as in other sections of this Appendix, each investigator should specifically assess the potential of this inducible system in a specific tissue culture or animal model. The tet-on-off systems were developed for naked DNA transfection systems or incorporated into viral vectors like the lentiviral vector (Wimerowicz and Trono, 2003).

The potential RNA silencing suppressor effect of the target

Prior to the recent unfolding of the gene silencing machinery, it was apparent that many plant viruses encode proteins, denoted RNA silencing suppressors that interfere with antiviral response (Ye *et al.*, 2003). Most reported cases originate from the plant viral world. However, recent reports also illustrate that numerous human pathogens encounter RNA silencing suppressor effects. One such case is the NS1 protein of human influenza A virus (Bucher *et al.*, 2004). The NS1 was shown to be capable of binding siRNAs. As such, the NS1 holds a counteracting innate antiviral response by sequestering siRNAs. Such an effect should be taken into consideration upon designing an RNAi approach to viral infections (Nishitsuji *et al.*, 2004).

Development of resistance to RNAi

One major drawback for most antiviral approaches is the development of resistance. This is most apparent in cases where the fidelity of the viral polymerase is low, especially in viruses with an RNA genome. To overcome this hurdle, most antiviral therapeutic protocols harness a strategy that uses multiple drugs targeting different viral proteins or steps in its life cycle. One example where such resistance has developed *in vitro* is in the case of RNAi against the Nef gene of HIV (Das *et al.*, 2004). Others have reported similar findings (Boden *et al.*, 2003a). To overcome this problem, it may be advisable to utilize a multi-target RNAi approach, possibly in combination with additional antiviral modalities.

Off-target effects

An unresolved issue which could encounter deleterious side-effects upon the use of RNAi is the issue of off-target effects (Jackson *et al.*, 2004). Numerous reports have shown major alterations in off-target genes. One notable report showed that the targeting of the MEN1 gene had a dramatic and significant change in protein levels of p53 and p21. Again, upon planning a gene therapy approach, this issue should be taken into consideration.

Activation of the innate immune response

The role of interferon signaling in RNAi has given rise to a series of conflicting reports. Although most studies suggest that there is very little non-specific effect of siRNA, others have shown that the Jak–Stat pathway is activated following siRNA transfection. This effect is mediated by dsRNA-dependent protein kinase (PKR) (Sledz *et al.*, 2003). While I expect that this issue will foster additional debates, at this point, it would be important to impart cautious interpretations upon describing RNAi effects. In addition, recent studies have suggested that although upon entrance or *in situ* propagation inside the cells siRNA does not activate the intracellular interferon machinery in mammalian cells, if they are shorter than 30 nt dsRNA, there is a non-specific innate immune response depicted by cytokine production (Sioud and Sorensen, 2003). Furthermore, this effect is dependent on the Toll-like receptor 3 (TLR3) that senses dsRNA and serves as its receptor. TLR3 is located intracellularly on the endosome membrane and signals through NFκB nuclear translocation for the production of inflammatory cytokines. Incubation of immune cells with siRNA induces the activation of cells (Kariko *et al.*, 2004). All of these effects could be dependent on the concentration of siRNA.

Literature for the Appendix

1. An, D. S., Xie, Y. M., Mao, S. H., Morizono, K., Kung, S. K. P. & Chen, I. S. Y. (2003). Efficient lentiviral vectors for short hairpin RNA delivery into human cells. *Human Gene Therapy* **14**, 1207–1212.
2. Anderson, J., Banerjea, A., Planelles, V. & Akkina, R. (2003). Potent suppression of HIV type 1 infection by a short hairpin anti-CXCR4 siRNA. *Aids Research and Human Retroviruses* **19**, 699–706.
3. Arad, U., Axelrod, J., Ben-nun-Shaul, O., Oppenheim, A. & Galun, E. (2004). Hepatitis B virus enhances transduction of human hepatocytes by SV40-based vectors. *J. Hepatology* **40**, 520–526.
4. Arts, G. J., Langemeijer, E., Tissingh, R., Ma, L. B., Pavliska, H., Dokic, K., Dooijes, R., Misic, E., Clasen, R., Michiels, F., van der Schueren, J., Lambrecht, M., Herman, S., Brys, R., Thys, K., Hoffmann, M., Tomme, P. & van Es, H. (2003). Adenoviral vectors expressing siRNAs for discovery and validation of gene function. *Genome Research* **13**, 2325–2332.
5. Banerjea, A., Li, M. J., Bauer, G., Remling, L., Lee, N. S., Rossi, J. & Akkina, R. (2003). Inhibition of HIV-1 by lentiviral vector-transduced siRNAs in T lymphocytes differentiated in SCID-hu mice and CD34(+) progenitor cell-derived macrophages. *Molecular Therapy* **8**, 62–71.
6. Boden, D., Pusch, O., Lee, F., Tucker, L. & Ramratnam, B. (2003a). Human immunodeficiency virus type 1 escape from RNA interference. *J. Virology* **77**, 11531–11535.
7. Boden, D., Pusch, O., Lee, F., Tucker, L. & Ramratnam, B. (2004a). Efficient gene transfer of HIV-1-specific short hairpin RNA into human lymphocytic cells using recombinant adeno-associated virus vectors. *Molecular Therapy* **9**, 396–402.

8. Boden, D., Pusch, O., Lee, F., Tucker, L., Shank, P. R. & Ramratnam, B. (2003b). Promoter choice affects the potency of HIV-1 specific RNA interference. *Nucleic Acids Research* **31**, 5033–5038.

9. Boden, D., Pusch, O., Silbermann, R., Lee, F., Tucker, L. & Ramratnam, B. (2004b). Enhanced gene silencing of HIV-1 specific siRNA using microRNA designed hairpins. *Nucleic Acids Research* **32**, 1154–1158.

10. Braasch, D. A., Jensen, S., Liu, Y. H., Kaur, K., Arar, K., White, M. A. & Corey, D. R. (2003). RNA interference in mammalian cells by chemically-modified RNA. *Biochemistry* **42**, 7967–7975.

11. Bucher, E., Hemmes, H., de Haan, P., Goldbach, R. & Prins, M. (2004). The influenza A virus NS1 protein binds small interfering RNAs and suppresses RNA silencing in plants. *J. General Virology* **85**, 983–991.

12. Butz, K., Ristriani, T., Hengstermann, A., Denk, C., Scheffner, M. & Hoppe-Seyler, F. (2003). siRNA targeting of the viral E6 oncogene efficiently kills human papillomavirus-positive cancer cells. *Oncogene* **22**, 5938–5945.

13. Carette, J. E., Overmeer, R. M., Schagen, F. H. E., Alemany, R., Barski, O. A., Gerritsen, W. R. & van Beusechem, V. W. (2004). Conditionally replicating adenoviruses expressing short hairpin RNAs silence the expression of a target gene in cancer cells. *Cancer Research* **64**, 2663–2667.

14. Chang, J. & Taylor, J. M. (2003). Susceptibility of human hepatitis delta virus RNAs to small interfering RNA action. *J. Virology* **77**, 9728–9731.

15. Chang, Y., Chang, S. S., Lee, H. H., Doong, S. L., Takada, K. & Tsai, C. H. (2004). Inhibition of the Epstein-Barr virus lytic cycle by Zta-targeted RNA interference. *J. General Virology* **85**, 1371–1379.

16. Chen, J., Wall, N. R., Kocher, K., Duclos, N., Fabbro, D., Neuberg, D., Griffin, J. D., Shi, Y. & Gilliland, D. G. (2004a). Stable expression of small interfering RNA sensitizes TEL-PDGF beta R to inhibition with imatinib or rapamycin. *J. Clinical Invest.* **113**, 1784–1791.

17. Chen, W., Yan, W., Du, Q., Fei, L., Liu, M., Ni, Z., Sheng, Z. & Zheng, Z. (2004b). RNA interference targeting VP1 inhibits foot-and-mouth disease virus replication in BHK-21 cells and suckling mice. *J. Virology* **78**, 6900–6907.

18. Chen, Y. C., Stamatoyannopoulos, G. & Song, C. Z. (2003). Down-regulation of CXCR4 by inducible small interfering RNA inhibits breast cancer cell invasion *in vitro. Cancer Research* **63**, 4801–4804.

19. Condiotti, R., Curran, M. A., Nolan, G. P., Giladi, H., Ketzinel-Gilad, M., Gross, E. & Galun, E. (2004). Prolonged liver-specific transgene expression by a non-primate lentiviral vector. *BBRC* **320**, 998–1006.

20. Cui, W. W., Ning, J. C., Naik, U. P. & Duncan, M. K. (2004). OptiRNAi, an RNAi design tool. *Computer Methods and Programs in Biomedicine* **75**, 67–73.

21. Das, A. T., Brummelkamp, T. R., Westerhout, E. M., Vink, M., Madiredjo, M., Bernards, R. & Berkhout, B. (2004). Human immunodeficiency virus type 1 escapes from RNA interference-mediated inhibition. *J. Virology* **78**, 2601–2605.

22. Ding, H. L., Schwarz, D. S., Keene, A., Affar, E. B., Fenton, L., Xia, X. A., Shi, Y., Zamore, P. D. & Xu, Z. S. (2003). Selective silencing by RNAi of a dominant allele that causes amyotrophic lateral sclerosis. *Aging Cell* **2**, 209–217.

23. Dorn, G., Patel, S., Wotherspoon, G., Hemmings-Mieszczak, M., Barclay, J., Natt, F. J. C., Martin, P., Bevan, S., Fox, A., Ganju, P., Wishart, W. & Hall, J. (2004). siRNA relieves chronic neuropathic pain. *Nucleic Acids Research* **32**, e49.

24. Dudek, P. & Picard, D. (2004). TROD: T7 RNAi Oligo Designer. *Nucleic Acids Research* **32**, W121–W123.

25. Edelstein, M. L., Abedi, M. R., Wixon, J. & Edelstein, R. M. (2004). Gene therapy clinical trials worldwide 1989–2004 — An overview. *J. Gene Medicine* **6**, 597–602.

26. Eliyahu, H., Servel, N., Domb, A. J. & Barenholz, Y. (2002). Lipoplex-induced hemagglutination: Potential involvement in intravenous gene delivery. *Gene Therapy* **9**, 850–858.

27. Fish, R. J. & Kruithof, E. K. (2004). Short-term cytotoxic effects and long-term instability of RNAi delivered using lentiviral vectors. *Molecular Biology* **3**, 9.

28. Galimi, F. & Verma, I. M. (2002). Opportunities for the use of lentiviral vectors in human gene therapy. *Lentiviral Vectors* **261**, 245–254.

29. Gao, G. P., Lebherz, C., Weiner, D. J., Grant, R., Calcedo, R., McCullough, B., Bagg, A., Zhang, Y. & Wilson, J. M. (2004). Erythropoietin gene therapy leads to autoimmune anemia in macaques. *Blood* **103**, 3300–3302.

30. Ge, Q., Filip, L., Bai, A. L., Nguyen, T., Eisen, H. N. & Chen, J. (2004). Inhibition of influenza virus production in virus-infected mice by RNA interference. *Proc. Natl. Acad. Sci. USA* **101**, 8676–8681.

31. Giladi, H., Ketzinel-Gilad, M., Rivkin, L., Felig, Y., Nussbaum, O. & Galun, E. (2003). Small interfering RNA inhibits hepatitis B virus replication in mice. *Molecular Therapy* **8**, 769–776.

32. Goodall, J., Wellbrock, C., Dexter, T. J., Roberts, K., Marais, R. & Goding, C. R. (2004). The Brn-2 transcription factor links activated BRAF to melanoma proliferation. *Molecular and Cellular Biology* **24**, 2923–2931.

33. Gupta, S., Schoer, R. A., Egan, J. E., Hannon, G. J. & Mittal, V. (2004). Inducible, reversible, and stable RNA interference in mammalian cells. *Proc. Natl. Acad. Sci. USA* **101**, 1927–1932.

34. Hacein-Bey-Abina, S., Le Deist, F., Carlier, F., Bouneaud, C., Hue, C., De Villartay, J., Thrasher, A. J., Wulffraat, N., Sorensen, R., Dupuis-Girod, S., Fischer, A., Cavazzana-Calvo, M., Davies, E. G., Kuis, W., Lundlaan, W. H. K. & Leiva, L. (2002). Sustained correction of X-linked severe combined immunodeficiency by *ex vivo* gene therapy. *New England J. Medicine* **346**, 1185–1193.

35. Hacein-Bey-Abina, S., Von Kalle, C., Schmidt, M., McCcormack, M. P., Wulffraat, N., Leboulch, P., Lim, A., Osborne, C. S., Pawliuk, R., Morillon, E., Sorensen, R., Forster, A., Fraser, P., Cohen, J. I., de Saint Basile, G., Alexander, I., Wintergerst, U., Frebourg, T., Aurias, A., Stoppa-Lyonnet, D., Romana, S., Radford-Weiss, I., Gross, F., Valensi, F., Delabesse, E., Macintyre, E., Sigaux, F., Soulier, J., Leiva, L. E., Wissler, M., Prinz, C., Rabbitts, T. H., Le Deist, F., Fischer, A. & Cavazzana-Calvo, M. (2003). LMO2-associated clonal T cell proliferation in two patients after gene therapy for SCID-X1. *Science* **302**, 415–419.

36. Hall, A. H. S. & Alexander, K. A. (2003). RNA interference of human papillomavirus type 18 E6 and E7 induces senescence in HeLa cells. *J. Virology* **77**, 6066–6069.

37. Hamasaki, K., Nakao, K., Matsumoto, K., Ichikawa, T., Ishikawa, H. & Eguchi, K. (2003). Short interfering RNA-directed inhibition of hepatitis B virus replication. *FEBS Letters* **543**, 51–54.

38. Heggestad, A. D., Notterpek, L. & Fletcher, B. S. (2004). Transposon-based RNAi delivery system for generating knockdown cell lines. *BBRC* **316**, 643–650.

39. Henschel, A., Buchholz, F. & Habermann, B. (2004). DEQOR: A web-based tool for the design and quality control of siRNAs. *Nucleic Acids Research* **32**, W113–W120.

40. Hill, J. A., Ichim, T. E., Kusznieruk, K. P., Li, M., Huang, X. Y., Yan, X. T., Zhong, R., Cairns, E., Bell, D. A. & Min, W. P. (2003). Immune modulation by silencing IL-12 production in dendritic cells using small interfering RNA. *J. Immunology* **171**, 3303–3303.

41. Hommel, J. D., Sears, R. M., Georgescu, D., Simmons, D. L. & DiLeone, R. J. (2003). Local gene knockdown in the brain using viral-mediated RNA interference. *Nature Medicine* **9**, 1539–1544.

42. Hosono, T., Mizuguchi, H., Katayama, K., Xu, Z. L., Sakurai, F., Ishii-Watabe, A., Kawabata, K., Yamaguchi, T., Nakagawa, S., Mayumi, T. & Hayakawa, T. (2004). Adenovirus vector-mediated doxycycline-inducible RNA interference. *Human Gene Therapy* **15**, 813–819.

43. Hosseinkhani, H., Azzam, T., Tabata, Y. & Domb, A. J. (2004). Dextran-spermine polycation: An efficient nonviral vector for *in vitro* and *in vivo* gene transfection. *Gene Therapy* **11**, 194–203.

44. Howard, K. (2003). First Parkinson gene therapy trial launches. *Nature Biotechnology* **21**, 1117–1118.

45. Jackson, A. L., Bartz, S. R., Schelter, J., Kobayashi, S. V., Burchard, J., Mao, M., Li, B., Cavet, G. & Linsley, P. S. (2003). Expression profiling reveals off-target gene regulation by RNAi. *Nature Biotechnology* **21**, 635–637.

46. Kaiser, J. (2004). Gene therapy — Side effects sideline hemophilia trial. *Science* **304**, 1423–1425.

47. Kanda, T., Kusov, Y., Yokosuka, O. & Gauss Muller, V. (2004). Interference of hepatitis A virus replication by small interfering RNAs. *BBRC* **318**, 341–345.

48. Kariko, K., Bhuyan, P., Capodici, J. & Weissman, D. (2004). Small interfering RNAs mediate sequence-independent gene suppression and induce immune activation by signaling through toll-like receptor 3. *J. Immunology* **172**, 6545–6549.

49. Karlas, A., Kurth, R. & Denner, J. (2004). Inhibition of porcine endogenous retroviruses by RNA interference: Increasing the safety of xenotransplantation. *Virology* **325**, 18–23.

50. Kasim, V., Miyagishi, M. & Taira, K. (2004). Control of siRNA expression using the Cre-loxP recombination system. *Nucleic Acids Research* **32**, e66.

51. Kawasaki, H. & Taira, K. (2003). Short hairpin type of dsRNAs that are controlled by tRNA(Val) promoter significantly induce RNAi-mediated gene silencing in the cytoplasm of human cells. *Nucleic Acids Research* **31**, 700–707.

52. Kawasaki, H. & Taira, K. (2004). Induction of DNA methylation and gene silencing by short interfering RNAs in human cells. *Nature* **431**, 211–217.

53. Kobayashi, N., Matsui, Y., Kawase, A., Hirata, K., Miyagishi, M., Taira, K., Nishikawa, M. & Takakura, Y. (2004). Vector-based *in vivo* RNA

interference: Dose- and time-dependent suppression of transgene expression. *J. Pharm. and Exp. Therapeutics* **308**, 688–693.

54. Konishi, M., Wu, C. H. & Wu, G. Y. (2003). Inhibition of HBV replication by siRNA in a stable HBV-producing cell line. *Hepatology* **38**, 842–850.

55. Kronke, J., Kittler, R., Buchholz, F., Windisch, M. P., Pietschmann, T., Bartenschlager, R. & Frese, M. (2004). Alternative approaches for efficient inhibition of hepatitis C virus RNA replication by small interfering RNAs. *J. Virology* **78**, 3436–3446.

56. Kumar, R., Conklin, D. S. & Mittal, V. (2003). High-throughput selection of effective RNAi probes for gene silencing. *Genome Research* **13**, 2333–2340.

57. Lee, M. T. M., Coburn, G. A., McClure, M. O. & Cullen, B. R. (2003). Inhibition of human immunodeficiency virus type 1 replication in primary macrophages by using Tat- or CCR5-specific small interfering RNAs expressed from a lentivirus vector. *J. Virology* **77**, 11964–11972.

58. Lewis, D. L., Hagstrom, J. E., Loomis, A. G., Wolff, J. A. & Herweijer, H. (2002). Efficient delivery of siRNA for inhibition of gene expression in postnatal mice. *Nature Genetics* **32**, 107–108.

59. Li, W. X., Li, H. W., Lu, R., Li, F., Dus, M., Atkinson, P., Brydon, E. W. A., Johnson, K. L., Garcia-Sastre, A., Ball, L. A., Palese, P. & Ding, S. W. (2004a). Interferon antagonist proteins of influenza and vaccinia viruses are suppressors of RNA silencing. *Proc. Natl. Acad. Sci. USA* **101**, 1350–1355.

60. Li, X. P., Li, G., Peng, Y., Kung, H. F. & Lin, M. C. (2004b). Suppression of Epstein-Barr virus-encoded latent membrane protein-1 by RNA interference inhibits the metastatic potential of nasopharyngeal carcinoma cells. *BBRC* **315**, 212–218.

61. Lu, A. L., Zhang, H. Q., Zhang, X. Y., Wang, H. X., Hu, Q. K., Shen, L., Schaffhausen, B. S., Hou, W. M. & Li, L. S. (2004). Attenuation of SARS coronavirus by a short hairpin RNA expression plasmid targeting RNA-dependent RNA polymerase. *Virology* **324**, 84–89.

62. Matsuda, T. & Cepko, C. L. (2004). Electroporation and RNA interference in the rodent retina *in vivo* and *in vitro*. *Proc. Natl. Acad. Sci. USA* **101**, 16–22.

63. McCaffrey, A. P., Meuse, L., Pham, T. T. T., Conklin, D. S., Hannon, G. J. & Kay, M. A. (2002). Gene expression — RNA interference in adult mice. *Nature* **418**, 38–39.

64. McCaffrey, A. P., Nakai, H., Pandey, K., Huang, Z., Salazar, F. H., Xu, H., Wieland, S. F., Marion, P. L. & Kay, M. A. (2003). Inhibition of hepatitis B virus in mice by RNA interference. *Nature Biotechnology* **21**, 639–644.

65. McCown, M., Diamond, M. S. & Pekosz, A. (2003). The utility of siRNA transcripts produced by RNA polymerase I in down regulating viral gene expression and replication of negative- and positive-strand RNA viruses. *Virology* **313**, 514–524.

66. Miller, D. G., Petek, L. M. & Russell, D. W. (2004a). Adeno-associated virus vectors integrate at chromosome breakage sites. *Nature Genetics* **36**, 767–773.

67. Miller, V. M., Gouvion, C. M., Davidson, B. L. & Paulson, H. L. (2004b). Targeting Alzheimer's disease genes with RNA interference: An efficient strategy for silencing mutant alleles. *Nucleic Acids Research* **32**, 661–668.

68. Minakuchi, Y., Takeshita, F., Kosaka, N., Sasaki, H., Yamamoto, Y., Kouno, M., Honma, K., Nagahara, S., Hanai, K., Sano, A., Kato, T., Terada, M. & Ochiya, T.

(2004). Atelocollagen-mediated syntehtic small interfering RNA delivery for effective gene silencing *in vitro* and *in vivo*. *Nucleic Acids Research* **22**, 109.

69. Mittal, V. (2004). Improving the efficiency of RNA interference in mammals. *Nature Reviews Genetics* **5**, 355–365.

70. Miyagishi, M., Sumimoto, H., Miyoshi, H., Kawakami, Y. & Taira, K. (2004). Optimization of an siRNA-expression system with an improved hairpin and its significant suppressive effects in mammalian cells. *J. Gene Medicine* **6**, 715–723.

71. Miyagishi, M. & Taira, K. (2002). Development and application of siRNA expression vector. *Nucleic Acids Research Suppl.* **2**, 113–114.

72. Morral, N., O'Neal, W., Rice, K., Leland, M., Kaplan, J., Piedra, P. A., Zhou, H. S., Parks, R. J., Velji, R., Aguilar-Cordova, E., Wadsworth, S., Graham, F. L., Kochanek, S., Carey, K. D. & Beaudet, A. L. (1999). Administration of helper-dependent adenoviral vectors and sequential delivery of different vector serotype for long-term liver-directed gene transfer in baboons. *Proc. Natl. Acad. Sci. USA* **96**, 12816–12821.

73. Naito, Y., Yamada, T., Ui-Tei, K., Morishita, S. & Saigo, K. (2004). siDirect: Highly effective, target-specific siRNA design software for mammalian RNA interference. *Nucleic Acids Research* **32**, W124–W129.

74. Nishitsuji, H., Ikeda, T., Miyoshi, H., Ohashi, T., Kannagi, M. & Masuda, T. (2004). Expression of small hairpin RNA by lentivirus-based vector confers efficient and stable gene-suppression of HIV-1 on human cells including primary non-dividing cells. *Microbes and Infection* **6**, 76–85.

75. Orba, Y., Sawa, H., Iwata, H., Tanaka, S. & Nagashima, K. (2004). Inhibition of virus production in JC virus-infected cells by postinfection RNA interference. *J. Virology* **78**, 7270–7273.

76. Park, W. S., Hayafune, M., Miyano-Kurosaki, N. & Takaku, H. (2003). Specific HIV-1 env gene silencing by small interfering RNAs in human peripheral blood mononuclear cells. *Gene Therapy* **10**, 2046–2050.

77. Park, W. S., Miyano-Kurosaki, N., Nakajima, E. & Takaku, H. (2001). Specific inhibition of HIV-1 gene expression by double-stranded RNA. *Nucleic Acids Research Suppl.* **1**, 219–220.

78. Paroo, Z. & Corey, D. R. (2004). Challenges for RNAi *in vivo*. *Trends Biotechnol.* **22**, 390–394.

79. Pebernard, S. & Iggo, R. D. (2004). Determinants of interferon-stimulated gene induction by RNAi vectors. *Differentiation* **72**, 103–111.

80. Phipps, K. M., Martinez, A., Lu, J., Heinz, B. A. & Zhao, G. S. (2004). Small interfering RNA molecules as potential anti-human rhinovirus agents: *In vitro* potency, specificity, and mechanism. *Antiviral Research* **61**, 49–55.

81. Poeschla, E. M. (2003). Non-primate lentiviral vectors. *Curr. Opin. Molecular Therapeutics* **5**, 529–540.

82. Reynolds, A., Leake, D., Boese, Q., Scaringe, S., Marshall, W. S. & Khvorova, A. (2004). Rational siRNA design for RNA interference. *Nature Biotechnology* **22**, 326–330.

83. Rund, D., Dagan, M., Dalyot-Herman, N., Kimchi-Sarfaty, C., Schoenlein, P. V., Gottesman, M. M. & Oppenheim, A. (1998). Efficient transduction of human hematopoietic cells with the human multidrug resistance gene 1 via SV40 pseudovirions. *Human Gene Therapy* **9**, 649–657.

84. Sago, N., Omi, K., Tamura, Y., Kunugi, H., Toyo-Oka, T., Tokunaga, K. & Hohjoh, H. (2004). RNAi induction and activation in mammalian muscle cells where Dicer and eIF2C translation initiation factors are barely expressed. *BBRC* **319**, 50–57.

85. Sakamoto, T., Miyazaki, E., Aramaki, Y., Arima, H., Takahashi, M., Kato, Y., Koga, M. & Tsuchiya, S. (2004). Improvement of dermatitis by iontophoretically delivered antisense oligonucleotides for interleukin-10 in NC/Nga mice. *Gene Therapy* **11**, 317–324.

86. Salvi, A., Arici, B., De Petro, G. & Barlati, S. (2004). Small interfering RNA urokinase silencing inhibits invasion and migration of human hepatocellular carcinoma cells. *Molecular Cancer Therapeutics* **3**, 671–678.

87. Sen, A., Steele, R., Ghosh, A. K., Basu, A., Ray, R. & Ray, R. B. (2003). Inhibition of hepatitis C virus protein expression by RNA interference. *Virus Research* **96**, 27–35.

88. Shlomai, A. & Shaul, Y. (2003). Inhibition of hepatitis B virus expression and replication by RNA interference. *Hepatology* **37**, 764–770.

89. Sioud, M. & Sorensen, D. R. (2003). Cationic liposome-mediated delivery of siRNAs in adult mice. *BBRC* **312**, 1220–1225.

90. Sledz, C. A., Holko, M., de Veer, M. J., Silverman, R. H. & Williams, B. R. G. (2003). Activation of the interferon system by short-interfering RNAs. *Nature Cell Biology* **5**, 834–839.

91. Song, E. W., Lee, S. K., Wang, J., Ince, N., Ouyang, N., Min, J., Chen, J. S., Shankar, P. & Lieberman, J. (2003). RNA interference targeting Fas protects mice from fulminant hepatitis. *Nature Medicine* **9**, 347–351.

92. Spankuch, B., Matthess, Y., Knecht, R., Zimmer, B., Kaufmann, M. & Strebhardt, K. (2004). Cancer inhibition in nude mice after systemic application of U6 promoter-driven short hairpin RNAs against PLK1. *J. National Cancer Institute* **96**, 862–872.

93. Sumimoto, H., Miyagishi, M., Miyoshi, H., Yamagata, S., Shimizu, A., Taira, K. & Kawakami, Y. (2004). Inhibition of growth and invasive ability of melanoma by inactivation of mutated BRAF with lentivirus-mediated RNA interference. *Oncogene* **23**, 6031–6039.

94. Takei, Y., Kadomatsu, K., Yuzawa, Y., Matsuo, S. & Muramatsu, T. (2004). A small interfering RNA targeting vascular endothelial growth factor as cancer therapeutics. *Cancer Research* **64**, 3365–3370.

95. Tiscornia, G., Tergaonkar, V., Galimi, F. & Verma, I. M. (2004). CRE recombinase-inducible RNA interference mediated by lentiviral vectors. *Proc. Natl. Acad. Sci. USA* **101**, 7347–7351.

96. Tompkins, S. M., Lo, C. Y., Tumpey, T. M. & Epstein, S. L. (2004). Protection against lethal influenza virus challenge by RNA interference *in vivo*. *Proc. Natl. Acad. Sci. USA* **101**, 8682–8686.

97. Tranchant, I., Thompson, B., Nicolazzi, C., Mignet, N. & Scherman, D. (2004). Physicochemical optimisation of plasmid delivery by cationic lipids. *J. Gene Medicine* **6**, S24–S35.

98. Van den Haute, C., Eggermont, K., Nuttin, B., Debyser, Z. & Baekelandt, V. (2003). Lentiviral vector-mediated delivery of short hairpin RNA results in

persistent knockdown of gene expression in mouse brain. *Human Gene Therapy* **14**, 1799–1807.

99. Ventura, A., Meissner, A., Dillon, C. P., McManus, M., Sharp, P. A., Van Parijs, L., Jaenisch, R. & Jacks, T. (2004). Cre-lox-regulated conditional RNA interference from transgenes. *Proc. Natl. Acad. Sci. USA* **101**, 10380–10385.

100. Wang, Z., Ren, L. L., Zhao, X. G., Hung, T., Meng, A., Wang, J. W. & Chen, Y. G. (2004). Inhibition of severe acute respiratory syndrome virus replication by small interfering RNAs in mammalian cells. *J. Virology* **78**, 7523–7527.

101. Wimerowicz, M. & Trono, D. (2003). Conditional suppression of cellular genes: Lentivirus vector-mediated drug-inducible RNA interference. *J. Virology* **77**, 8957–8961.

102. Xia, H. B., Mao, Q. W., Eliason, S. L., Harper, S. Q., Martins, I. H., Orr, H. T., Paulson, H. L., Yang, L., Kotin, R. M. & Davidson, B. L. (2004). RNAi suppresses polyglutamine-induced neurodegeneration in a model of spinocerebellar ataxia. *Nature Medicine* **10**, 816–820.

103. Xia, X. G., Zhou, H. S., Ding, H. L., Affar, El Bashiz, B., Shi, Y. & Xu, Z. (2003). An enhanced U6 promoter for synthesis of short hairpin RNA. *Nucleic Acids Research* **31**, e100.

104. Yague, E., Higgins, C. F. & Raguz, S. (2004). Complete reversal of multidrug resistance by stable expression of small interfering RNAs targeting MDR1. *Gene Therapy* **11**, 1170–1174.

105. Ye, K. Q., Malinina, L. & Patel, D. J. (2003). Recognition of small interfering RNA by a viral suppressor of RNA silencing. *Nature* **426**, 874–878.

106. Ying, C. X., De Clercq, E. & Neyts, J. (2003). Selective inhibition of hepatitis B virus replication by RNA interference. *BBRC* **309**, 482–484.

107. Yokota, T., Sakamoto, N., Enomoto, N., Tanabe, Y., Miyagishi, M., Maekawa, S., Yi, L., Kurosaki, M., Taira, K., Watanabe, M. & Mizusawa, H. (2003). Inhibition of intracellular hepatitis C virus replication by synthetic and vector-derived small interfering RNAs. *EMBO Reports* **4**, 602–608.

108. Zender, L., Hutker, S., Liedtke, C., Tillmann, H. L., Zender, S., Mundt, B., Waltemathe, M., Gosling, T., Flemming, P., Malek, N. P., Trautwein, C., Manns, M. P., Kuhnel, F. & Kubicka, S. (2003). Caspase 8 small interfering RNA prevents acute liver failure in mice. In *Proc. Natl. Acad. Sci. USA* **100**, 7797–7802.

109. Zhang, Y., Zhang, Y. F., Bryant, J., Charles, A., Boado, R. J. & Pardridge, W. M. (2004b). Intravenous RNA interference gene therapy targeting the human epidermal growth factor receptor prolongs survival in intracranial brain cancer. *Clinical Cancer Research* **10**, 3667–3677.

110. Zhang, Y. J., Li, T. S., Fu, L., Yu, C. M., Li, Y. H., Xu, X. L., Wang, Y. Y., Ning, H. X., Zhang, S. P., Chen, W., Babiuk, L. A. & Chang, Z. J. (2004a). Silencing SARS-CoV Spike protein expression in cultured cells by RNA interference. *FEBS Letters* **560**, 141–146.

111. Zhao, L. J., Jian, H. & Zhu, H. H. (2003). Specific gene inhibition by adenovirus-mediated expression of small interfering RNA. *Gene* **316**, 137–141.

INDEX

lymphomagenesis, 224
lysine, 312, 313

M. musculus, 338
Müller glial cells, 220
Macho, 353
macronucleus, 155
MADS-box, 349–351, 358, 359
Magnaporthe grisea (M. grisea), 41, 42
Maimonides, 2
maize, 288
malaria disease, 151
male sterility (ms), 278
Maleficent, 222
mammalian brain development, 275
mammalian cells, 200, 205
mammalian diseases, 192, 224
mammalian embryonal differentiation, 266
mammals, 194, 273, 289
man, 149
manos, 92
Manufacture of Medical and Health Products by Transgenic Plants, ix
MAP, 165
MAP2 (microtubule-associated protein 2), 207, 258
Marchantia polymorpha, 331
mature miRNA, 241, 242, 330
McClintock, Barbara, viii
Medea, 372
MEF2A, 207
meiosis, 45
meiotic arrest, 195
meiotic division, 195
meiotic silencing by unpaired DNA (MSUD), 45
melanocytes, 287
Melanoma, 420
MEN1 (multiple endocrine neoplasma type 1), 230, 429
Mendel, Gregor, viii, 3
Mendelian genetics, 174
merozoites, 151
mesodermal cell, 68
mesophyll protoplast, 348
messenger RNA, 193

metamorphosis, 91, 363
metaphase, 195, 196
metastasis, 421
Metathron, 295
metazoa, 244
Methanococcus vannielii, 131
methionine, 312, 313
methylation, 10, 260
methylation-induced premeiotically, MIP, 36
7-methylguanosine (m^{7}G), 218
miRNA, 4
micro-electroporation, 189
microinjection, 189
micronucleus, 155, 156
microRNA (miRNA), 33, 45, 71, 361, 240
microtubules, 135
miR numbers, 71
mir-100, 127, 128
mir-125, 127, 128
mir-14, 127
mir-34, 127, 128
miRNA recognition elements, 270
miRNA seed, 262
miRNA*, 340
miRNA:miRNA* duplex, 241, 248, 340
miRNAs in mammals, 240
miRNP, 340
miRNP complex, 245, 268
miRNP particles, 250
miRNP, 269
MiRscan program, 74
miRx, 427
Mishna, 373
mitochondrial, 347
mitochondrion, 131
mitogen-activated protein (MAP), 196
mode of silencing, 254
molecular parasitology, 132
Mollusca, 173
Moloney murine leukemia virus (Mo-MuLV), 423
molting stages, 68
monkey COS7 cells, 200
monoclonal antibodies, 419
monogenetic maladies, 407
Morgan, Thomas Hunt, 174